Systems Biology

Volume II: Networks, Models, and Applications

SYS BIO

Series in Systems Biology

Edited by Dennis Shasha, New York University

Systems Biology describes the discipline that seeks to understand biological phenomena on a large scale: the association of gene with function, the detailed modeling of the interaction among proteins and metabolites, and the function of cells. Systems Biology has wide-ranging application, as it is informed by several underlying disciplines, including biology, computer science, mathematics, physics, chemistry, and the social sciences. The goal of the series is to help practitioners and researchers understand the ideas and technologies underlying Systems Biology. The series volumes will combine biological insight with principles and methods of computational data analysis.

Cellular Computing, edited by Martyn Amos
Systems Biology, Volume I: Genomics, edited by Isidore Rigoutsos and Gregory Stephanopoulos
Systems Biology, Volume II: Networks, Models, and Applications, edited by Isidore Rigoutsos and Gregory Stephanopoulos

Systems Biology

Volume II:
Networks, Models, and Applications

Edited by
Isidore Rigoutsos
&
Gregory Stephanopoulos

OXFORD
UNIVERSITY PRESS
2007

OXFORD
UNIVERSITY PRESS

Oxford University Press, Inc., publishes works that further
Oxford University's objective of excellence
in research, scholarship, and education.

Oxford New York
Auckland Cape Town Dar es Salaam Hong Kong Karachi
Kuala Lumpur Madrid Melbourne Mexico City Nairobi
New Delhi Shanghai Taipei Toronto

With offices in
Argentina Austria Brazil Chile Czech Republic France Greece
Guatemala Hungary Italy Japan Poland Portugal Singapore
South Korea Switzerland Thailand Turkey Ukraine Vietnam

Published by Oxford University Press, Inc.
198 Madison Avenue, New York, New York 10016

www.oup.com

Library of Congress Cataloging-in-Publication Data
Systems biology/edited by Isidore Rigoutsos and Gregory Stephanopoulos.
v. ; cm.—(Series in systems biology)
Includes bibliographical references and indexes.
Contents: 1. Genomics—2. Networks, models, and applications.
ISBN-13: 978-0-19-530081-9 (v. 1)
ISBN 0-19-530081-5 (v. 1)
ISBN-13: 978-0-19-530080-2 (v. 2)
ISBN 0-19-530080-7 (v. 2)
1. Computational biology. 2. Genomics. 3. Bioinformatics. I. Rigoutsos, Isidore.
II. Stephanopoulos, G. III. Series.
[DNLM: 1. Genomics. 2. Computational Biology. 3. Models, Genetic. 4. Systems Biology.
QU58.5 S995 2006]
QH324.2.S97 2006
570—dc22 2005031826

9 8 7 6 5 4 3 2 1

Printed in the United States of America
on acid-free paper

To our mothers

Acknowledgments

First and foremost, we wish to thank all the authors who contributed the chapters of these two books. In addition to the professionalism with which they handled all aspects of production, they also applied the highest standards in authoring pieces of work of the highest quality. For their willingness to share their unique expertise on the many facets of systems biology and the energy they devoted to the preparation of their chapters, we are profoundly grateful. Next, we wish to thank Dennis Shasha, the series editor, and Peter Prescott, Senior Editor for Life Sciences, Oxford University Press, for embracing the project from the very first day that we presented the idea to them. Peter deserves special mention for it was his continuous efforts that helped remove a great number of obstacles along the way.

We also wish to thank Adrian Fay, who coordinated several aspects of the review process and provided input that improved the flow of several chapters, as well as our many reviewers, Alice McHardy, Aristotelis Tsirigos, Christos Ouzounis, Costas Maranas, Daniel Beard, Daniel Platt, Jeremy Rice, Joel Moxley, Kevin Miranda, Lily Tong, Masaru Nonaka, Michael MacCoss, Michael Pitman, Nikos Kyrpides, Rich Jorgensen, Roderic Guigo, Rosaria De Santis, Ruhong Zhou, Serafim Batzoglou, Steven Gygi, Takis Benos, Tetsuo Shibuya, and Yannis Kaznessis, for providing helpful and detailed feedback on the early versions of the chapters; without their help the books would not have been possible. We are also indebted to Kaity Cheng for helping with all of the administrative aspects of this project. And, finally, our thanks go to our spouses whose understanding and patience throughout the duration of the project cannot be overstated.

Contents

Contributors

C. A. FLOUDAS
Department of Chemical Engineering
Princeton University
Princeton, New Jersey
floudas@titan.princeton.edu

H. K. FUNG
Department of Chemical Engineering
Princeton University
Princeton, New Jersey
hfung@princeton.edu

ROBERT S. GERMAIN
IBM T.J. Watson Research Center
Yorktown Heights, New York
germainr@us.ibm.com

ERWIN P. GIANCHANDANI
Department of Biomedical Engineering
University of Virginia
Charlottesville, Virginia
epg5g@virginia.edu

JOSEPH. L. GREENSTEIN
Department of Biomedical Engineering
Johns Hopkins University
Baltimore, Maryland
Jgreenst@bme.jhu.edu

G. A. HELD
IBM T.J. Watson Research Center
Yorktown Heights, New York
gaheld@us.ibm.com

PATRICK A. HELM
Robert M. Berne Cardiovascular Research Center
Charlottesville, Virginia
Ph7t@virginia.edu

MARKUS J. HERRGÅRD
Department of Bioengineering
University of California, San Diego
La Jolla, California
mherrgar@ucsd.edu

BYUNG HUN KIM
Department of BioSystems and Bioinformatics Research Center
Korea Advanced Institute of Science and Technology
Daejeon, Korea

TAE YONG KIM
Department of BioSystems and Bioinformatics Research Center
Korea Advanced Institute of Science and Technology
Daejeon, Korea

JOHN D. LAMBRIS
Department of Pathology and Laboratory Medicine
University of Pennsylvania Biomedical Graduate Studies
Philadelphia, Pennsylvania
lambris@mail.med.upenn.edu

DONG-YUP LEE
Department of BioSystems and Bioinformatics Research Center
Korea Advanced Institute of Science and Technology
Daejeon, Korea

SANG JUN LEE
Department of BioSystems and Bioinformatics Research Center
Korea Advanced Institute of Science and Technology
Daejeon, Korea

SANG YUP LEE
Department of BioSystems and Bioinformatics Research Center
Korea Advanced Institute of Science and Technology
Daejeon, Korea
leesy@kaist.ac.kr

BING LIM
Department of Stem Cell and Developmental Biology
Genome Institute of Singapore
Singapore
limb1@gis.a-star.edu.sg

DIMITRIOS MASTELLOS
Department of Pathology and Laboratory Medicine
University of Pennsylvania Biomedical Graduate Studies
Philadelphia, Pennsylvania
dimitri7@mail.med.upenn.edu

HUCK HUI NG
Department of Stem Cell and Developmental Biology
Genome Institute of Singapore
Singapore

HASAN H. OTU
Harvard Institutes of Medicine
Harvard Medical School
Boston, Massachusetts

BERNHARD Ø. PALSSON
Department of Bioengineering
University of California, San Diego
La Jolla, California
palsson@ucsd.edu

JASON A. PAPIN
Department of Biomedical Engineering
University of Virginia
Charlottesville, Virginia
papin@virginia.edu

JED W. PITERA
Almaden Research Center
San Jose, California
pitera@us.ibm.com

PAUL ROBSON
Department of Stem Cell and Developmental Biology
Genome Institute of Singapore
Singapore

CHRISTIAN I. RUSE
Department of Cell Biology
The Scripps Research Institute
La Jolla, California
cruse@scripps.edu

GUSTAVO STOLOVITZKY
IBM T.J. Watson Research Center
Yorktown Heights, New York
gustavo@us.ibm.com

SHANKAR SUBRAMANIAM
Department of Bioengineering
University of California, San Diego
La Jolla, California
shankar@sdsc.edu

WILLIAM C. SWOPE
Almaden Research Center
San Jose, California
swope@almaden.ibm.com

ANDREW M. THOMSON
Department of Stem Cell and Developmental Biology
Genome Institute of Singapore
Singapore

YUHAI TU
IBM T.J. Watson Research Center
Yorktown Heights, New York
yuhai@us.ibm.com

RAIMOND L. WINSLOW
Department of Biomedical Engineering
Johns Hopkins University
Baltimore, Maryland
rwinslow@bme.jhu.edu

JOHN R. YATES III
Department of Cell Biology
The Scripps Research Institute
La Jolla, California
jyates@scripps.edu

Systems Biology: A Perspective

As recently as a decade ago, the core paradigm of biological research followed an established path: beginning with the generation of a specific hypothesis a concise experiment would be designed that typically focused on studying a small number of genes. Such experiments generally measured a few macromolecules, and, perhaps, small metabolites of the target system.

The advent of genome sequencing and associated technologies greatly improved scientists' ability to measure important classes of biological molecules and their interactions. This, in turn, expanded our view of cells with a bevy of previously unavailable data and made possible genome-wide and cell-wide analyses. These newly found lenses revealed that hundreds (sometimes thousands) of molecules and interactions, which were outside the focus of the original study, varied significantly in the course of the experiment.

The term *systems biology* was coined to describe the field of scientific inquiry which takes a global approach to the understanding of cells and the elucidation of biological processes and mechanisms. In many respects, this is also what physiology (from the Greek *physis* = nature and *logos* = word-knowledge) focused on for the most part of the twentieth century. Indeed, physiology's goal has been the study of function and characteristics of living organisms and their parts and of the underlying physiochemical phenomena. Unlike physiology, systems biology attempts to interpret and contextualize the large and diverse sets of biological measurements that have become visible through our genomic-scale window on cellular processes by taking a holistic approach and bringing to bear theoretical, computational, and experimental advances in several fields. Indeed, there is considerable excitement that, through this integrative perspective, systems biology will succeed in elucidating the mechanisms that underlie complex phenomena and which would have otherwise remained undiscovered.

For the purposes of our discussion, we will be making use of the following definition: "Systems biology is an integrated approach that brings together and leverages theoretical, experimental, and computational approaches in order to establish connections among important molecules or groups of molecules in order to aid the eventual mechanistic explanation of cellular processes and systems." More specifically, we view systems biology as a field that aims to uncover concrete molecular relationships for targeted analysis through the interpretation of cellular phenotype in terms of integrated biomolecular networks. The fidelity and breadth of our network and state characterization are intimately related to the degree of our understanding of the system under study. As the readers will find, this view permeates the treatises that are found in these two books.

Cells have always been viewed as elegant systems of immense complexity that are, nevertheless, well coordinated and optimized for a particular purpose. This apparent complexity led scientists to take a reductionist approach to research which, in turn, contributed to a rigorous understanding of low-level processes in a piecemeal fashion. Nowadays, completed genomic sequences and systems-level probing hold the potential to accelerate the discovery of unknown molecular mechanisms and to organize the existing knowledge in a broader context of high-level cellular understanding. Arguably, this is a formidable task. In order to improve the chances of success, we believe that one must anchor systems biology analyses to specific questions and build upon the existing core infrastructure that the earlier, targeted research studies have allowed us to generate.

The diversity of molecules and reactions participating in the various cellular functions can be viewed as an impediment to the pursuit of a more complete understanding of cellular function. However, it actually represents a great opportunity as it provides countless possibilities for modifying the cellular machinery and commandeering it toward a specific goal. In this context, we distinguish two broad categories of questions that can guide the direction of systems biology research. The first category encompasses topics of medical importance and is typically characterized by forward-engineering approaches that focus on preventing or combating disease. The second category includes problems of industrial interest, such as the genetic engineering of microbes so as to maximize product formation, the creation of robust-production strains, and so on. The applications of the second category comprise an important reverse-engineering component whereby microbes with attractive properties are scrutinized for the purpose of transferring any insights learned from their functions to the further improvement and optimization of production strains.

PRIOR WORK

As already mentioned, and although the term *systems biology* did not enter the popular lexicon until recently, some of the activities it encompasses have been practiced for several decades. As we cannot possibly be exhaustive, we present a few illustrative examples of approaches that have been developed in recent years and successfully applied to relatively small systems. These examples can serve as useful guides in our attempt to tackle increasingly larger challenges.

Metabolic Control Analysis (MCA)

Metabolic pathways and, in general, networks of reactions are characterized by substantial stoichiometric and (mostly) kinetic complexity in their own right. The commonly applied assumption of a single rate-limiting step leads to great simplification of the reaction network and often yields analytical expressions for the conversion rates. However, this assumption is not justified for most biological systems where kinetic control is not concentrated in a single step but rather is distributed among several enzymatic steps. Consequently, kinetics and flux control of a bioreaction network represent properties of the entire system and can be determined from the characteristics of individual reactions in a bottom-up approach or from the response of the overall system in a top-down approach. The concepts of MCA and distribution of kinetic control in a reaction pathway have had a profound impact on the identification of target enzymes whose genetic modification permitted the amplification of the product flux through a pathway.

Signaling Pathways

Signal transduction is the process by which cells communicate with each other and their environment and involves a multitude of proteins that can be in active or inactive states. In their active (phosphorylated) state they act as catalysts for the activation of subsequent steps in the signaling cascade. The end result is the activation of a transcription factor which, in turn, initiates a gene transcription event. Until recently, and even though several of the known proteins participate in more than one signaling cascade, such systems were being studied in isolation from one another. A natural outcome of this approach was of course the ability to link a single gene with a single ligand in a causal relationship whereby the ligand activates the gene. However, such findings are not representative in light of the fact that signaling pathways branch and interact with one another creating a rather intricate and complex signaling network. Consequently, more tools, computational as well as experimental, are required if we are to improve our understanding of signal transduction. Developing such tools is among the goals of the recently formed Alliance for Cellular Signaling, an NIH-funded project involving several laboratories and research centers (www.signaling-gateway.org).

Reconstruction of Flux Maps

Metabolic pathway fluxes are defined as the actual rates of metabolite interconversion in a metabolic network and represent most informative measures of the actual physiological state of cells and organisms. Their dependence on enzymatic activities and metabolite concentrations makes them an accurate representation of carbon and energy flows through the various pathway branches. Additionally, they are very important in identifying critical reaction steps that impact flux control for the entire pathway. Thus, flux determination is an essential component of strain evaluation and metabolic engineering. Intracellular flux determination requires the enumeration and satisfaction of all intracellular metabolite balances along with the use of sufficient measurements typically derived from the introduction of isotopic tracers and metabolite and mass isotopomer measurement by gas chromatography–mass spectrometry. It is essentially a problem of constrained parameter estimation in overdetermined systems with overdetermination providing the requisite redundancy for reliable flux estimation. These approaches are basically methods of network reconstruction whereas the obtained fluxes represent properties of the entire system. As such, the fluxes accurately reflect changes introduced through genetic or environmental modifications and, thus, can be used to assess the impact of such modifications on cell physiology and product formation, and to guide the next round of cell modifications.

Metabolic Engineering

Metabolic engineering is the field of study whose goal is the improvement of microbial strains with the help of modern genetic tools. The strains are modified by introducing specific transport, conversion, or deregulation changes that lead to flux redistribution and the improvement of product yield. Such modifications rely to a significant extent on modern methods from molecular biology. Consequently, the following central question arises: "What is the real difference between genetic engineering and metabolic engineering?" We submit that the main difference is that metabolic engineering is concerned with the entire metabolic system whereas genetic engineering specifically focuses on a particular gene or a small collection of genes. It should be noted that over- or underexpression of a single gene or a few genes may have little or no impact on the attempt to alter cell physiology. On the other hand, by examining the properties of the metabolic network as a whole, metabolic engineering attempts to identify targets for amplification as well as rationally assess the effect that such changes will incur on the properties of the overall network. As such, metabolic engineering can be viewed as a precursor to functional genomics and systems biology in the sense that it represents the first organized effort to reconstruct and modify pathways using genomic tools while being guided by the information of postgenomic developments.

WORDS OF CAUTION

In light of the many exciting possibilities, there are high expectations for the field of systems biology. However, as we move forward, we should not lose sight of the fact that the field is trying to tackle a problem of considerable magnitude. Consequently, any expectations of immediate returns on the scientific investment should be appropriately tempered. As we set out to forecast future developments in this field, it is important to keep in mind several points.

Despite the wealth of available genomic data, there are still a lot of regions in the genomes of interest that are functional and which have not been identified as such. In order to practice systems biology, lists of "parts" and "relationships" that are as complete as possible are needed. In the absence of such complete lists, one generally hopes to derive at best an approximate description of the actual system's behavior. A prevalent misconception among scientists states that nearly complete lists of *parts* are already in place. Unfortunately, this is not the case—the currently available parts lists are incomplete as evidenced by the fact that genomic maps are continuously updated through the addition of removal of (occasionally substantial amounts of) genes, by the discovery of more regions that code for RNA genes, and so on.

Despite the wealth of available genomic data, knowledge about existing optimal solutions to important problems continues to elude us. The current efforts in systems biology are largely shaped by the available knowledge. Consequently, optimal solutions that are implemented by metabolic pathways that are unknown or not yet understood are beyond our reach. A characteristic case in point is the recent discovery, in sludge microbial communities, of a *Rhodocyclus*-like polyphosphate-accumulating organism that exhibits enhanced biological phosphorus removal abilities. Clearly, this microbe is a great candidate to be part of a biological treatment solution to the problem of phosphorus removal from wastewater. Alas, this is not yet an option as virtually nothing is known about the metabolic pathways that confer phosphorus removal ability to this organism.

Despite the wealth of available genomic data, there are still a lot of important molecular interactions of whose existence we are unaware. Continuing on our parts and relationships comment from above, it is worth noting another prevalent misconception among scientists: it states that nearly complete lists of *relationships* are already in place. For many years, pathway analysis and modeling has been characterized by protein-centric views that comprised concrete collections of proteins participating in well-understood interactions. Even for well-studied pathways, new important protein interactions are continuously discovered. Moreover, accumulating experimental evidence shows that numerous important interactions are in fact effected by the action of RNA molecules on DNA molecules and by extension on proteins. Arguably, posttranscriptional gene silencing and RNA interference represent one area of research activity with the potential to substantially revise our current understanding of cellular processes. In fact, the already accumulated knowledge suggests that the traditional protein-centric views of the systems of interest are likely incomplete and need to be augmented appropriately. This in turn has direct consequences on the modeling and simulation efforts and on our understanding of the cell from an integrated perspective.

Constructing biomolecular networks for new systems will require significant resources and expertise. Biomolecular networks incorporate a multitude of relationships that involve numerous components. For example, constructing gene interaction maps requires large experimental investments and computational analysis. As for global protein–protein interaction maps, these exist for only a handful of model species. But even reconstructing well-studied and well-documented networks such as metabolic pathways in a genomic context can prove a daunting task. The magnitude of such activities has already grown beyond the capabilities of a single investigator or a single laboratory.

Even when one works with a biomolecular network database, the system picture may be incomplete or only partially accurate. In the postgenomic era, the effort to uncover the structure and function of genetic regulatory networks has led to the creation of many databases of biological knowledge. Each of these databases attempts to distill the most salient features from incomplete, and at times flawed, knowledge. As an example, several databases exist that enumerate protein interactions for the yeast genome and have been compiled using the yeast two-hybrid screen. These databases currently document in excess of 80,000 putative protein–protein interactions; however, the knowledge content of these databases has only a small overlap, suggesting a strong dependence of the results on the procedures used and great variability in the criteria that were applied before an interaction could be entered in the corresponding knowledge repository. As one might have expected, the situation is less encouraging for those organisms with lower levels of direct interaction experimentation and scrutiny (e.g., *Escherichia coli*) or which possess larger protein interaction spaces (e.g., mouse and human); in such cases, the available databases capture only a minuscule fraction of the knowledge spectrum.

Carrying out the necessary measurements requires significant resources and expertise. Presently, the only broadly available tool for measuring gene expression is the DNA chip (in its various incarnations). Conducting a large-scale transcriptional experiment will incur unavoidable significant costs and require that the involved scientists be trained appropriately. Going a step further, in order to measure protein levels, protein states, regulatory elements, and metabolites, one needs access to complex and specialized equipment. Practicing systems biology will necessitate the creation of partnerships and the collaboration of faculty members across disciplines. Biologists, engineers, chemists, physicists, mathematicians, and computer scientists will need to learn to speak one another's language and to work together.

It is unlikely that a single/complex microarray experiment will shed light on the interactions that a practitioner seeks to understand. Even leaving aside the large amounts of available data and the presence of noise, many of the relevant interactions will simply not incur any large or direct transcriptional changes. And, of course, one should remain mindful of the fact that transcript levels do not necessarily correlate with protein levels, and that protein levels do not correlate well with activity level. The situation is accentuated further if one considers that both transcript and protein levels are under the control of agents such as microRNAs that were discovered only recently—the action of such agents may also vary temporally contributing to variations across repetitions of the same experiment.

Patience, patience, and patience: the hypotheses that are derived from systems-based approaches are more complex than before and disproportionately harder to validate. For a small system, it is possible to design experiments that will test a particular hypothesis. However, it is not obvious how this can be done when the system under consideration encompasses numerous molecular players. Typically, the experiments that have been designed to date strove to keep most parameters constant while studying the effect of a small number of changes introduced to the system in a controlled manner. This conventional approach will need to be reevaluated since now the number of involved parameters is dramatically higher and the demands on system controls may exceed the limits of present experimentation.

ABOUT THIS BOOK

From the above, it should be clear that the systems biology field comprises multifaceted research work across several disciplines. It is also hierarchical in nature with single molecules at one end of the hierarchy and complete, interacting organisms at the other. At each level of the hierarchy, one can distinguish "parts" or active agents with concrete static characteristics and dynamic behavior. The active agents form "relationships" by interacting among themselves within each level, but can also be involved in inter-level interactions (e.g., a transcription factor, which is an agent associated with the proteomic level, interacts at specific sites with the DNA molecule, an agent associated with the genomic level of the hierarchy).

Clearly, intimate knowledge and understanding of the specifics at each level will greatly facilitate the undertaking of systems biology activities. Experts are needed at all levels of the hierarchy who will continue to generate results with an eye toward the longer-term goal of the eventual mechanistic explanation of cellular processes and systems.

The two books that we have edited try to reflect the hierarchical nature of the problem as well as this need for experts. Each chapter is contributed by authors who have been active in the respective domains for many years and who have gone to great lengths to ensure that their presentations serve the following two purposes: first, they provide a very extensive overview of the domain's main activities by describing their own and their colleagues' research efforts; and second, they enumerate currently open questions that interested scientists should consider tackling. The chapters are organized into a "Genomics" and a "Networks, Models, and Applications" volume, and are presented in an order that corresponds roughly to a "bottom-up" traversal of the systems biology hierarchy.

The "Genomics" volume begins with a chapter on prebiotic chemistry on the primitive Earth. Written by Stanley Miller and James Cleaves, it explores and discusses several geochemically reasonable mechanisms that may have led to chemical self-organization and the origin of life. The second chapter is contributed by Antonio Lazcano and examines possible events that may have led to the appearance of encapsulated replicative systems, the evolution of the genetic code, and protein synthesis. In the third chapter, Granger Sutton and Ian Dew present and discuss algorithmic techniques for the problem of fragment assembly which, combined with the shotgun approach to DNA sequencing, allowed for significant advances in the field of genomics. John Besemer and Mark Borodovsky review, in chapter 4, all of the major approaches in the development of gene-finding algorithms. In the fifth chapter, Temple Smith, through a personal account, covers approximately twenty years of work in biological sequence alignment algorithms that culminated in the development of the Smith–Waterman algorithm. In chapter 6, Michael Galperin and Eugene Koonin discuss the state of the art in the field of functional annotation of complete genomes and review the challenges that proteins of unknown function pose for systems biology. The state of the art of protein structure prediction is discussed by Jeffrey Skolnick and Yang Zhang in chapter 7, with an emphasis on knowledge-based comparative modeling and threading approaches. In chapter 8, Gary Stormo presents and discusses experimental and computational approaches that allow the determination of the specificity of a transcription factor and the discovery of regulatory sites in DNA. Michael Syvanen presents and discusses the phenomenon of horizontal gene transfer in chapter 9 and also presents computational questions that relate to the phenomenon. The first volume concludes with a chapter by John Mattick on non-protein-coding RNA and its involvement in regulatory networks that are responsible for the various developmental stages of multicellular organisms.

The "Networks, Models, and Applications" volume continues our ascent of the systems biology hierarchy. The first chapter, which is written by Cristian Ruse and John Yates III, introduces mass spectrometry and discusses its numerous uses as an analytical tool for the analysis of biological molecules. In chapter 2, Chris Floudas and Ho Ki Fung review mathematical modeling and optimization methods for the de novo design of peptides and proteins. Chapter 3, written by William Swope, Jed Pitera, and Robert Germain, describes molecular modeling and simulation techniques and their use in modeling and studying biological systems. In chapter 4, Glen Held, Gustavo Stolovitzky, and Yuhai Tu discuss methods that can be used to estimate the statistical significance of changes in the expression levels that are measured with the help of global expression assays. The state of the art in high-throughput technologies for interrogating cellular signaling networks is discussed in chapter 5 by Jason Papin, Erwin Gianchandani, and Shankar Subramaniam, who also examine schemes by which one can generate genotype–phenotype relationships given the available data. In chapter 6, Dimitrios Mastellos and John Lambris use the complement system as a platform to describe systems approaches that can help elucidate gene regulatory networks and innate immune pathway associations, and eventually develop effective therapeutics. Chapter 7, written by Sang Yup Lee, Dong-Yup Lee, Tae Yong Kim, Byung Hun Kim, and Sang Jun Lee, discusses how computational and "-omics" approaches can be combined in order to appropriately engineer "improved" versions of microbes for industrial applications. In chapter 8, Markus Herrgård and Bernhard Palsson discuss the design of metabolic and regulatory network models for complete genomes and their use in exploring the operational principles of biochemical networks. Raimond Winslow, Joseph Greenstein, and Patrick Helm review and discuss the current state of the art in the integrative modeling of the cardiovascular system in chapter 9. The volume concludes with a chapter on embryonic stem cells and their uses in testing and validating systems biology approaches, written by Andrew Thomson, Paul Robson, Huck Hui Ng, Hasan Otu, and Bing Lim.

The companion website for *Systems Biology* Volumes I and II provides color versions of several figures reproduced in black and white in print. Please refer to http://www.oup.com/us/sysbio to view these figures in color:

Volume I: Figures 7.5 and 7.6
Volume II: Figures 3.10, 5.1, 7.4 and 9.8

Systems Biology

Volume II: Networks, Models, and Applications

1

Mass Spectrometry in Systems Biology

Cristian I. Ruse & John R. Yates III

The cell is the fundamental entity of a living organism. Four main types of chemical compounds are essential for any living organism: amino acids, fatty acids, nucleic acids, and sugars [1]. Fatty acids form membranes (composed mainly of phospholipids) that in turn define the physical space of a cell and mediate its interaction with the environment. Hereditary information is stored in genes (made up by nucleic acids). Proteins, which constitute more than 50% of the dry weight of the cell, give the functional form of the genes. The cell draws energy from sugar that can subsequently be stored both in fatty acids and polysaccharides.

Omics technologies aim to indiscriminately analyze and characterize genes (genomics/transcriptomics), proteins (proteomics), and metabolites (metabolomics). Systems biology uses these large-scale strategies to portray interactions among genes, gene products, and molecules involved in interactions with their environment (i.e., metabolites, lipids, carbohydrates, hormones, etc.). As the biological system reaches another level of complexity by assembly of multicellular systems into organs, metabonomics links the high-throughput technologies with tissue histology by analyzing biological fluids and tissues [2].

Understanding how these components assemble in a living cell (or any other physiological entity), communicate, and function requires the ability to formulate and test hypotheses. Analytical chemistry provides the measurement tools and computational modeling supplies the necessary framework to produce knowledge from the integration of these measurements [3]. Systems biology analysis integrates experimental data with a computational model; it uses the model to test how predictions fit experimental observations and eventually allows for generation of new hypotheses [4].

Mass spectrometry (MS), one of many tools used in analytical chemistry, has the ability to perform unbiased global analysis of the structure and the quantity of almost all molecular components of a biological system [5]. The goal of a large-scale MS experiment is to provide input data for any type of computational model in systems biology, from abstract (high-level) models to very detailed (low-level) ones [3]. MS can systematically analyze compounds in all three major subdomains of systems biology: genomics, proteomics, and metabolomics (including its branches lipidomics, glycomics, etc.). Proteomics (analysis and characterization of large collections of proteins) analysis by MS is by far the most advanced [6].

In this chapter, we will outline the basic principles of MS. We will focus on analytical contribution of MS to proteomics and to a lesser extent to lipidomics. State-of-the-art technology is available for proteomics and is emerging for lipidomics. Proteomics and lipidomics experiments employ different facets of analytical methods that are coupled with or based on MS. Integration of large-scale MS data for proteomics, particularly with mRNA expression profiles generated from microarray studies, and from some lipidomics studies, will be highlighted. Finally, we will see the promise of MS for systems biology.

BASIC PRINCIPLES OF MASS SPECTROMETRY

Mass spectrometers are composed of three major units: an ion source, a mass analyzer, and a detector. Together, these three units allow mass spectrometers to measure a fundamental physicochemical property of a molecule: the mass-to-charge (m/z) ratio of its gas-phase ions.

Ionization

Biological sample introduction to a mass spectrometer requires production of volatile, gas-phase ions of biomolecules. Biomolecular MS relies mostly on two processes for robust and reasonably unbiased ionization of bioanalytes: electrospray ionizaton (ESI) and matrix-assisted laser desorption ionization (MALDI). Underscoring the importance of soft ionization processes, discovery of ESI and MALDI was recognized with the Nobel Prize for Chemistry in 2002 [7,8].

In ESI, a high electric field (2–5 kV) applied between the tip of the capillary (anode) and the inlet of the mass spectrometer (cathode) forms electrically charged liquid droplets that are then further evaporated. Gas-phase ions are then transferred to the mass spectrometer down both a potential and a pressure gradient. A significant advantage of ESI is the ease of interfacing separation techniques with mass spectrometers.

In MALDI, biomolecules are cocrystallized with an energy-absorbent substance (matrix) [9]. Matrix molecules rapidly absorb the energy of irradiation (from laser pulses) and promote vaporization of matrix–analyte complexes followed by ionization of the analytes. For both MALDI and ESI, the efficiency of the ionization process is an important determinant of the sensitivity of detection, that is, the response of the instrument for a given concentration of analyte.

Peptides are readily ionized in acidic buffers. Depending on the polarity of the molecule one could use either positive or negative ionization (e.g., anionic lipids, peptides carrying negative charges—phosphopeptides, etc.). Ionization of nonpolar compounds could be promoted by formation of metal clusters (e.g., lithium adducts of triacylglycerols) [10]. In global analysis of lipids, sensitivity of analysis depends on the concentration of lipids; lipids tend to form aggregates at higher concentration, therefore linearity of the instrument response is obtained only at low concentration of lipids [10]. For extremely hydrophilic compounds, such as carbohydrates, chemical derivatization might be used for better ionization [11].

Mass Spectrometers

Several types of mass analyzers are employed in analysis of bioanalytes, including time-of-flight, quadrupole, quadrupole ion trap, and ion cyclotron resonance trap. We will briefly introduce these mass analyzers below.

Three parameters characterize the performance of the mass analyzer [5]:

- *resolution* or the ratio of m/z value of a bioanalyte to the width of the m/z peak, that is, $m/z/\Delta m/z$;
- *mass accuracy*, which defines how close the measurement of the mass is to the actual molecular weight of the bioanalyte (expressed in parts per million, ppm); and
- *scan speed* or how fast a mass spectrometer collects data.

Time-of-flight (TOF) analyzer determines the m/z of an ion from its measured flight time in field-free region. Ionized molecules receive a predetermined amount of kinetic energy from an acceleration voltage, therefore kinetic energy = Uz, where U is the accelerating voltage and z is charge. Therefore, the velocity of an ion is given by $v = (2Uz/m)^{1/2}$, where m is the mass of the ionized molecule. In other words, for the same distance ions with low m/z will travel faster than ions with high m/z.

Sources that generate ions in repetitive pulses are the natural choice for TOF analyzers (for example, MALDI). Ideally, acceleration voltage would be applied to a population of ions with unique initial velocities per each m/z value. However, vaporization and ionization distribute initial kinetic energy on the ions, therefore resulting in different initial velocities of ions with the same m/z values. Ions with the same m/z value travel the field-free tube accelerated at different initial velocities and reach the detector over a time interval (Δt). Consequently, Δt eventually results in $\Delta m/z$, therefore affecting the resolution of mass measurement. By using longer flight tubes, one could minimize the effects of initial velocities. An alternative solution is to use a reflectron (figure 1.1a). A reflectron focuses ions of same m/z but different initial velocities by reversing their flight path, with lower initial velocities

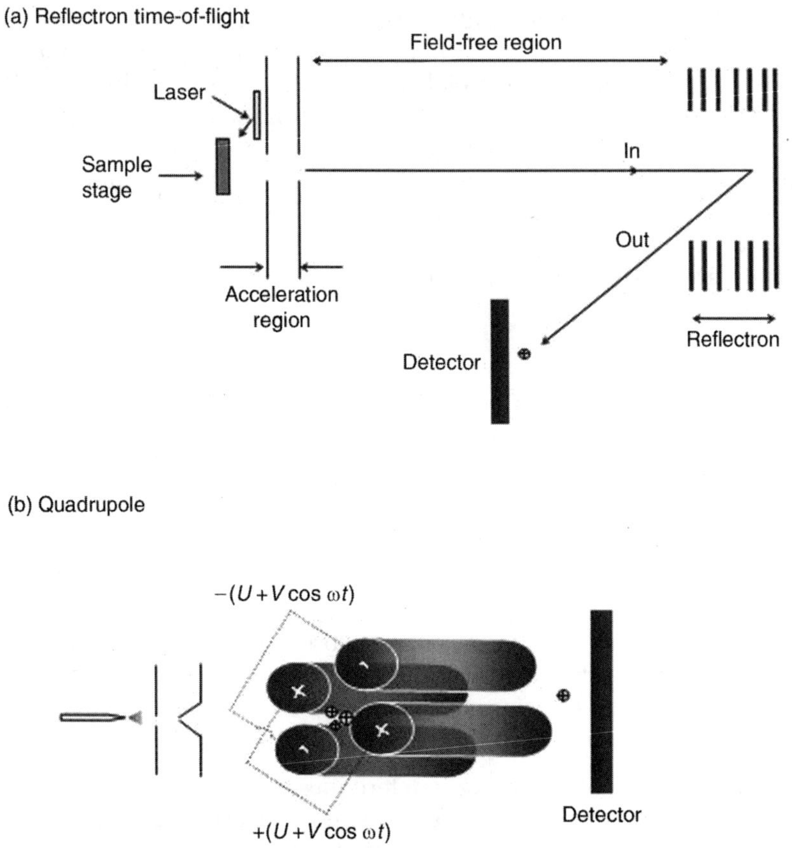

Figure 1.1 Schematic diagram of single-stage mass spectrometers built from basic mass analyzers: (a) Reflectron time-of-flight. A mass spectrum is generated by the time-of-flight of ions down a field-free tube. The reflectron mirror corrects for the spreading of energy values of ions. (b). Opposite electrodes have RF potentials of opposing polarity to form an electric field that constrains the motion of ions. Superimposed on the radio frequency fields is a direct DC potential. By varying the ratio of RF amplitudes and DC potentials, quadrupole can selectively analyze light ions (low m/z) or heavy ions (high m/z).

ions being reflected faster. Reflectron TOF resolution is 10,000 and its mass accuracy 10–20 ppm. Considering an accelerating voltage of +20 kV to +30 kV, a TOF analyzer can acquire an m/z range of 500–4000 in tens of microseconds, no scanning being required. However, acquisition rates also depend on the time required to produce ions as well as other factors.

Quadrupole mass analyzer is formed by four rods (figure 1.1b). An electric field with radio frequency and DC components is applied to one pair of rods. The second pair of rods has potential of

(c) Quadrupole (electric) ion trap

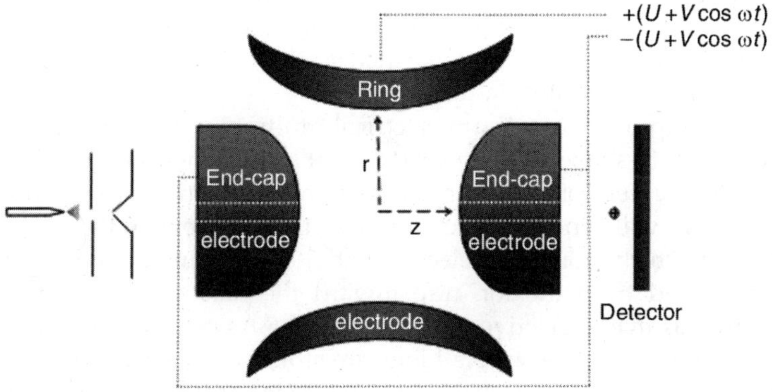

(d) ICR (magnetic) trap

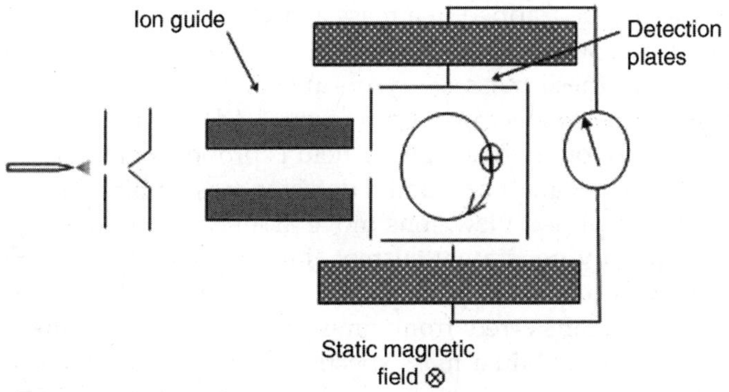

Figure 1.1 (*continued*) (c) Quadrupole ion trap constrains ions into the mass analyzer that are then ejected to a detector based on their m/z values. Essentially a single mass analyzer, ion trap can also function as a tandem instrument by successively concentrating, fragmenting, and analyzing ions (a process called tandem-in-time) in the physical mass analyzer. (d) In an ion cyclotron resonance cell, trapped ions rotate with a specific frequency that is proportional to their mass-to-charge ratio. (Reprinted, with permission, from the *Annual Review of Biophysics and Biomolecular Structure,* vol. 33, © 2004 Annual Reviews, www.annualreviews.org.)

opposite polarity. Quadrupole mass filter operates in a mass-selective stability mode: by varying the amplitudes of RF and DC potentials, quadrupole selects only the m/z value that will have a stable trajectory through the filter and arrive at the detector, while all the other m/z values will have unstable trajectories and be removed from the filter. Usually, quadrupole mass analyzer has a mass accuracy of 300 ppm.

It is worth noting that for quadrupole filters unit resolution is used rather that resolution. Unit resolution defines the ability of the quadrupole mass filter to resolve m/z signals that are 1 unit apart. Recent instrument developments have increased quadrupole mass accuracy to 5 ppm and its resolution up to 5000 [5].

Ion trap instruments are electrical multipoles, formed by rods and "doughnut"-shaped electrodes (toroidal arrangement) (figure 1.1c). The RF electric field of an ion trap constrains the motion of ions in the XY direction while end-cap electrodes constrain the ions in the Z direction. Upon perturbation of the electric field, ions with increasing m/z values are ejected from the ion trap toward the detector, in what became known as mass-selective instability mode. As opposed to quadrupole mass filter, in an ion trap all ions are stored except the ones with m/z value matching the ejecting potential. Mass accuracy and unit resolution are similar in quadrupole mass filters, 300 ppm and 2000. However, because of highly efficient use of all ions produced, ion traps are more sensitive than quadrupole mass filters.

Ions can also be trapped in a mass analyzer by static high magnetic fields. Ions trapped within a high magnetic field assume a cyclotronic motion giving these mass analyzers their name, ion cyclotron resonance (ICR) mass spectrometers (figure 1.1d). In ICR, the rotation frequency of an ion in the magnetic field is proportional to its mass-to-charge ratio. Mass analyzers and a detector are combined in ICR trap. In a highly simplified view, ions move at a specific frequency in ICR and consequently generate a current (image current) that is detected over time by the detection plates (electrodes) of ICR cells. The recorded signal is then transferred from time to frequency domain (Fourier transform, FT) to obtain a mass spectrum, since ion frequency is proportional to m/z. FT-ICR MS has high mass accuracy (1–2 ppm) and high resolution (>100,000).

Mass-to-charge ratios can be resolved and measured accurately but they do not describe the covalent structure of biomolecules. Mass spectrometers were developed to measure both molecular weight and covalent structure using the following sequence of events: (i) mass measurement of molecular ions, (ii) fragmentation reaction, and (iii) mass measurement of fragments of molecular ions. This sequence of events describes the operation of tandem mass spectrometer (MS/MS) that is commonly used in the structural analysis of biomolecules.

One example of a tandem mass spectrometer is the serial arrangement of two quadrupole mass filters interspaced by a collision cell, also known as a triple quadrupole system (figure 1.2a). Mass measurement in the first mass filter is followed by collision-activated dissociation (CAD) with noble gases such as argon or helium and analysis of the fragment ions in the second mass analyzer. These devices can also be used for highly specific and accurate quantitation of ions by monitoring

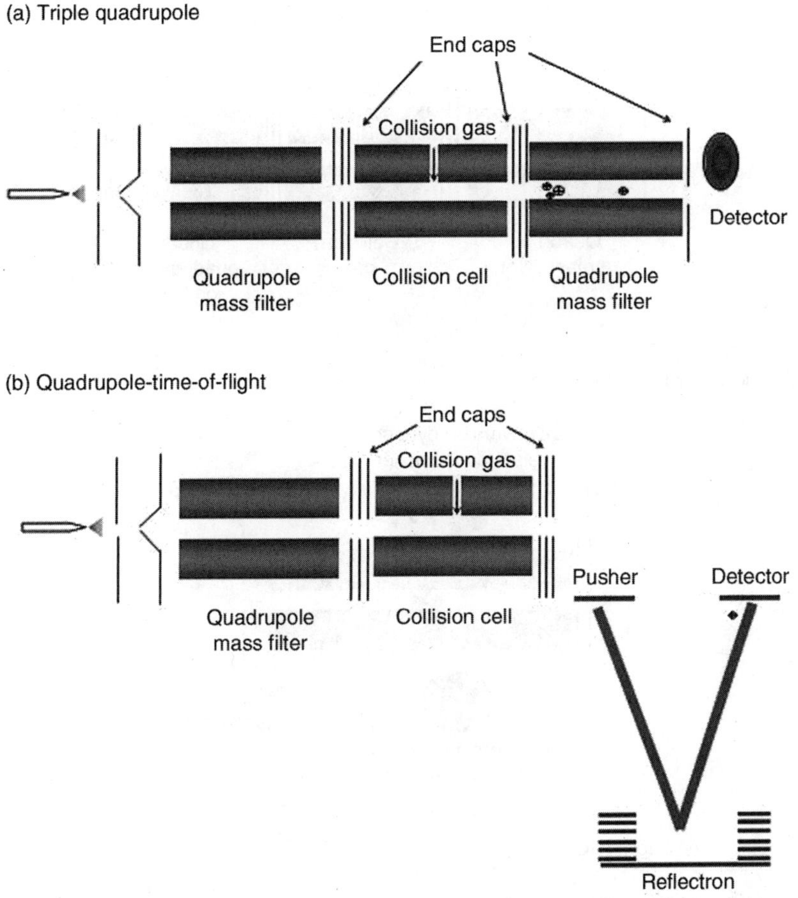

Figure 1.2 Schematic diagram of multistage mass spectrometers built from different type of mass analyzers. (a) Triple quadrupole. (b) Q-TOF. (Reprinted, with permission, from the *Annual Review of Biophysics and Biomolecular Structure*, vol. 33, © 2004 Annual Reviews, www.annualreviews.org.)

both the precursor ion and the production of one fragment ion (multiple ion reaction monitoring). Triple quadrupoles are often used in metabolomics.

A quadruople-time-of-flight (Q-TOF) (figure 1.2b) uses the configuration of triple quadrupole in which the third segment is a TOF analyzer rather than a quadrupole, therefore increasing both resolution and mass accuracy [12]. In a quadrupole linear ion trap (figure 1.2c), the TOF analyzer is replaced by a linear ion trap [13]. Sequencing experiments (MS/MS) take place in the linear ion trap. This instrument combines selectivity of the quadrupole mass filter with high-sensitivity analysis by the linear ion trap.

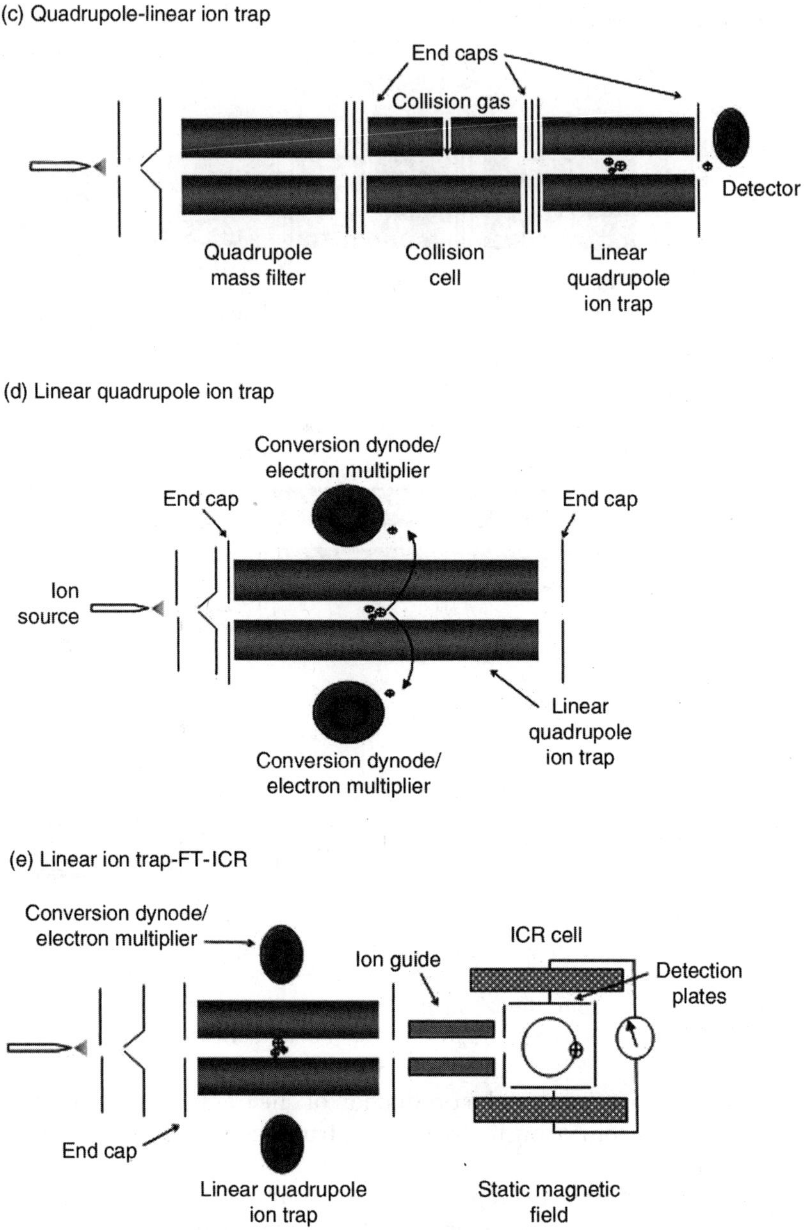

Figure 1.2 (*continued*) (c) Quadrupole linear trap. (d) Linear quadrupole ion trap. (e) Linear ion trap-Fourier transform mass spectrometer.

A simple yet powerful setup is the linear ion trap (LIT) as a stand-alone instrument (figure 1.2d). The end-cap electrodes control the injection of ions into the linear ion trap and when ions are ejected from the trap, they exit to detectors on both sides of the trap. A benefit to this configuration is that there is no loss of ions during detection. Linear ion traps are fast-scanning instruments and can acquire at a rate of 5 spectra/s [14]. MS/MS is performed in the same physical location in the following sequence: isolate the ion of interest, perform CAD (fragmentation), and measure the products of the reaction. In general, ion traps can perform this cycle (isolation, fragmentation, and mass analysis) for many (n) stages (MS^n).

Fourier transform mass spectrometry (FTMS) is unparalleled in its mass measurement accuracy and resolution, but is not very effective at performing multistage mass spectrometry experiments. However, hybrid FTMS instruments that combine a linear ion trap as the first mass analyzer for FTMS allows MS/MS experiments to be performed at high scanning speed (in linear ion trap) in advance of high mass accuracy and resolution measurements (figure 1.2e).

Operation of mass spectrometers is directed by an instrument control system. For example, the instrument control system monitors mass analyzers, detectors, and a portion of ion source under proper vacuum conditions. Besides maintaining operational parameters, the software for instrument control is the interface for designing unique mass spectrometry experiments. One such example is data-dependent acquisition, illustrated in figure 1.3. The figure legend contains detailed information about the instrument routine.

Most of the current large-scale analysis of peptides is based on data-dependent acquisition performed under the control of a computer, where measurement of m/z of biomolecules (full scan mass spectrum) is followed by tandem mass spectra of the most abundant ions from full scan MS. However, beside data-dependent acquisition, more specialized experiments (i.e., precursor ion scan, product ion scan, neutral loss ion scan, presented in refs. [15] and [16]) describe fine structural properties of biomolecules. For example, precursor ion scanning has been used for profiling classes of lipids [10].

Analysis of biological samples by large-scale MS experiments represents a compromise between the acquisition rate of the instrument (scan speed) and how informative the collected data are. To summarize, we extract information from an experiment only under proper sensitivity conditions (see previous section) and appropriate resolution and mass accuracy. Moreover, in analyzing a mixture of compounds, fragmentation spectra are necessary for complete identification. When designing an experiment for comprehensive characterization of a biological entity, the choice of one instrument over another should take into consideration how any of the above factors will influence the

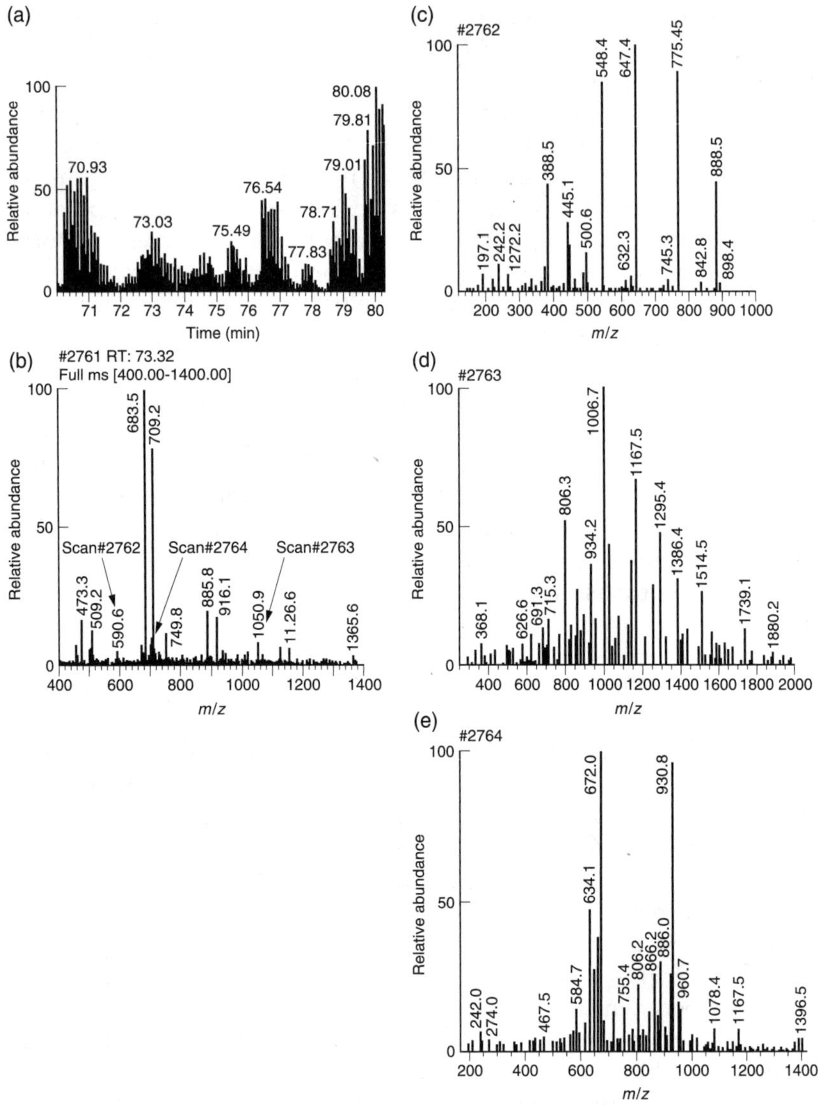

Figure 1.3 Data-dependent analysis of a standard protein mixture digested with three different proteases (illustration adapted from ref. [16] with data from authors' laboratory). The panel presents a time window of 10 minutes of an LC-MS/MS analysis (a). The chromatogram has a jagged appearance as the instrument alternates between recording a full scan MS and three consecutive MS/MS scans. Full scan MS spectrum at retention time 73.32 (on chromatography experiments a minute is divided into 100 equal seconds) is recorded as scan # 2761 (b). A range of 1000 m/z units is surveyed, from m/z 400 to m/z 1400. The three most abundant ions are selected for further MS/MS analysis. Once an ion has been selected for MS/MS fragmentation, it is

expected outcome. This chapter is centered on one aspect of large-scale analysis by MS, tandem MS or sequencing of biomolecules by MS/MS.

Sequencing of Bioanalytes by Mass Spectrometry

Mass measurement alone does not elucidate the covalent structure of ions, particularly for peptides and proteins, thus complete physicochemical characterization of a bioanalyte requires additional analysis. Tandem MS analyzes complex mixtures of biomolecules by first selecting a precursor ion for fragmentation to obtain an MS/MS spectrum that contains a specific signature (pattern) for each biomolecule. We compiled below the nomenclature and salient features of MS/MS (sequencing) spectra of peptides and lipids. Sequencing spectra together with appropriate multistage MS experiments could describe the structure as well as the profile changes of biomolecules.

MS/MS Spectra of Peptides. Fragmentation of peptide bonds produces two major types of ions: b ions (charge on peptide N-terminus) and y ions (charge on peptide C-terminus) [17]. MS/MS spectra of peptides are readouts of their amino acid sequence. The appearance of a peptide MS/MS spectrum resembles a histogram, where the bars (fragment ions) often represent differences corresponding to the molecular weight of an amino acid. Database searching of theoretical spectra or library matching of previously collected MS/MS spectra identifies the peptides [18]. Large-scale analysis of proteins involves searching hundreds of thousands of MS/MS spectra. Consequently, results of database searches should be statistically evaluated in order to provide cutoff filters for protein identification [19,20].

Peptides are polymers of repeating monomeric units. Fragmentation along the backbone of the peptide produces a mass ladder, that is, a double fingerprint that sometimes is complemented with fragmentation of chemical units branching from the backbone (figure 1.4).

MS/MS Spectra of Lipids. Annotation of lipid MS/MS spectra is given by XX:Y (XX is the total number of carbons of the fatty acids, Y being

---

included on an exclusion list for limited duration. Therefore, after evaluating scan #2761, the instrument proceeds with MS/MS sequencing of the ions that are not already in the exclusion list. Consequently, the fragmentation spectra of m/z 509.2, 1050.9, and 701.3 are recorded in scan #2762 (c), 2763 (d), and 2764 (e), respectively. Starting with scan #2765 the instrument repeats the cycle of acquisition. Using the Sequest database search algorithm, the following peptide sequences were identified: #2762 KIKVYLPR (doubly charged) from ovalbumin, #2763 LKEC^{am}C^{am}DKPLLEKSHC^{am}IA (doubly charged) from albumin, #2764 LKEC^{am}C^{am}DKPLLEKSHC^{am}IA (triply charged) from albumin. Cysteines present in the albumin peptide are carbamidomethylated.

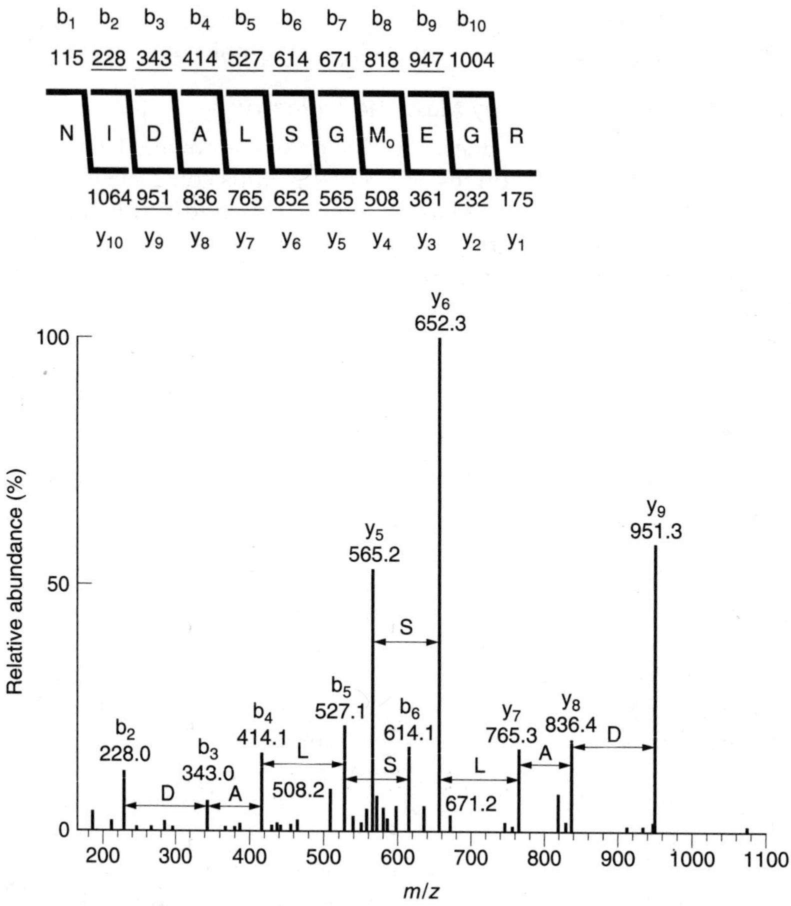

Figure 1.4 Fragmentation spectrum of a peptide is a double fingerprint: peptide sequence can be read from the amino or carboxy terminus. Fragment ions (vertical bars) define differences in amino acids masses that make up the peptide sequence. As an example, a peptide from human cardiac troponin I is presented. The upper part of the figure summarizes in a mass ladder the information derived from the MS/MS spectrum. Highlighted on the MS/MS spectrum are fragment ions showing a difference equal to the molecular weight of an amino acid.

the number of double bonds) followed by description of the headgroup [21]. Generally, MS/MS spectra of lipids are less complex than those of peptides [22]. The presence of ions corresponding to a lipid head group, easily highlighted in an MS/MS spectrum, together with appropriate scanning routine is sufficient to profile classes of lipids [23]. However, lipids identification by MS/MS sometimes requires synthetic analogs for complete structural characterization. Identification by comparison to a MS/MS spectrum of a pure synthetic analog is also applied to metabolites analysis because of their considerable chemical diversity [24].

Elucidation of bioanalyte structure to fine details, particularly when comparing differences in structure, might require more than MS/MS fragmentation. Recent developments in mass spectrometry instrumentation have introduced routine analysis of bioanalytes structures by multiple stages (e.g., MS^3 and higher) of fragmentation in combination with neutral loss detection in a data-dependent manner (see Mass spectrometers section above).

Separation of Bioanalytes

Separation of bioanalytes prior to presentation to mass spectrometry analysis is a key element for global analysis of biomolecules regardless of their nature: proteins, peptides, nucleic acids, lipids, carbohydrates, pharmaceuticals, organic molecules, and so on. Efficient separation is required before presenting biomolecules to the mass spectrometer. Application of MS to biological molecules was facilitated by the development of separation techniques, particularly of capillary chromatography interfaced online with ESI that reduced the flow rate of analysis and increased its sensitivity. High-resolution separations of complex biological mixtures are important to decrease ion suppression during ionization and to increase dynamic range. For peptides, capillary liquid chromatography (LC) simplifies complicated mixtures before they reach the ionization source. Alternatively, for lower molecular weight compounds (i.e., metabolites), capillary gas chromatography (GC) separates complex mixtures if bioanalytes are volatile or can be derivatized to volatile compounds. GC coupled online with MS (GC-MS) is the mainstay of metabolomics analysis. In this section, we will emphasize LC applications to large-scale analysis of peptides for proteomics.

Proteomics is a biological assay [25], where the identification of proteins is made either by reconstituting sequence information from shotgun analysis of peptides (bottom-up approach) [26–28] or by direct molecular weight measurements and subsequent fragmentation of proteins (top-down approach) [29–31].

In bottom-up proteomics, proteolytically digested proteins (either in solution or in gel) are sequenced by tandem mass spectrometry. Samples are loaded on capillary columns interfaced online with a mass spectrometer. A pressure vessel is used to load volumes of up to 1 ml in the capillary columns. Microcolumn/capillary chromatography is the workhorse of current research in proteomics (the inner diameter of a capillary column is 75–100 μm) [32]. The reader is referred to ref. [33] for a comprehensive tutorial on building and assembling nanoflow-HPLC columns and micro-ESI as well as for sample loading on capillary columns.

MS analysis of complex peptide mixtures can be improved by combinations of orthogonal chromatographic phases (i.e., based on complementary selectivity for peptides) in which each fraction from the upstream column is further separated on the downstream column [26,34,35]. The most common arrangement includes separation by ion exchange on the upstream column followed by analysis of each fraction in the reversed-phase downstream (analytical) column. Online integration of multidimensional chromatography and tandem mass spectrometry (figure 1.5) provides a robust automated platform for global proteome analysis by shotgun sequencing [27,28,33]. Together with the Sequest database searching algorithm, online multidimensional chromatography-tandem mass spectrometry forms the core of multidimensional protein identification technology (MudPIT) platform (figure 1.5).

Shotgun sequencing has been performed using one-dimensional chromatography together with high-accuracy measurement of mass (by FTICR) to assign accurate mass tags for peptides [36]. In subsequent analyses, accurate mass tag and retention times during 1D chromatography of the peptide form a unique identifier which can then be used for large-scale proteomic experiments [37].

More narrowly focused proteomic approaches are being used to monitor subpopulations of peptides with distinct physicochemical properties. Most of these approaches use a number of affinity methods in a preparative rather than online analytical fashion. The most notable examples are phosphopeptides enriched by immobilized metal affinity chromatography (IMAC) and glycosylated peptides isolated by lectin chromatography. Amino acid specific selection has been demonstrated by diagonal chromatography where a sample is run twice: before and after a chemical reaction specific to an amino acid of interest [38]. However, most of these methods focus on certain classes of peptides while discarding the rest of the sample. Some combination of affinity techniques with multidimensional separations will eventually provide an integrated analysis of specific modified peptides in the context of their parent proteins.

LC techniques offer a wide variety of choices for large-scale peptide analysis, but only few tailored methods for specific classes of lipids. Gross and colleagues introduced intrasource separation by using the electrical properties of lipids. In this method, the polarity of ESI (positive/negative ion) has been used to separately analyze classes of lipids that may ionize preferentially or with stronger signals as positive ions (neutral polar species) or negative ions (anionic species) [10,23]. 2D ESI has also been applied to multidimensional analysis of phosphopeptides, where the negative ion polarity is used for improved detection of phosphopeptides that are further sequenced by MS/MS in positive ion polarity [39].

Capillary electrophoresis (CE) separates bioanalytes based on size-to-charge ratio and constitutes yet one more possible dimension of

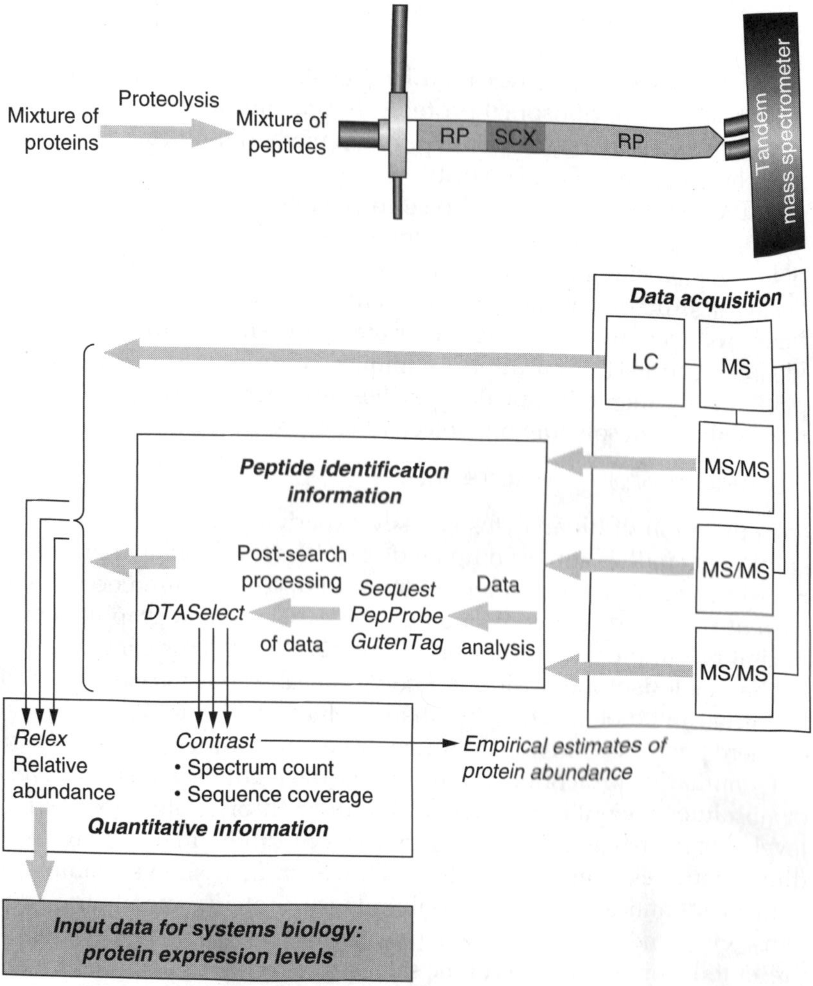

Figure 1.5 MudPIT platform is constituted of three major units: (i) sample preparation, (ii) online liquid chromatography-tandem mass spectrometry, and (iii) data analysis. (i) Protein samples are digested chemically with CNBr and specific proteases (Endo-Lys C, trypsin) [27]. Alternatively, samples are digested by a combination of nonspecific and specific proteases (elastase, subtilisin, trypsin) and then pooled for LC-MS/MS analysis [28] or by a single nonspecific protease (proteinase K) [80]. Using nonspecific proteases increases the sequence coverage of posttranslationally modified regions of proteins [28, 80]. (ii) The sample is then loaded onto a narrow capillary column and peptides are alternatively eluted from SCX and RP phases [26,27]. Peptides are ionized by ESI and introduced in a tandem mass spectrometer. (iii) Data analysis is performed by a suite of software packages. Peptides are identified by Sequest [18], PepProbe [67], or GutenTag [122]. Peptide information is assembled in protein identification [68]. Using DTASelect [78], results are further inspected based on criteria of interest. Quantitative experiments using stable isotopically labeled samples use Relex software for obtaining the relative protein abundances [59,60]. Statistical modeling of the data [57] or experimental parameters [77] can be used to empirically estimate amounts of proteins without stable isotope labeling.

separation in proteomics research. Tong et al. [40] interfaced RP and CE for the analysis of ribosomal proteins. Beside multidimensionality, CE might be used as a stand-alone separation method for metabolites, possibly by providing faster separation than GC.

GC-MS is the method of choice for analysis of metabolites, either with a quadrupole or a TOF detector. An attractive feature of GC is the ability to predict relative retention time of bioanalytes. For a predefined chemical structure, mass measurement (m/z) and relative retention time produced by a GC-MS experiment are enough for compound identification [41]. Separation techniques contribute to MS analytical platforms by increasing peak capacities, in other words by increasing the number of resolvable biomolecular components.

QUANTIFICATION OF BIOANALYTES BY MS

Quantification of bioanalytes is a key experiment in systems biology studies. Quantification of compounds by MS is traditionally performed by using a stable isotope internal standard to correct differences in ion current profiles due to ionization efficiencies, chromatographic reproducibility, sample handling errors, or matrix effects. If the concentration of the stable isotope analog is known, an absolute measurement of quantity can be obtained rather than a relative measure. We will present here predominantly quantitative proteomics studies.

Quantitative proteomics follows either the relative expression levels or absolute amounts of proteins. Quantitation of relative expression levels for proteome analysis has its roots in differential display two-dimensional electrophoresis (2DE). However, 2DE shows limitations both in identification (e.g., low-abundance proteins, membrane proteins, etc.) and in quantitation (i.e., posttranslational modifications presented as multiple spots of the same protein) experiments. Solution-phase proteomics methods using LC-MS/MS are more sensitive and specific than traditional 2DE separations, but relative quantities cannot be achieved without some form of stable isotope labeling or well-controlled chromatography (see below).

Bottom-up proteomics emerged as the method of choice for analysis of relative expression levels of proteins by stitching quantitative information at the peptide level back to the parent proteins. Two different strategies offer high-throughput capabilities for quantification of peptide levels by mass spectrometry (i) isotope labeling and (ii) stable isotope free.

Isotope labeling methods are composed of a growing number of variants. When manipulation of the biological system is possible, metabolic labeling provides an accurate measurement of relative quantities of peptides by one of the following strategies:

1. Incorporation of ^{15}N into all amino acids [42,43]. ^{15}N incorporation has been applied to organism level by metabolic labeling of *C. elegans* [44] and rats [45].

2. Enrichment of amino acids with ^{13}C [46].
3. Incorporation of only a select number of isotopically enriched amino acids in cell culture, SILAC [47].

Sometimes, manipulation of biological samples is not possible. In such situations, chemical labeling is an alternative. There are a number of possible chemistries for stable isotope labeling of peptides/proteins (for review see ref. [48]). Developed by Aebersold and colleagues, ICAT (isotope-coded affinity tag reagents) is one of the most successful strategies for chemical labeling of peptides [49]. ICAT reagent contains three units: an affinity tag (usually biotin), a linker (that is isotope-coded), and a thiol-specific reactive group. A variety of ICAT reagents have been developed over the years (for review see ref. [50]). ICAT strategies have a particular advantage, which is that it simplifies the protein digestion mixture by affinity enrichment of a subpopulation of peptides. However, because most of the ICAT reagent targets only cysteine residues, its applicability is limited to cysteine-containing peptides. As such, identification of posttranslational modified regions of proteins is reduced to only those present in cysteine-containing peptides [51].

One of the most stringent requirements for quantitative analyses is that the standard compound is to be added at the exact starting point of the experiment. Therefore, requirements for quantitative analysis of a whole class of biological samples, in particular human tissue, emphasize the need for quantitative proteomics without stable isotope labeling. Proof of principle experiments have been performed for profiling of protein complexes [52] and quantification relative to a native reference peptide (from the protein of interest) of site-specific phosphorylation [53]. Western blotted quantities of nitrated proteins paralleled quantities derived from site-specific nitration relative to native reference peptide [54]. Absolute amounts (mole of modifications/mole of protein) can be obtained by constructing a response curve with external standards. Alternatively, by spiking the protein solution with a native peptide (isotopically labeled) that is concomitantly processed by digestion, one could obtain absolute quantification of proteins levels while accounting for digestion efficiency [55]. Absolute quantitation can also be obtained by addition of an isotope label internal standard of the peptide of interest (already processed by proteolysis), followed by a single reaction monitoring experiment (one of the most sensitive MS routines for quantification) that detects both the analyte and its labeled standard (a method known as AQUA [56]). Practical considerations for protein profiling when performing large-scale analysis of peptides by MudPIT led Liu et al. [57] to statistically evaluate the sampling of peptide ions by data-dependent acquisition (as defined above). Their model showed that spectral sampling is an accurate estimate of relative protein abundance over two orders of magnitude.

Quantitative proteomics experiments produce input data for systems biology. From this prospective, protein expression levels should be statistically accounted for at less than two-fold difference in order to be useful in the description of a complex system [58]. Recently, a suite of methods using MudPIT as core technology has been developed to accurately quantify small differences in protein expression levels en masse, by using metabolically labeled organisms [45,59] and sophisticated software for quantitative proteomics [59,60].

Global quantitation of cellular lipids employs a different strategy as compared to stable isotope labeling. Because ESI efficiency depends mostly on lipid polar head, classes of lipids can be quantified by a single internal standard [23]. This represents a practical advantage as it would be close to impossible to synthesize stable-isotope internal standards for thousands of lipids. In spite of recent advances in lipidomics, methodological developments have to mature in widespread technologies. However, as we will see in the next section, proteomics technologies already fully benefit from the individual analytical aspects of both LC and MS as well as their use in concerted manner as LC-MS.

Platforms for Proteomics

Online assembly of LC and MS constitutes a continuum that generates multidimensional data: chromatospectrograms. Chromatospectrograms are the usual chromatographic profiles that embed mass spectral information, MS and/or MS/MS spectra. Combinations of chromatographic profiles and m/z intensities are used for extracting quantitative information without using stable isotope labeling [61].

Three parameters completely describe a biomolecule in an LC-MS experiment: a series of m/z values (present in MS and MS/MS spectra), its retention time (RT), and its corresponding elution profile (i.e., the area under the curve that is proportional with the amount). The data generated from an experiment can be re-inspected using any of these parameters and information about its value creates the opportunity for hypothesis-driven mass spectrometry. For example, a comparison of two states can highlight only those ions showing a change in abundance and those ions can then be analyzed using a tandem mass spectrometry experiment to determine the identity of the molecule. Instruments used for this strategy combine MALDI with quadrupole ion traps or Q-TOF detection. By its nature, MALDI would allow for an analysis to be repeated until the sample is consumed.

In hypothesis-driven mass spectrometry there are two stages of analysis: (a) survey of the sample in high-throughput manner to generate a hypothesis (usually by MALDI-MS or MS/MS) followed by (b) testing of the hypothesis with a different strategy (that could either use MALDI MS^n or LC-MS/MS). MALDI quadrupole ion trap allows for in-depth interrogation of proteomic samples based on the prediction of an m/z value of a precursor ion of low abundance [62]. Alternatively, MALDI–MS/MS could identify the presence of a neutral loss (for example, neutral loss of H_3PO_4 from Ser/Thr phosphorylated peptides). Then, one could employ targeted MS^3 for structural elucidation of the peptides showing the neutral loss [63]. Correspondingly, in an LC-MS/MS arrangement, data-dependent neutral loss on an ESI linear ion trap performs the same experiment for hypothesis-free analysis. Either evaluation of differential quantities of peptides, MS/MS spectra, or neutral loss scans can be used as criteria for hypothesis-driven mass spectrometry. Graber et al. [64] proposed an integrated workflow based on the use of MALDI to make decisions on follow-up experiments using ESI. The workflow relies on a relational database that organizes the acquired data for further result-driven mass spectrometry.

Recently, Venable et al. [59] developed a data-independent strategy for the acquisition of tandem mass spectrometry data for large-scale quantitative proteome analysis using fast scanning instruments like the linear ion trap. Data-independent tandem MS method indiscriminately collects MS/MS spectra without MS survey of the data as required by data-dependent acquisition.

Ultimately, the basic function of any proteomic platform (hyphothesis-driven or hypothesis free) is elucidation of structures of peptides as well as precise amino acid localization of posttranslational modifications as a function of cellular events. From these data, we can then reassemble the scaffold of proteins by describing their sequence coverage, modifications, and relative expression level, parameters that will eventually describe the biological system.

Table 1.1 summarizes the different proteomics platforms, separation techniques, and MS instrumentation used in papers cited in this chapter. The first three sections of the table summarize analytical techniques used for analysis of proteome and large-scale identifications of modified peptides (phosphorylated and/or other modifications). Analysis of modified peptides requires in some cases preparative protocols of protein biochemistry (column I). After proteolytic digestion of proteins, the peptide mixture is further simplified by preparative offline chromatography for peptide quantification or identification of modifications (column II). Column III summarizes the combinations of LC-MS used. For proteome analysis, MudPIT presents widespread applicability with no other sample preparation other than digestion of lysates or cellular organelles. In order for single LC coupled with MS/MS to analyze samples of the same complexity, one needs to rely either on extensive sample preparation (column I or II) and/or employ MS instruments with increased mass accuracy and resolving power (Q-TOF, FTICR). On the other hand, column IV emphasizes the role of ion trap MS as the workhorse of proteomics studies, particularly its scanning capabilities for large-scale analysis of peptides. Analysis of modified peptides

Table 1.1 High-throughput platforms and mass spectrometry instrumentation in systems biology

Separation of proteins[a]	Offline separation of peptides[b]	Platform	MS	References
Proteome				
		MudPIT	IT	26–28,33,43, 45,57,59,60,67, 68,71,76–78, 80,120,122
		LC-ESI-MS/MS	Q-TOF	91,92
		LC-ESI-MS/MS followed by LC-ESI-MS (AMT)	IT FTICR	37,72
		LC-ESI-MS/MS	IT	81
	SCX, avidin affinity (ICAT)	LC-ESI-MS/MS	IT	74,75,89
	CE	LC-MS/MS	IT	40
Phosphorylated Proteins				
		LC-ESI(–)-MS followed by LC-ESI(+)-MS/MS	TQ	39
	IMAC	LC-ESI-MS/MS	IT	97
	IMAC, SAX	LC-ESI-MS/MS	Q-TOF	103
	Avidin affinity	LC-ESI-MS/MS	IT, Q-TOF	104
		LC-ESI-MS/MS	IT	56
SDS-Page	IMAC	MALDI-MS followed by LC-ESI-MS/MS MALDI-MS/MS	MALDI-TOF Q-TOF MALDI-Q-TOF	103
SDS-Page	SCX	LC- ESI-MS/MS	IT	98
IP	IMAC	LC-ESI-MS/MS	IT	99
	IMAC	LC-ESI-MS/MS	IT	83
SDS-Page		LC-ESI-MS, LC-ESI-MS/MS	IT	53,54
	IMAC, SCX	LC-ESI-MS/MS	IT	119
		MALDI-MS/MS	MALDI-IT	62,63
Other Posttranslational Modifications of Proteins				
		MudPIT	IT	28,80,106
	RP,RP	LC-MS/MS	Q-TOF	38
Lectin affinity	AE, lectin affinity	LC-MS/MS	Q-TOF	108
IP	SCX	LC-ESI-MS/MS	IT	107
	SCX/avidin affinity	LC-ESI-MS^n	LIT	110

Table 1.1 *(continued)*

	Platform	MS	References
Profiling	MALDI-TOF/MS	MALDI-TOF	113,114
	SELDI-TOF/MS	SELDI-TOF	115–118
	LC-MS/MS	IT	119
Sample preparation[c]	**Platform**	**MS**	**References**
Metabolomics	GC-ESI/MS	TOF-MS	93
	ESI (+/−)-MS	TQ	10,22
	ESI-MS/MS	TQ	21

[a] Subcellular fractionation was not considered.
[b] Cleanup steps (like solid phase extraction) are not listed.
[c] Sample preparation for metabolomics includes the following steps: quenching, extraction, and concentration.
ESI(+/−), electrospray positive/negative ion mode; LIT, linear ion trap; MS^n, multiple stages MS/MS; IP, immunoprecipitation; IT, ion trap; RP, reversed-phase; SAX, strong anion exchange; SCX, strong cation exchange; SDS-PAGE, sodium dodecyl sulfate–polyacrylamide gel electrophoresis; TQ, triple quadrupole.

draws upon the complementary nature of ESI and MALDI for efficient ionization of modified peptides. Moreover, analysis of modified peptides requires increased information content through better mass accuracy and resolution of TOF instruments, specific detection of neutral loss ions by triple quadrupole as well as multiple stages of fragmentation by triggered neutral losses in LIT. Even so, ion trap is still the leading instrument whose scanning abilities are often enough to complement enrichment techniques and LC separation in identifying modified peptides.

The last two sections of table 1.1 (profiling and metabolomics) underscore the power of mass measurement as mostly single-stage MS is performed. Profiling techniques emphasize the utility of solid-phase support for certain clinical samples.

MASS SPECTROMETRY DATA IN SYSTEMS BIOLOGY

The availability of large data sets such as the global protein expression data, protein–protein interaction data, and localization data that are available for *S. cerevisiae* can be used to extend and integrate observations made from large-scale MS analysis of proteins and metabolites.

Results of large-scale analysis of protein collections provide a statistical estimate of the sample composition [65]. Large-scale analysis of peptides by MS-based platforms is no exception. In a global analysis of complex mixture of peptides, their MS/MS spectra are matched against peptide sequences in a protein database [18]. These matches have a calculated probability of being correct [66,67]. Then, considering the total number of spectra collected by MudPIT and the size of the database, a confidence score for protein identification is derived [68]. Clearly, bottom-up proteomic analysis necessitates complex statistical analysis of data after database matching processes per single biological sample. In contrast, for metabolome profiling, evaluation of chromatospectrogram traces is sometimes sufficient to distinguish between treated and untreated samples, often at the expense of identification of the metabolites of interest. Identification of metabolites of interest is then usually performed in a targeted analysis.

In this section, we will see how mass spectrometry data (i) can help describe a biological system and (ii) can be used in a correlative analysis of different classes of compounds for a broader understanding of biological system.

Description of the System

The first stage in describing a biological system is to define its components (i.e., nucleic acids, proteins, and metabolites). We present here MS utilization in proteomics and metabolomics because analyses of these data were followed by their correlation to data generated by other high-throughput assays.

PROTEOMICS

There are two major categories of proteomics: functional proteomics and interaction proteomics. In addition to the identification of proteins, both functional and interaction proteomics analyze either one or all of the three major parameters of proteins: the amount of proteins (quantity), the modification state of proteins, and subcellular distribution. However, functional versus interaction proteomics differs in at least two major ways: (i) sample preparation and (ii) in refining their specific set of data against other global investigations.

Many proteins organize in macromolecular complexes, transmit information through signaling pathways, and in doing so usually reside at a specific cellular location. Proteins found to interact (via MS) provide information on interaction and sometimes on localization. In the context of protein–protein interaction, intracellular membranes of various organelles are also docking structures where proteins can localize or transiently associate in order to perform their functions. Thus, analyzing proteins of specific organelles by mass spectrometry can provide a wealth of biological information. Functional proteomics can address specific, targeted proteins, follow changes of entire cellular pathways, or dissect cellular organelles for their bona fide proteins (for review see refs. [69] and [70]).

An example of integrated data from a targeted MS approach to help answer questions about specific proteins was demonstrated by Hazbun et al. [71]. This targeted approach to assign function to 100 uncharacterized ORFs included MS, classical biochemical techniques (affinity purification), cell biology protocols (two-hybrid analysis, fluorescence microscopy), and bioinformatics analysis (structure prediction, Gene Ontology annotation). Affinity purification followed by mass spectrometry emerged as one of the techniques capable of assigning function and biological processes to the uncharacterized ORFs [71]. In another example, integration of MS data from proteomic analysis of mitochondria with publicly available data sets and functional genomics methods led to construction of a predictive score of protein localization in mitochondria [72]. Both these studies formatted the results relative to Gene Ontology (GO) annotation [73].

Quantitative proteomic analysis using ICAT and MS complemented with traditional biochemical assays outlined changes in cellular pathways [74,75]. Depending on the biological question, one could filter these results to reveal novel functions of targeted proteins, as with Myc oncoprotein in Shiio et al. [74], or expand knowledge on cellular pathways by ascribing interactions of proteins showing changes in expression in response to an agonist, for example, the response of human liver cells to interferon [75].

Recently, tissue-specific internal standards have become available by metabolic labeling of rats [45]. Using this method, changes in proteins involved in xenobiotic metabolism and protein-folding machinery of the endoplasmic reticulum between animals treated with cycloheximide (a protein synthesis blocker) and controls were measured.

An alternative to relative quantitation by stable isotope labeling for shotgun proteomics is to use information available in MudPIT experiments for data clustering based on raw data, statistical modeling [57], or subtractive proteomics. Protein expression patterns from *Plasmodium falciparum* were clustered based on sequence coverage to distinguish among stage-specific proteins of malaria parasite [76]. Subtractive proteomics (which is the identification of proteins in certain samples or fractions) [77] identified previously unknown associations of nuclear membrane proteins with dystrophies [78].

LIPIDOMICS

Global analysis of complex mixtures of cellular lipids is the analytical goal of the lipidomics. Cellular lipids can be classified in three major groups: polar, nonpolar, and metabolites. Lipid metabolites are of significant interest when comparing disease versus control samples in pathological conditions [23]. In this section, we will discuss analytical MS techniques applied to a subgroup of polar lipids, phospholipids [79].

The plasma membrane controls the flow of information from and into the cell by transport of solutes and signaling mechanisms. Beside proteins, various concentrations and compositions of lipids confer versatility and plasticity to (i) membranous structures regulating the communication between the cell and its environment and (ii) membranes of different organelles. Phospholipids, major components of the membrane bilayer, could be described by two main classes, choline glycerophospholipids and ethanolamine glycerophospholipids, which usually localize on the inner and outer leaflet of the membrane respectively. Beside their structural role, phospholipids at the plasma membrane constitute a reserve of precursors for second messengers in cell signaling.

Analysis of plasma membrane structures illustrates one of the major challenges of omics technologies: sample preparation. In order to understand how membrane sheets integrate proteins and lipids in a symbiotic manner, we need to simultaneously extract and conserve these two classes of bioanalytes for further characterization by mass spectrometry. Two preparative methods hold the promise for an integrated proteomics–lipidomics analysis. Wu et al. developed a proteomics assay to characterize membrane proteins from nonsolubilized membrane samples [70,80]. This strategy produces two sets of bioanalytes: transmembrane domains embedded in lipid bilayer and the proteolytically digested soluble domains of membrane proteins. Potentially, lipids could also be analyzed as well as transmembrane domains of proteins. Such an approach is one possible way to integrate proteomics and lipidomics in one analysis. Chloroform/methanol extraction of lipids complemented by analysis of membrane proteins (the method proposed by Blonder et al. [81]) provides a different strategy. Though similar to the above-mentioned strategy, this approach will destroy the topology of the membrane and will hamper analysis of posttranslational modifications [70].

The Alliance for Cellular Signaling (AfCS) outlined a framework for global quantitative analysis of signaling pathways in B lymphocytes and cardiac myocytes [82]. Recently, AfCS released two data sets for one of their two model systems, the WHEI-231 B lymphoma cell line. Shu et al. [83] identified 193 phosphorylation sites in 107 proteins using IMAC followed by LC-MS/MS in B lymphoma cell line treated with calyculin A (a Ser/Thr phosphatase inhibitor). As part of the AfCS project, Forrester et al. [21], using a combination of mass spectrometry techniques (positive ion mode, negative ion mode, MS/MS), generated a time profile of lipids following the changes in glycerophospholipids upon stimulation with anti-IgM. It is worth noting that these authors relied on ESI source as the only separation tool for analysis of complex mixtures of lipids. Difficulties in coupling MS with separating techniques for lipids and lack of stable isotope labeled internal standards were compensated by laborious mathematical analysis. Full scan mass spectra for each time point were processed for data normalization much like in microarray data normalization. The authors then used control charts to extract statistically significant changes in the time profile of phospholipids. The results summarize the direction of change for each lipid (1 for increase in concentration, 0 for no change, and –1 for decrease in concentration). The authors speculated on monitoring the activity of phospholipases by quantitating the amounts of different classes of lipids over time.

While the plasma membrane is the physical barrier of the cell and coordinates its overall exchange of information with the extracellular milieu, localized regions in the membrane perform more specific functions by aggregating proteins and lipids in microdomains [84]. One such structure is lipid rafts. Various laboratories are beginning to carry out proteomics experiments targeting these lipid microdomains. For example, Foster et al. [85] used stable isotope labeling for quantitative proteomics to describe proteins specific to lipid rafts, based on the cholesterol dependence of composition of rafts.

In some instances, new technologies have enhanced the ability to analyze lipid raft proteomes. Development of a new proteomics platform, online LC-MALDI-MS/MS, combined the advantage of chromatographic separation with robustness of MALDI ionization [86]. A recent application investigated lipid rafts proteome based on SDS solubilization and analysis of raft proteins. Analysis of this sample would have been more challenging with an ESI platform because of the interference of SDS on ESI.

Lipidomics and proteomics are only two examples of large-scale analysis of biomolecules. Analytical technologies based on MS are widely used in proteomics and are rapidly developing for lipidomics. Collection of proteomic data sets for different biological points is only the first step in a complex experiment. Identification of proteins requires established database algorithms and statistical evaluation of identified proteins [19]. Quantitative proteomics provides analytical validated techniques for comparison of protein levels [45,74,75]. Alternatively, isotope-free quantitative techniques [57] and integration of proteomic data sets with annotated protein databases [73] could be used to provide new knowledge on the biological systems. Lipidomics data sets identify profiles of cellular lipids [23] and trends in concentration of lipids upon applying a stimulus [21]. Eventually, biological validation is required for results of both proteomics and lipidomics.

Integration of Data Sets

As mentioned in previous sections, integration of information (in this case mostly MS data) strengthens our knowledge of the biological system and lays the foundation for its systematic analysis. Generating new hypotheses involves prediction of how a system would respond to a certain perturbation. Experiments can then be performed by applying this very stimulus, monitoring responses, integrating data from different omics approaches, and comparing the outcome with the initial prediction in order to refine the hypothesis [4].

Eventually, computational modeling should monitor the system response to perturbation and output. Experiments should identify components that participate in biological process, quantify their amounts, and describe their interactions through perturbation, response, and output. In others words, we need to define each particular chemical structure (primary sequence for proteins), its quantity, and what portion of that chemical structure transmits information through interaction. High-throughput genomics and large-scale proteomics and metabolomics experiments are capable of performing these tasks and their data could be further processed to calculate the level of correlation of trends and changes in quantitaties across different data sets.

We summarize here some examples of correlative studies between mRNA and protein expression levels. Figure 1.6 illustrates the current state on integration of omics data from the prospective mass spectrometry. Two approaches characterize correlative studies of mRNA and protein expression levels. First and most stringent is the derivation of a correlation coefficient that is based on statistics of two global measurements: quantitative proteomics (provided by stable isotope analogs followed by mass spectrometry measurements) and microarray or SAGE experiments (an interesting discussion on integration of high throughput omics data is presented in ref. [87]). The second approach uses an RNA/protein concordance test based on detection of quantities of these two bioanalytes by mass spectrometry and microarrays. Eventually, in both cases, one uses a defined metric (for definition and applications of metrics see ref. [88]) to cluster mRNA/protein data in pathways, complexes, and so on, with the goal of extracting specific features of the system under analysis.

In one mRNA/protein correlative study, Ideker et al. [89] measured the perturbation applied to wild-type yeast and mutants by the absence of galactose. Quantitative proteomics (ICAT, MS) and microarray experiments showed a weak but significant correlation ($r = 0.61$) of mRNA/protein expression levels for 289 genes. Correlation of the observed gene products was superimposed to data describing protein–protein interactions in yeast [90]. In general, gene products known to interact from two-hybrid experiments showed increased correlation as compared to the whole data set.

Washburn et al. [43] compared yeast grown in rich and minimal media (enriched in ^{15}N). Quantitative MudPIT and oligonucleotide arrays provided data for mRNA/protein expression levels plotted on a log-log graph. Analysis of the graph showed a weak correlation

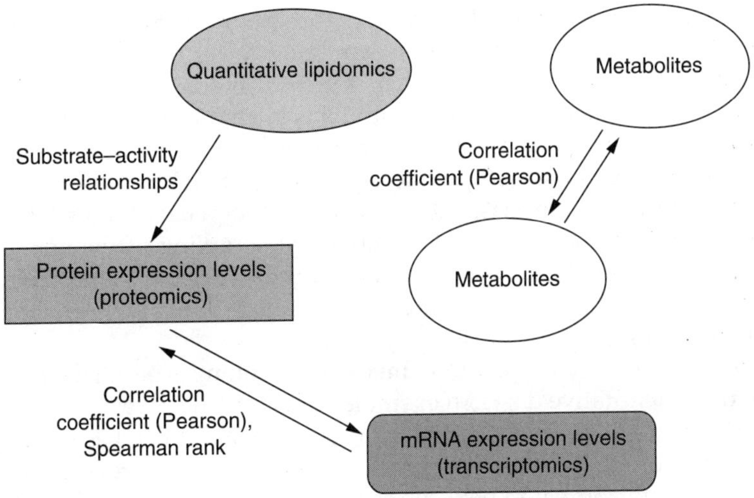

Figure 1.6 Mass spectrometry data in systems biology. Mass spectrometry is the technique of choice for identification and quantification of biological molecules except mRNA. Data from proteomics and transcriptomics experiments (usually performed by microarray) can be correlated using statistical coefficients [43,74] for a comprehensive description of the biological system. Statistical correlation of metabolite pairs (measured by mass spectrometry) is another way to follow biological perturbations [93]. Integration of data acquired at mRNA, protein, and metabolite level will involve high-throughput data generated by mass spectrometry and microarray experiments together with statistical methods for correlation of these data sets. Color code: dark gray, data are presented in an array format, ready for complex statistical analysis (i.e., clustering); light gray, data are presented in an array format that describes the trend in changing of concentrations, which might be potentially used for monitoring enzyme activites; white, data are derived from profiling experiments for further comparison with metabolites determined in the same set of experiments rather than comparison with other large omics data sets.

overall (Spearman rank 0.45) for 678 loci. Close examination of the log-log plot indicated analysis of data both by loci and by biological pathway. Better correlation has been found across loci involved in the same biological pathway.

In another study involving the integrated analysis of mitochondria, Mootha et al. [91] used a significance test rather than a correlation coefficient. In this study, genes at the transcript level (mRNA) from four tissues (brain, heart, kidney, and liver), that is, from multicellular environments, were compared with protein identification from proteomics experiments. Statistics of these two data sets (mRNAs and proteins) across four tissues tested the null hypothesis of no correlation between detection of protein and mRNA expression level and rejected it for 426 out of 569 gene pairs. Further statistical modeling estimated the specificity of mitochondrial gene products in all four tissues. While the results of this study complemented the atlas of mitochondrial proteins, the proteomics platform (based on single separation by reversed-phase chromatography) emphasized the need for adequately tailored multidimensional chromatography for unbiased sampling of the analytes. Furthermore, the authors moved beyond mitochondrial gene products and defined a metric (called the neighborhood index [92]) to explore coregulated genes.

In addition to proteomics, many metabolomics studies include relative quantitative data. An example of such a study was reported by Weckwerth et al. [93]. In this study, using a robust GC-TOF/MS system, more than 1000 metabolites were monitored by relative quantification and compared between three different plant lines [93]. Three parameters—m/z values, RT and chromatographic profile (that is, a chromatospectrogram as defined under Platforms for Proteomics), and in some cases MS/MS spectra of reference compounds—were used for identification and quantification of plant metabolites without stable isotope labeling. Processing of the data included deconvolution of peaks, peak identification, and statistics of metabolite peaks. Based on a large pool of metabolite pairs showing medium ($r > 0.60$) to strong correlation ($r > 0.80$), the authors were able to distinguish between different plant lines or phenotypes.

These studies clearly outline the potential for integration of MS data into systems biology approaches that include mass spectrometry data. Washburn et al. [43] proved that integrative approaches can be successful in building knowledge from protein pathways. Integration of individual protein pathways (i.e., signaling, trafficking, etc.) in a computational model is limited only by our ability to test them experimentally [94]. In this framework, signaling pathways detach themselves as realistic and intermediate steps in modeling a whole biological system.

Signaling Pathways

Cellular signaling usually translates information from extracellular stimuli to the cell nucleus through specific pathways. Compensation of signaling pathways (through interaction or crosstalk) defines the biological response of the cell [1]. Proteins assemble in dynamic macromolecular complexes regulating the specificity of individual signaling pathways (i.e., usually on the same scaffolding protein). Protein phosphorylation plays a prominent role in many pathways. Cascades of kinase phosphorylation and phosphatase dephosphorylation act as on/off switches in formation and dissociation of protein–protein interaction and in functional and enzymatic activity and consequently in signal transduction. Therefore, global analysis of phosphorylation along a signaling pathway or of the whole cell is of paramount importance for understanding the cellular response to particular signals. While many methodologies (Western blot, Edman sequencing, etc.) have contributed to our knowledge of protein phosphorylation, clearly mass spectrometry approaches compressed the time associated with discovery and characterization of novel phosphorylation events. Collectively, mass spectrometry platforms for analysis of global protein phosphorylation are grouped under the name phosphoproteomics [95,96].

The ubiquitous role of phosphorylation in cell signaling does not equate in highly abundant phosphoprotein species. In contrast, infinitesimal amounts of phosphorylation per site can confer the initiation of a signaling event and often involves proteins relatively low in abundance. Global analysis of phosphorylation sites in a sea of more abundant proteins requires highly specific tools and enrichment methods (see Separation of bioanalytes, above). Ficarro et al. [97] provided proof of principle for the global analysis of phosphorylation sites in yeast, by using IMAC for the enrichment of phosphopeptides followed by analysis by LC-MS/MS. Recently, a HeLa cell nuclear phosphoprotein analysis identified about 1700 phosphopeptides using a different chromatography method to enrich for phosphopeptides [98]. In both cases, the amount of sample used was practically unlimited (replenished from cell culture) and very few species identified were multiply phosphorylated peptides. Data analysis of these large collections of phosphopeptides still remains a challenge, particularly for Ser/Thr-containing phosphopeptides.

A combination of IMAC and immunoprecipitation allowed identification of about 60 tyrosine phosphorylated sites (either from Gleevec/STI571 treatment of chronic myelogenous leukemia cells or T-cell activation) and followed some of the identified species through three time points [99].

Description of site-specific protein phosphorylation has benefited over the years from computational tools able to predict both phosphorylation sites and the kinases responsible [100,101]. Some investigators have attempted to use both mass spectrometry and prediction algorithms in one experimental platform. In such a pl... , Hjerrild et al. [102] described four potential substrates of c...-dependent protein kinase A by using algorithm prediction ... /MS. However, the specificity level of neural networks ... n was low (41%) and phosphorylation sites were derived ... ro kinase reactions, where PKA promiscuously phosphor... Thr sites.

In another study, comb... gy of the membrane and phosphoproteomics, Nuhs... defined phosphorylation sites for about 200 *Arabidop*... mbrane proteins. Further integration of information ... predicting transmembrane domains of proteins resulted in localization of phosphorylation sites in dynamically flexible intradomain loops.

Ultimately, phosphoproteomic studies aim to reveal how information is transmitted in a signaling pathway, and in some instances the extent of phosphorylation may be of critical importance. For that purpose, several methods have been developed for quantification of site-specific phosphorylation: relative quantification using stable isotope labeling (PhIAT) [104] or a native reference peptide from the modified protein [53] as well as absolute quantification [53,56] by external addition of an isotope analog of the phosphopeptide [56]. Because of its extreme importance in many biological processes, quantification of site-specific protein phosphorylation continues to be an area of active research [105].

OTHER POSTTRANSLATIONAL MODIFICATIONS

Besides phosphorylation, a host of other posttranslational modifications can regulate the activity of proteins. Because of the immense interest in these modification events, numerous proteomics-based methods have been devised to identify proteins containing modifications. Global analyses of sumoylation [106], ubiquitination [107], and N-linked glycoproteins [108] have been published recently.

Eventually, global strategies of analysis will address detection of posttranslational modifications that dynamically regulate each other. For example, β-linked N-acetylglucosamine (O-GlcNAc) is a dynamic posttranslational modification that might be involved in antagonizing the functional effect of phosphorylation [109]. Khidekel et al. [110] devised a method that will selectively label and enrich O-GlcNAc glycosylated proteins for further purification from brain tissue. Though this method will help study O-GlcNAc modification, clearly, new methods are needed to address the dynamic interplay between posttranslational modifications at the global level.

Mass Spectrometry in Medicine

Molecular medicine is one of the major disciplines where systems biology can be applied [111]. Two classes of specimens are available for molecular medicine studies: body fluids and human tissues, often in very limited amounts (for example, obtained from biopsies). Organization in cellular environments of tissues brings another level of complexity [illegible] parameters detectable by omics technologies and in particular mass spectrometry. However, recent developments of mass spectrometry support [illegible]tion of these technologies in routine clinical analysis with tremendous value for medical diagnostics [112].

Caprioli and colleagues [illegible] direct tissue profiling by mass spectrometry by adapting MALDI technology. Imaging of tissue sections comprises two steps: (i) interrogation of a grid of spots over a tissue surface by laser irradiation and MS detection over a 500 Da to 80 kDa range and (ii) data analysis by color coding to indicate the amount of sample detected (obtained by data integration of each peak) [113]. Comparison of an optical image and MS image integrates histopathological and biochemical features of the tissue sample for better medical assessment. Further clustering of mass spectral data could distinguish over different histological groups [114].

In a different strategy for profiling serum to identify ovarian cancer, Petricoin et al. [115] used SELDI-TOF to identify potential biomarkers. In SELDI (surface-enhanced laser desorption ionization) the MALDI target is replaced by a chromatographic surface. Consequently, a subset of proteins/peptides is fractionated on the target plate based on the affinity for certain chemistries [116]. The initial high-throughput appeal of this protein chip platform has been hampered by two important issues: (i) analytical validation of the data and (ii) correct bioinformatics analysis. Analytical validation of this platform was recently performed by Zhou et al. [117] using a high-resolution TOF instrument. At the same time, new methods for data analysis in SELDI-TOF are emerging [118]. Thus, it appears that this technology will make a major impact in this area of research.

A systems biology approach can be applied in the clinical setting. Beside clinical evaluation of tissue specimens or body fluids, mass spectrometry can serve as a hypothesis-free tool for proteomics platforms, as we have seen in the above examples of systems biology studies. For example, multiple analyses of a sample using different technologies can be integrated in a single database. A database infrastructure for high-throughput description of a given tissue should organize (i) clinical information on patients, (ii) data available in public domains (for example, databases of microarray experiments) together with functional annotation of gene products, and (iii) results from experiments. Studies on nonfailing human hearts integrated in a relational database transcriptomics and proteomics data with validated phosphoproteome analysis, with mRNAs and proteins being extracted sequentially from the same preparation [119]. Many more databases, similar to the one described above, exist and are in the process of being developed. Integrated databases are the templates in systems biology research for comparative studies that aim for better diagnostic capabilities and eventually to unravel the cause of disease.

CONCLUSION

Mass spectrometry holds the promise of transforming results of biological experiments from qualitatively described models to quantitative models, in the era of "sequencing the primordial soup" (a phrase coined by Shabanowitz et al. [32]). A large-scale MS experiment performs identification and quantification of biomolecules in the following sequence: separated biomolecules are introduced into mass spectrometers by soft ionization (ESI or MALDI) and detected at a predetermined scan rate in the mass range of mass analyzers (i.e., quadrupole, ion traps, time-of-flight, ICR) with their specific mass accuracy and resolving power. Then, biomolecules are further sequenced by tandem mass spectrometry. New generation technologies of mass spectrometry for proteomics will increasingly address analysis of transient structures [120]. Development of methodologies will address the necessity to monitor multiple time points in a quantitative manner [121].

Development of MS instruments with better sensitivity and higher mass accuracy and resolving power is required for better understanding of complex biological systems. Tandem MS as well as MS^n together with biomolecule identification through database searches and statistical evaluation of the results are currently a research area driven by requirements of studies of biological systems. In this context, separation technologies need to be improved for specific samples and quantitation methods should provide solutions for any biological system. Metabolomics will need efficient sample preparation protocols to take full advantage of analytical MS. Assembly of the results from large-scale (omics) experiments of different families of biomolecules will be a critical step for transforming information into knowledge.

ACKNOWLEDGMENTS

The authors acknowledge support from NIH 5R01MH067880-02 and NIH RR 11823-08. The authors would like to thank Drs. Greg Cantin, Claire Delahunty, and John Venable for critical reading of the manuscript.

REFERENCES

1. Alberts, B., A. Johnson, J. Lewis, et al. *Molecular Biology of the Cell*, 4th ed. Garland Science Publishing, New York, 2002.
2. Nicholson, J. K., J. Connelly, J. C. Lindon, et al. Metabonomics: a platform for studying drug toxicity and gene function. *Nature Reviews Drug Discovery*, 1(2):153–61, 2002.
3. Ideker, T. and D. A. Lauffenburger. Building with a scaffold: emerging strategies for high- to low-level cellular modeling. *Trends in Biotechnology*, 21(6):255–62, 2003.
4. Ideker, T., T. Galitski and L. Hood. A new approach to decoding life: systems biology. *Annual Review of Genomics and Human Genetics*, 2:343–72, 2001.
5. Yates III, J. R. Mass spectrometry as an emerging tool for systems biology. *Biotechniques*, 36(6):917–19, 2004.
6. Yates III, J. R. Mass spectral analysis in proteomics. *Annual Review of Biophysics and Biomolecular Structure*, 33:297–316, 2004.
7. Fenn, J. B. Electrospray wings for molecular elephants (Nobel Lecture). *Angewandte Chemie International Edition*, 42(33):3871–94, 2003.

8. Tanaka, K. The origin of macromolecule ionization by laser irradiation (Nobel Lecture). *Angewandte Chemie International Edition*, 42(33):3860–70, 2003.
9. Karas, M. and F. Hillenkamp. Laser desorption ionization of proteins with molecular masses exceeding 10,000 daltons. *Analytical Chemistry*, 60(20): 2299–301, 1988.
10. Han, X. and R. W. Gross. Global analyses of cellular lipidomes directly from crude extracts of biological samples by ESI mass spectrometry: a bridge to lipidomics. *Journal of Lipid Research*, 44(6):1071–9, 2003.
11. Zaia, J. Mass spectrometry of oligosaccharides. *Mass Spectrometry Reviews*, 23(3):161–227, 2004.
12. Morris, H. R., T. Paxton, M. Panico, et al. A novel geometry mass spectrometer, the Q-TOF, for low-femtomole/attomole-range biopolymer sequencing. *Journal of Protein Chemistry*, 16(5):469–79, 1997.
13. Le Blanc, J. C., J. W. Hager, A. M. Ilisiu, et al. Unique scanning capabilities of a new hybrid linear ion trap mass spectrometer (Q TRAP) used for high sensitivity proteomics applications. *Proteomics*, 3(6):859–69, 2003.
14. Schwartz, J. C., M. W.Senko and J. E. Syka. A two-dimensional quadrupole ion trap mass spectrometer. *Journal of the American Society of Mass Spectrometry*, 13(6):659–69, 2002.
15. Griffiths, W. J., A. P. Jonsson, S. Liu, et al. Electrospray and tandem mass spectrometry in biochemistry. *Biochemical Journal*, 355(3):545–61, 2001.
16. Kinter, M. and N. E. Sherman. *Protein Sequencing and Identification Using Tandem Mass Spectrometry*. Wiley-Interscience, New York, 2000.
17. Roepstorff, P. and J. Fohlman. Proposal for a common nomenclature for sequence ions in mass spectra of peptides. *Biomedical Mass Spectrometry*, 11(11):601, 1984.
18. Eng, J. K., A. L. McCormack and J. R. Yates. An approach to correlate tandem mass-spectral data of peptides with amino-acid-sequences in a protein database. *Journal of the American Society for Mass Spectrometry*, 5(11):976–89, 1994.
19. Sadygov, R. G., D. Cociorva and J. R. Yates. Large scale database searching using tandem mass spectra: looking up the answer in the back of the book. *Nature Methods*, 1(3):195–202, 2004.
20. MacCoss, MJ. Computational analysis of shotgun proteomics data. *Current Opinion in Chemical Biology*, 9(1):88–94, 2005.
21. Forrester, J. S., S. B. Milne, P. T. Ivanova, et al. Computational lipidomics: a multiplexed analysis of dynamic changes in membrane lipid composition during signal transduction. *Molecular Pharmacology*, 65(4):813–21, 2004.
22. Han, X. and R. W. Gross. Electrospray ionization mass spectroscopic analysis of human erythrocyte plasma membrane phospholipids. *Proceedings of the National Academy of Sciences USA*, 91(22):10635–9, 1994.
23. Han, X. and R. W. Gross. Shotgun lipidomics: electrospray ionization mass spectrometric analysis and quantitation of cellular lipidomes directly from crude extracts of biological samples. *Mass Spectrometry Reviews*, 24(3):367–412, 2005.
24. Villas-Boas, S. G., S. Mas, M. Akesson, et al. Mass spectrometry in metabolome analysis. *Mass Spectrometry Reviews*, 24(5):613–46, 2005.
25. Patterson, S. D. and R. H. Aebersold. Proteomics: the first decade and beyond. *Nature Genetics*, 33(Suppl):311–23, 2003.

26. Link, A. J., J. Eng, D. M. Schieltz, et al. Direct analysis of protein complexes using mass spectrometry. *Nature Biotechnology*, 17(7):676–82, 1999.
27. Washburn, M. P., D. Wolters and J. R. Yates III. Large-scale analysis of the yeast proteome by multidimensional protein identification technology. *Nature Biotechnology*, 19(3):242–7, 2001.
28. MacCoss, M. J., W. H. McDonald, A. Saraf, et al. Shotgun identification of protein modifications from protein complexes and lens tissue. *Proceedings of the National Academy of Sciences USA*, 99(12):7900–5, 2002.
29. Fridriksson, E. K., A. Beavil, D. Holowka, et al. Heterogeneous glycosylation of immunoglobulin E constructs characterized by top-down high-resolution 2-D mass spectrometry. *Biochemistry*, 39(12):3369–76, 2000.
30. Reid, G. E. and S. A. McLuckey. "Top down" protein characterization via tandem mass spectrometry. *Journal of Mass Spectrometry*, 37(7):663–75, 2002.
31. Kelleher, N. L. Top-down proteomics. *Analytical Chemistry*, 76(11): 197A–203A, 2004.
32. Shabanowitz, J., R. E. Settlage, J. A. Marto, et al. Sequencing the primordial soup. In A. L. Burlingame, S. A. Carr and M. A. Baldwin (Eds.), *Mass Spectrometry in Biology and Medicine* (pp. 163–179). Humana Press, Totowa, N.J., 2000.
33. Delahunty, C. and J. R. Yates III. Protein identification using 2D-LC-MS/MS. *Methods*, 35(3):248–55, 2005.
34. Opiteck, G. J. and J. W. Jorgenson. Two-dimensional SEC/RPLC coupled to mass spectrometry for the analysis of peptides. *Analytical Chemistry*, 69(13):2283–91, 1997.
35. Link, A. J. Multidimensional peptide separations in proteomics. *Trends in Biotechnology*, 20(12 Suppl):S8–13, 2002.
36. Conrads, T. P., G. A. Anderson, T. D. Veenstra, et al. Utility of accurate mass tags for proteome-wide protein identification. *Analytical Chemistry*, 72(14):3349–54, 2000.
37. Lipton, M. S., L. Pasa-Tolic, G. A. Anderson, et al. Global analysis of the *Deinococcus radiodurans* proteome by using accurate mass tags. *Proceedings of the National Academy of Sciences USA*, 99(17):11049–54, 2002.
38. Gevaert, K., M. Goethals, L. Martens, et al. Exploring proteomes and analyzing protein processing by mass spectrometric identification of sorted N-terminal peptides. *Nature Biotechnology*, 21(5):566–9, 2003.
39. Annan, R. S., M. J. Huddleston, R. Verma, et al. A multidimensional electrospray MS-based approach to phosphopeptide mapping. *Analytical Chemistry*, 73(3):393–404, 2001.
40. Tong, W., A. Link, J. K. Eng, et al. Identification of proteins in complexes by solid-phase microextraction/multistep elution/capillary electrophoresis/tandem mass spectrometry. *Analytical Chemistry*, 71(13):2270–8, 1999.
41. Weckwerth, W. Metabolomics in systems biology. *Annual Review of Plant Biology*, 54:669–89, 2003.
42. Oda, Y., K. Huang, F. R. Cross, et al. Accurate quantitation of protein expression and site-specific phosphorylation. *Proceedings of the National Academy of Sciences USA*, 96(12):6591–6, 1999.
43. Washburn, M. P., A. Koller, G. Oshiro, et al. Protein pathway and complex clustering of correlated mRNA and protein expression analyses in

Saccharomyces cerevisiae. *Proceedings of the National Academy of Sciences USA*, 100(6):3107–12, 2003.
44. Krijgsveld, J., R. F. Ketting, T. Mahmoudi, et al. Metabolic labeling of *C. elegans* and *D. melanogaster* for quantitative proteomics. *Nature Biotechnology*, 21(8):927–31, 2003.
45. Wu, C. C., M. J. MacCoss, K. E. Howell, et al. Metabolic labeling of mammalian organisms with stable isotopes for quantitative proteomic analysis. *Analytical Chemistry*, 76(17):4951–9, 2004.
46. Stocklin, R., J. F. Arrighi, K. Hoang-Van, et al. Positive and negative labeling of human proinsulin, insulin, and C-peptide with stable isotopes. New tools for in vivo pharmacokinetic and metabolic studies. *Methods in Molecular Biology*, 146:293–315, 2000.
47. Ong, S. E., L. J. Foster and M. Mann. Mass spectrometric-based approaches in quantitative proteomics. *Methods*, 29(2):124–30, 2003.
48. Lill, J. Proteomic tools for quantitation by mass spectrometry. *Mass Spectrometry Reviews*, 22(3):182–94, 2003.
49. Gygi, S. P., B. Rist, S. A. Gerber, et al. Quantitative analysis of complex protein mixtures using isotope-coded affinity tags. *Nature Biotechnology*, 17(10):994–9, 1999.
50. Zhang, H., W. Yan and R. Aebersold. Chemical probes and tandem mass spectrometry: a strategy for the quantitative analysis of proteomes and subproteomes. *Current Opinion in Chemical Biology*, 8(1):66–75, 2004.
51. MacCoss, M. J. and J. R. Yates III. Proteomics: analytical tools and techniques. *Current Opinion in Clinical Nutrition and Metabolic Care*, 4(5):369–75, 2001.
52. Bondarenko, P. V., D. Chelius and T. A. Shaler. Identification and relative quantitation of protein mixtures by enzymatic digestion followed by capillary reversed-phase liquid chromatography-tandem mass spectrometry. *Analytical Chemistry*, 74(18):4741–9, 2002.
53. Ruse, C. I., B. Willard, J. P. Jin, et al. Quantitative dynamics of site-specific protein phosphorylation determined using liquid chromatography electrospray ionization mass spectrometry. *Analytical Chemistry*, 74(7):1658–64, 2002.
54. Willard, B. B., C. I. Ruse, J. A. Keightley, et al. Site-specific quantitation of protein nitration using liquid chromatography/tandem mass spectrometry. *Analytical Chemistry*, 75(10):2370–6, 2003.
55. Barnidge, D. R., E. A. Dratz, T. Martin, et al. Absolute quantification of the G protein-coupled receptor rhodopsin by LC/MS/MS using proteolysis product peptides and synthetic peptide standards. *Analytical Chemistry*, 75(3):445–51, 2003.
56. Gerber, S. A., J. Rush, O. Stemman, et al. Absolute quantification of proteins and phosphoproteins from cell lysates by tandem MS. *Proceedings of the National Academy of Sciences USA*, 100(12):6940–5, 2003.
57. Liu, H., R. G. Sadygov and J. R. Yates III. A model for random sampling and estimation of relative protein abundance in shotgun proteomics. *Analytical Chemistry*, 76(14):4193–201, 2004.
58. Prudhomme, W., G. Q. Daley, P. Zandstra, et al. Multivariate proteomic analysis of murine embryonic stem cell self-renewal versus

differentiation signaling. *Proceedings of the National Academy of Sciences USA*, 101(9):2900–5, 2004.

59. Venable, J., M. Q. Dong, J. Wohlschlegel, et al. Automated approach for quantitative analysis of complex peptide mixtures from tandem mass spectra, *Nature Methods*, 1(1):39–45, 2004.
60. MacCoss, M. J., C. C. Wu, H. Liu, et al. A correlation algorithm for the automated quantitative analysis of shotgun proteomics data. *Analytical Chemistry*, 75(24):6912–21, 2003.
61. Radulovic D., Jelveh S., Ryu S., et al. Informatics platform for global proteomic profiling and biomarker discovery using liquid chromatography-tandem mass spectrometry. *Molecular and Cellular Proteomics*, 3(10):984–97, 2004.
62. Kalkum, M., G. J. Lyon and B. T. Chait. Detection of secreted peptides by using hypothesis-driven multistage mass spectrometry. *Proceedings of the National Academy of Sciences USA*, 100(5):2795–800, 2003.
63. Chang, E. J., V. Archambault, D. T. McLachlin, et al. Analysis of protein phosphorylation by hypothesis-driven multiple-stage mass spectrometry. *Analytical Chemistry*, 76(15):4472–83, 2004.
64. Graber, A., P. S. Juhasz, N. Khainovski, et al. Result-driven strategies for protein identification and quantitation—a way to optimize experimental design and derive reliable results. *Proteomics*, 4(2):474–89, 2004.
65. Zhu, H., M. Bilgin and M. Snyder. Proteomics. *Annual Review of Biochemistry*, 72:783–812, 2003.
66. Keller, A., A. I. Nesvizhskii, E. Kolker, et al. Empirical statistical model to estimate the accuracy of peptide identifications made by MS/MS and database search. *Analytical Chemistry*, 74(20):5383–92, 2002.
67. Sadygov, R. G. and J. R. Yates III. A hypergeometric probability model for protein identification and validation using tandem mass spectral data and protein sequence databases. *Analytical Chemistry*, 75(15):3792–8, 2003.
68. Sadygov, R. G., H. Liu and J. R. Yates III. Statistical models for protein validation using tandem mass spectral data and protein amino acid sequence databases. *Analytical Chemistry*, 76(6):1664–71, 2004.
69. Warnock, D. E., E. Fahy and S. W. Taylor. Identification of protein associations in organelles, using mass spectrometry-based proteomics. *Mass Spectrometry Reviews*, 23(4):259–80, 2004.
70. Wu, C. C. and J. R. Yates III. The application of mass spectrometry to membrane proteomics. *Nature Biotechnology*, 21(3):262–7, 2003.
71. Hazbun, T. R., L. Malmstrom, S. Anderson, et al. Assigning function to yeast proteins by integration of technologies. *Molecular Cell*, 12(6):1353–65, 2003.
72. Prokisch, H., C. Scharfe, D. G. Camp II, et al. Integrative analysis of the mitochondrial proteome in yeast. *PLoS Biology*, 2(6):795–804, 2004.
73. Gene Ontology Consortium. The Gene Ontology (GO) database and informatics resource. *Nucleic Acids Research*, 32(Database issue):D258–61, 2004.
74. Shiio, Y., S. Donohoe, E. C. Yi, et al. Quantitative proteomic analysis of Myc oncoprotein function. *EMBO Journal*, 21(19):5088–96, 2002.
75. Yan, W., H. Lee, E. C. Yi, et al. System-based proteomic analysis of the interferon response in human liver cells. *Genome Biology*, 5(8):R54, 2004.

76. Florens, L., M. P. Washburn, J. D. Raine, et al. A proteomic view of the *Plasmodium falciparum* life cycle. *Nature*, 419(6906):520–6, 2002.
77. Tabb, D. L., W. H. McDonald and J. R. Yates III. DTASelect and Contrast: tools for assembling and comparing protein identifications from shotgun proteomics. *Journal of Proteome Research*, 1(1):21–6, 2002.
78. Schirmer, E. C., L. Florens, T. Guan, et al. Nuclear membrane proteins with potential disease links found by subtractive proteomics. *Science*, 301(5638):1380–2, 2003.
79. Ivanova PT, Milne SB, Forrester JS, et al. LIPID arrays: new tools in the understanding of membrane dynamics and lipid signaling. *Molecular Interventions*, 4(2):86–96, 2004.
80. Wu, C. C., M. J. MacCoss, K. E. Howell, et al. A method for the comprehensive proteomic analysis of membrane proteins. *Nature Biotechnology*, 21(5):532–8, 2003.
81. Blonder, J., M. B. Goshe, R. J. Moore, et al. Enrichment of integral membrane proteins for proteomic analysis using liquid chromatography-tandem mass spectrometry. *Journal of Proteome Research*, 1(4):351–60, 2002.
82. Overview of the Alliance for Cellular Signaling. *Nature*, 420(6916):703–6, 2002.
83. Shu, H., S. Chen, Q. Bi, et al. Identification of phosphoproteins and their phosphorylation sites in the WEHI-231 B lymphoma cell line. *Molecular and Cellular Proteomics*, 3(3):279–86, 2004.
84. Maxfield, F. R. Plasma membrane microdomains. *Current Opinion in Cell Biology*, 14(4):483–7, 2002.
85. Foster, L. J., C. L. De Hoog and M. Mann. Unbiased quantitative proteomics of lipid rafts reveals high specificity for signaling factors. *Proceedings of the National Academy of Sciences USA*, 100(10):5813–18, 2003.
86. Li, N., A. R. Shaw, N. Zhang, et al. Lipid raft proteomics: analysis of in-solution digest of sodium dodecyl sulfate-solubilized lipid raft proteins by liquid chromatography-matrix-assisted laser desorption/ionization tandem mass spectrometry. *Proteomics*, 4(10):3156–66, 2004.
87. Ge, H, A. J. Walhout and M. Vidal. Integrating "omic" information: a bridge between genomics and systems biology. *Trends in Genetics*, 19(10):551–60, 2003.
88. Quackenbush, J. Computational analysis of microarray data. *Nature Reviews in Genetics*, 2(6):418–27, 2001.
89. Ideker, T., V. Thorsson, J. A. Ranish, et al. Integrated genomic and proteomic analyses of a systematically perturbed metabolic network. *Science*, 292(5518):929–34, 2001.
90. Schwikowski, B., P. Uetz and S. Fields. A network of protein-protein interactions in yeast. *Nature Biotechnology*, 18(12):1257–61, 2000.
91. Mootha, V. K., J. Bunkenborg, J. V. Olsen, et al. Integrated analysis of protein composition, tissue diversity, and gene regulation in mouse mitochondria. *Cell*, 115(5):629–40, 2003.
92. Mootha, V. K., P. Lepage, K. Miller, et al. Identification of a gene causing human cytochrome c oxidase deficiency by integrative genomics. *Proceedings of the National Academy of Sciences USA*, 100(2):605–10, 2003.

93. Weckwerth, W., M. E. Loureiro, L. Wenzel, et al. Differential metabolic networks unravel the effects of silent plant phenotypes. *Proceedings of the National Academy of Sciences USA*, 101(20):7809–14, 2004.
94. Wiley, H. S., S. Y. Shvartsman and D. A. Lauffenburger. Computational modeling of the EGF-receptor system: a paradigm for systems biology. *Trends in Cell Biology*, 13(1):43–50, 2003.
95. Kalume, D. E., H. Molina and A. Pandey. Tackling the phosphoproteome: tools and strategies. *Current Opinion in Chemical Biology*, 7(1):64–9, 2003.
96. Cantin, G. T. and J. R. Yates III. Strategies for shotgun identification of post-translational modifications by mass spectrometry. *Journal of Chromatography A*, 1053(1–2):7–14, 2004.
97. Ficarro, S. B., M. L. McCleland, P. T. Stukenberg, et al. Phosphoproteome analysis by mass spectrometry and its application to *Saccharomyces cerevisiae*. *Nature Biotechnology*, 20(3):301–5, 2002.
98. Beausoleil, S. A., M. Jedrychowski, D. Schwartz, et al. Large-scale characterization of HeLa cell nuclear phosphoproteins. *Proceedings of the National Academy of Sciences USA*, 101(33):12130–5, 2004.
99. Salomon, A. R., S. B. Ficarro, L. M. Brill, et al. Profiling of tyrosine phosphorylation pathways in human cells using mass spectrometry. *Proceedings of the National Academy of Sciences USA*, 100(2):443–8, 2003.
100. Blom, N., T. Sicheritz-Ponten, R. Gupta, et al. Prediction of post-translational glycosylation and phosphorylation of proteins from the amino acid sequence. *Proteomics*, 4(6):1633–49, 2004.
101. Obenauer, J. C., L. C. Cantley and M. B. Yaffe. Scansite 2.0: proteome-wide prediction of cell signaling interactions using short sequence motifs. *Nucleic Acids Research*, 31(13):3635–41, 2003.
102. Hjerrild, M., A. Stensballe, T. E. Rasmussen, et al. Identification of phosphorylation sites in protein kinase A substrates using artificial neural networks and mass spectrometry. *Journal of Proteome Research*, 3(3):426–33, 2004.
103. Nuhse, T. S., A. Stensballe, O. N. Jensen, et al. Phosphoproteomics of the *Arabidopsis* plasma membrane and a new phosphorylation site database. *Plant Cell*, 16(9):2394–405, 2004.
104. Goshe, M. B., T. D. Veenstra, E. A. Panisko, et al. Phosphoprotein isotope-coded affinity tags: application to the enrichment and identification of low-abundance phosphoproteins. *Analytical Chemistry*, 74(3):607–16, 2002.
105. Blagoev, B., S. E. Ong, I. Kratchmarova, et al. Temporal analysis of phosphotyrosine-dependent signaling networks by quantitative proteomics. *Nature Biotechnology*, 22(9):1139–45, 2004.
106. Wohlschlegel, J. A., E. S. Johnson, S. I. Reed, et al. Global analysis of protein sumoylation in *Saccharomyces cerevisiae*. *Journal of Biological Chemistry*, 279(44):45662–8, 2004.
107. Peng, J., D. Schwartz, J. E. Elias, et al. A proteomics approach to understanding protein ubiquitination. *Nature Biotechnology*, 21(8):921–6, 2003.
108. Kaji, H., H. Saito, Y. Yamauchi, et al. Lectin affinity capture, isotope-coded tagging and mass spectrometry to identify N-linked glycoproteins. *Nature Biotechnology*, 21(6):667–72, 2003.
109. Slawson, C. and G. W. Hart. Dynamic interplay between O-GlcNAc and O-phosphate: the sweet side of protein regulation. *Current Opinion in Structural Biology*, 13(5):631–6, 2003.

110. Khidekel, N., S. B. Ficarro, E. C. Peters, et al. Exploring the O-GlcNAc proteome: direct identification of O-GlcNAc-modified proteins from the brain. *Proceedings of the National Academy of Sciences USA*, 101(36):13132–7, 2004.
111. Kitano, H. Systems biology: a brief overview. *Science*, 295(5560):1662–4, 2002.
112. Kinter, M. Toward broader inclusion of liquid chromatography-mass spectrometry in the clinical laboratory. *Clinical Chemistry*, 50(9):1500–2, 2004.
113. Stoeckli, M., P. Chaurand, D. E. Hallahan, et al. Imaging mass spectrometry: a new technology for the analysis of protein expression in mammalian tissues. *Nature Medicine*, 7(4):493–6, 2001.
114. Chaurand, P., S. A. Schwartz and R. M. Caprioli. Assessing protein patterns in disease using imaging mass spectrometry. *Journal of Proteome Research*, 3(2):245–52, 2004.
115. Petricoin, E. F., A. M. Ardekani, B. A. Hitt, et al. Use of proteomic patterns in serum to identify ovarian cancer. *Lancet*, 359(9306):572–7, 2002.
116. Petricoin, E. F. and L. A. Liotta. SELDI-TOF-based serum proteomic pattern diagnostics for early detection of cancer. *Current Opinion in Biotechnology*, 15(1):24–30, 2004.
117. Zhou, M., D. A. Lucas, K. C. Chan, et al. An investigation into the human serum "interactome." *Electrophoresis*, 25(9):1289–98, 2004.
118. Conrads, T. P., V. A. Fusaro, S. Ross, et al. High-resolution serum proteomic features for ovarian cancer detection. *Endocrine Related Cancer*, 11(2): 163–78, 2004.
119. Ruse, C. I., F. L. Tan, M. Kinter, et al. Intregrated analysis of the human cardiac transcriptome, proteome and phosphoproteome. *Proteomics*, 4(5):1505–16, 2004.
120. Skop, A. R., H. Liu, J. R. Yates III, et al. Dissection of the mammalian midbody proteome reveals conserved cytokinesis mechanisms. *Science*, 305(5680):61–6, 2004.
121. Zhang Y., A. Wolf-Yadlin, P. L. Ross, et al. Time-resolved mass spectrometry of tyrosine phosphorylation sites in the epidermal growth factor receptor signaling network reveals dynamic modules. *Molecular and Cellular Proteomics*, 4(9):1240–50, 2005.
122. Tabb, D. L., A. Saraf and J. R. Yates III. GutenTag: high-throughput sequence tagging via an empirically derived fragmentation model. *Analytical Chemistry*, 75(23):6415–21, 2003.

2

Mathematical Modeling and Optimization Methods for De Novo Protein Design

C. A. Floudas & H. K. Fung

The de novo peptide and protein design, first suggested almost two decades ago, begins with a postulated or known flexible protein three-dimensional structure and aims at identifying amino acid sequence(s) compatible with this structure. Initially, the problem was denoted as the "inverse folding problem" [1,2] since protein design has intimate links to the well-known protein folding problem [3]. In contrast to the characteristic of protein folding to associate a given protein sequence with its own unique shape, the inverse folding problem exhibits high levels of degeneracy, that is, a large number of sequences will be compatible with a given protein structure, although the sequences will vary with respect to properties such as activity and stability.

In silico protein design allows for the screening of overwhelmingly large sectors of sequence space, with this sequence diversity subsequently leading to the possibility of a much broader range of properties and degrees of functionality among the selected sequences. Allowing for all 20 possible amino acids at each position of a small 50-residue protein results in 20^{50} combinations, or more than 10^{65} possible sequences. From this large number of sequences, the computational sequence selection process aims at selecting those sequences that will be compatible with a given structure using efficient optimization of energy functions that model the molecular interactions.

In an effort to make the difficult nature of the energy modeling and combinatorial optimization manageable, the first attempts at computational protein design focused only on a subset of core residues and explored steric van der Waals-based energy functions through exhaustive searches for compatible sequences [4,5]. Over time, the models have evolved to incorporate improved rotamer libraries in combination with detailed energy models and interaction potentials. Although the consideration of packing effects on structural specificity is sometimes sufficient, as shown through the design of compatible structures using backbone-dependent rotamer libraries with only van der Waals energy evaluations for a subset of hydrophobic residues [6,7], there has been extensive research to develop models including hydrogen bonding, solvent, and electrostatic effects [8–11]. These functional additions to the design models are especially important for full sequence design since packing interactions no longer dominate for noncore residues (e.g., surface and intermediate residues). The incorporation of these additional noncore residues increases the potential for diversity, and therefore enhances the probability for improving functionality when compared to the parent system. An additional complication is the need to account for changes in amino acid compositions and inherent propensities through the appropriate definition of a reference state [9,12–13].

TEMPLATE FLEXIBILITY

Many computational protein design efforts were based on the premise that the three-dimensional coordinates of the template or backbone were fixed. This assumption was first proposed by Ponder and Richards [4], and was appealing because it greatly reduced the search space, and the time required to converge to a solution for the minimum energy sequence, regardless of the kind of search method employed. However, the assumption was also highly questionable. Protein backbones had been observed to allow residues that would not have been permissible had the backbone been fixed [14]. In the Protein Data Bank, there exist numerous examples of proteins that exhibit multiple NMR structures. Though commonly assumed as rigid bodies as a first approximation, the secondary structures of α-helices and β-sheets actually display some twisting and bending in the protein fold, and Emberly et al. [15,16] had applied principal component analysis of database protein structures to quantify the degree and modes of their flexibility.

Su and Mayo [17] claimed that their ORBIT (Optimization of Rotamers By Iterative Techniques) computational protein design process was robust against 15% change in the backbone. Nevertheless, they found out on a later case study on T4 lysosome that core repacking to stabilize the fold was difficult to achieve without considering a flexible template [18]. Therefore, to ensure that good sequence solutions are not rejected, it is more desirable to assume backbone flexibility in de novo protein design.

Researchers have formulated several methods to incorporate template variability. First, backbone flexibility can simply be modeled by using smaller atomic radii in the van der Waals potential. One common practice has been to scale down the radii by 5–10% [6,19], thus permitting slight overlaps between atoms due to backbone movements. Key disadvantages of this simple approach include overestimation of the attractive forces and also the possibility of hydrophobic core overpacking.

Another way to allow for backbone flexibility is through considering a discrete set of templates by using genetic algorithms and

Monte Carlo sampling. This is the approach adopted by both Desjarlais and Handel [20] and Kraemer-Pecore et al. [21]. Under this approach an ensemble of related backbone conformations close to the template are generated at random. Then a sequence will be designed for each of them under the rigid backbone assumption, and finally the backbone sequence combination with the lowest energy will be selected. For symmetric proteins, backbone structure can actually be modeled by parametric fitting, and this should improve computational efficiency. However, the vast majority of protein structures are nonsymmetric, which make this parametric approach infeasible. Su and Mayo [17] overcame this difficulty by treating α-helices and β-sheets as rigid bodies and designing sequences for several template variations of the protein Gβ1. Farinas and Regan [22] considered a discrete set of templates when they designed the metal binding sites in Gβ1, and they identified varied residue positions that would have been missed if average three-dimensional coordinates had been used for calculations. Harbury et al. [23] incorporated template flexibility through an algebraic parameterization of the backbone, when they designed a family of α-helical bundle proteins with right-handed superhelical twist. They were able to achieve a root mean square coordinate deviation between the predicted structure and the actual structure of the de novo designed protein of around 0.2 Å.

One natural approach to incorporate backbone flexibility is to allow for variability in each position in the template. The deterministic in silico sequence selection method, recently proposed by Klepeis et al. [24,25] using the integer linear optimization technique, takes into account template flexibility via the introduction of a distance-dependent force field in the sequence selection stage. Pairwise amino acid interaction potential, which depends on both the types of the two amino acids and the distance between them, was used to calculate the total energy of a sequence. Instead of being a continuous function, the dependence of the interaction potential on distance is discretized into bins. With typical bin sizes of 0.5 to 1 Å, the overall protein design model that Klepeis et al. [24,25] developed implicitly incorporated backbone movements of roughly the same order of magnitude.

MATHEMATICAL MODELING AND OPTIMIZATION METHODS

Once an energy function has been defined, sequence selection is accomplished through an optimization-based search designed to minimize the energy objective. Both stochastic and deterministic methods have been applied to the computational protein design problem. The Self-Consistent Mean Field (SCMF) [26] and dead-end elimination (DEE) [27] are both good examples of deterministic methods.

The key limitations imposed on the SCMF and DEE are (i) the backbone/template is fixed, and (ii) sequence search is restricted to a discrete set of rotamers. In their application of the SCMF method, Koehl and Delarue [28–30] refined iteratively a conformational matrix whose element *CM(i,j)* gives the probability that side chain *i* of a protein takes on rotamer *j*. Hence *CM(i,j)* sums to unity over all possible rotamers for a given side chain *i*. With an initial guess for the conformational matrix, which is usually based on the assumption that all rotamers had the same probability, that is, for rotamer *k* of residue *i*:

$$CM(i,k) = \frac{1}{K_i} \quad (1)$$

for $k = 1, 2, \ldots, K_i$. The mean-field potential *E(i,k)* is calculated using [28]:

$$E(i,k) = U(x_{ikC}) + U(x_{ikC}, x_{0C}) + \sum_{j=1, j\neq i}^{N} \sum_{l=1}^{K_j} CM(j,l)U(x_{ikC}, x_{jlC}) \quad (2)$$

where x_{0C} corresponds to the coordinates of the atoms in the template, and x_{ikC} corresponds to the coordinates of the atoms of residue *i* whose conformation is described by rotamer *k*. Lennard-Jones (12-6) potential can be used for the potential energy *U* [28]. Energies of the K_i possible rotamers of residue *i* can subsequently be converted into probabilities using the Boltzmann law:

$$CM_1(i,k) = \frac{e^{\frac{-E(i,k)}{RT}}}{\sum_{l=1}^{K_i} e^{\frac{-E(i,l)}{RT}}} \quad (3)$$

$CM_1(i,k)$ provides an update on *CM(i,k)* which can be used to repeat the calculation of energies and another update on the conformational matrix until convergence is attained. The convergence criterion is usually set as 10^{-4} to define self-consistency [28]. In addition, oscillations during convergence could be removed by updating $CM_1(i,k)$ with a "memory" of the previous step [28]:

$$CM = \lambda CM_1 + (1 - \lambda)CM \quad (4)$$

where optimal step size λ was found to be 0.9 [28]. The main disadvantage of SCMF is that though it is deterministic in nature, it does not guarantee to yield a global minimum in energy [26].

In contrast, DEE ensures the convergence to a globally optimal solution consistently. DEE operates on the systematic elimination of rotamers that are not allowable to be parts of the sequence with the lowest energy. The energy function in DEE is written in the form of a sum of individual term (rotamer–template) and pairwise term (rotamer–rotamer):

$$E = \sum_{i=1}^{N} E(i_r) + \sum_{i=1}^{N-1} \sum_{j>i}^{N} E(i_r, j_s) \tag{5}$$

where $E(i_r)$ is the rotamer–template energy for rotamer i_r of amino acid i, $E(i_r,j_s)$ is the rotamer–rotamer energy of rotamer i_r and rotamer j_s of amino acids i and j respectively, and N is the total number of residues in the protein [31]. The pruning criterion in DEE is based on the concept that if the pairwise energy between rotamer i_r and rotamer j_s is higher than that between rotamers i_t and j_s for all j_s in a certain rotamer set $\{S\}$, then i_r cannot be the global energy minimum conformation (GMEC) and thus can be eliminated. Mathematically the idea can be expressed as the following inequality [32]:

$$E(i_r) + \sum_{j\neq i}^{N} E(i_r, j_s) > E(i_t) + \sum_{j\neq i}^{N} E(i_t, j_s) \qquad \forall\{S\} \tag{6}$$

So rotamer i_r can be pruned if the above holds true. Bounds implied by (6) can be utilized to generate the following computationally more tractable inequality [32]:

$$E(i_r) + \sum_{j\neq i}^{N} \min_{s} E(i_r, j_s) > E(i_t) + \sum_{j\neq i}^{N} \max_{s} E(i_t, j_s) \tag{7}$$

The above inequality can be extended to eliminate pairs of rotamers. This is done by determining a rotamer pair i_r and j_s which always contributes higher energies than rotamer pair i_u and j_v for all possible rotamer combinations. The analogous computationally tractable inequality is [32]:

$$\varepsilon(i_r, j_s) + \sum_{k\neq i,j}^{N} \min_{t} \varepsilon(i_r, j_s, k_t) > \varepsilon(i_u, j_v) + \sum_{k\neq i,j}^{N} \max_{t} \varepsilon(i_u, j_v, k_t) \tag{8}$$

where ε is the total energies of rotamer pairs:

$$\varepsilon(i_r, j_s) = E(i_r) + E(j_s) + E(i_r, j_s) \tag{9}$$

$$\varepsilon(i_r, j_s, k_t) = E(i_r, k_t) + E(j_s, k_t) \tag{10}$$

The Mayo group has pioneered the development of DEE and has applied the method to design a variety of proteins [18,33–36]. Goldstein [37] improved the original DEE criterion by stating that rotamer i_r can be pruned if the energy contribution is always reduced by an alternative rotamer i_t:

$$E(i_r) - E(i_t) + \sum_{j \neq i}^{N} \min_{s}[E(i_r, j_s) - E(i_t, j_s)] > 0 \tag{11}$$

For rotamer pair elimination, the corresponding inequality is [32]:

$$\varepsilon(i_r, j_s) - \varepsilon(i_u, j_v) + \sum_{k \neq i,j}^{N} \min_{t}[\varepsilon(i_r, j_s, k_t) - \varepsilon(i_u, j_v, k_t)] > 0 \tag{12}$$

In general, rotamer pair elimination is computationally more expensive than single rotamer elimination, and methods have been developed by Gordon and Mayo [38] to predict which doubles elimination inequalities are the strongest.

Pierce et al. [31] introduced Split DEE which split the conformational space into partitions and thus eliminated the dead-ending rotamers more efficiently:

$$E(i_r) - E(i_t) + \sum_{j, j \neq k \neq i}^{N} \{\min_{u}[E(i_r, j_u) - E(i_t, j_u)]\} + [E(i_r, k_v) - E(i_t, k_v)] > 0 \tag{13}$$

Further revisions and improvements on DEE had been performed by Wernisch et al. [13] and Gordon et al. [39].

The protein design problem has been proved to be NP-hard [40], which means that the time required to solve the problem varies exponentially according to n^m, where n is the average number of amino acids to be considered per position and m is the number of residues. Hence as the protein becomes big enough, deterministic methods may reach a plateau, and this is when stochastic methods come into play. Monte Carlo methods and genetic algorithms are the most commonly used stochastic methods for de novo protein design. In Monte Carlo methods, a mutation is performed at a certain position in the sequence, the Boltzmann probability is calculated from the energies before and after the mutation, and the temperature is compared to a random number. The mutation is allowed if the Boltzmann probability is higher than the random number, and rejected otherwise. Dantas et al.'s [41] protein design computer program, RosettaDesign, applied Monte Carlo optimization algorithms. In completely redesigning nine globular proteins, RosettaDesign yielded sequences of 70–80% identity as the final results of energy optimization when multiple runs were started with different random sequences [41]. Originated in genetics and evolution, genetic algorithms generate a multitude of random amino acid sequences and exchange for a fixed template. Sequences with low energies form hybrids with other sequences while those with high energies are eliminated in an iterative process which only terminates when a converged solution is attained [42]. Desjarlais and Handel [20] have applied a two-stage combination of Monte Carlo and genetic algorithms to design the hydrophobic core of protein 434cro. Both Monte Carlo methods and genetic algorithms can search larger combinatorial space compared to deterministic methods, but they share the common disadvantage of lacking consistency in finding the global minimum in energy.

Recent methods attempt to avoid the problem of optimizing residue interactions by manipulation of the shapes of free energy landscapes [43]. Another class of methods focus on a statistical theory for combinatorial protein libraries which provides probabilities for the selection of amino acids in each sequence position [44–46]. The set of site-specific amino acid probabilities obtained at the end actually represents the sequence with the maximum entropy subject to all of the constraints imposed [44,45,47]. This statistical computationally assisted design strategy (*scads*) has been employed to characterize the structure and functions of membrane protein KcsA and to enhance the catalytic activity of a protein with a dinuclear metal center [47]. It has also been used to calculate the identity probabilities of the varied positions in the immunoglobulin light chain-binding domain of protein L [45]. *Scads* serves as a useful framework for interpreting and designing protein combinatorial libraries, as it provides clues about the regions of the sequence space that are most likely to produce well-folded structures [48].

Several sequence selection approaches have been tested and validated by experiment, thereby firmly establishing the feasibility of computational protein design. The first computational design of a full sequence to be experimentally characterized was that of a stable zinc-finger fold ($\beta\beta\alpha$) using a combination of a backbone-dependent rotamer library with atomistic level modeling and a dead-end elimination-based algorithm [49]. Recently, Kuhlman et al. [50] introduced a computational framework that iterates between sequence design and structure prediction, designed a new fold for a 93-residue α/β protein, and validated its fold and stability experimentally. Despite these accomplishments, the development of a computational protein design technique to rigorously address the problems of fold stability and functional design remains a challenge. As mentioned earlier, one important reason for this is either the almost universal specification of a fixed backbone, or the use of a discrete set of backbones, which does not allow for the true flexibility that would afford more optimal sequences and more robust predictions of stability. Moreover, several models that attempt to incorporate backbone flexibility highlight a second difficulty, namely, inadequacies inherent to energy modeling [20]. The need for empirically derived weighting factors, and the dependence on specific heuristics, limit the generic nature of these computational protein design methods. Such modeling-based assumptions also raise issues regarding the appropriateness of the optimization method and underscore the question of whether it is sufficient to merely identify the globally optimal sequence or, more likely, a subset of low-lying energy sequences. An even more difficult problem relevant to both flexibility and energy modeling is to correctly model the interactions that control the functionality and activity of the designed sequences.

DE NOVO PROTEIN DESIGN FRAMEWORK

In Klepeis et al. [24,25], a novel two-stage computational peptide and protein design method is presented, not only to select and rank sequences for a particular fold, but also to validate the stability and specificity of the fold for these selected sequences. The sequence selection phase relies on a novel integer linear programming (ILP) model with several important constraint modifications that improve both the tractability of the problem and its deterministic convergence to the global minimum. In addition, a rank-ordered list of low-lying energy sequences is identified along with the global minimum energy sequence. Once such a subset of sequences has been identified, the fold validation stage is used to verify the stabilities and specificities of the designed sequences through a deterministic global optimization approach that allows for backbone flexibility. The selection of the best designed sequences is based on rigorous quantification of energy-based probabilities. In the following, we will discuss the two stages in detail.

In Silico Sequence Selection

To correctly select a sequence compatible with a given backbone template, an appropriate energy function must first be identified. Desirable properties of energy models for protein design include both accuracy and rapid evaluation. Moreover, the functions should not be overly sensitive to fixed backbone approximations. In certain cases, additional requirements, such as the pairwise decomposition of the potential for application of the dead-end elimination algorithm [27], may be necessary.

Instead of employing a detailed atomistic level model, which requires the empirical reweighting of energetic terms, the proposed sequence selection procedure is based on optimizing a pairwise *distance-dependent* interaction potential. Such a statistically based empirical energy function assigns energy values for interactions between amino acids in the protein based on the alpha-carbon separation distance for each pair of amino acids. Such structure-based pairwise potentials are fast to evaluate, and have been used in fold recognition and fold prediction [51]. One advantage of this approach is that there is no need to derive empirical weights to account for individual residue propensities. Moreover, the possibility that such interaction potentials lack sensitivity to local atomic structure is addressed within the context of the overall two-stage approach. In fact, the coarser nature of the energy function in the in silico sequence selection phase may prove beneficial in that it allows for an inherent flexibility to the backbone.

A number of different parameterizations for pairwise residue interaction potentials exist. The simplest approach is the development of a binary version of the model such that each contact between two amino acids is assigned according to the residues types and the requirement that a contact is defined as the separation between the side chains of two amino acids being less than 6.5 Å [52]. An improvement of this model is based on the incorporation of distance dependence for the energy of each amino acid interaction. Specifically, the alpha-carbon distances are discretized into a set of 13 bins to create a finite number of interactions, the parameters of which were derived from a linear optimization formulated to favor native folds over optimized decoy structures [53,54]. The use of a distance-dependent potential allows for the implicit inclusion of side chains and the specificity of amino acids. The resulting potential, which involves 2730 parameters, was shown to provide higher Z scores than other potentials and place native folds lower in energy [53,54].

The linearity of the resulting formulation based on this distance-dependent interaction potential [55] is also an attractive characteristic of the in silico sequence selection procedure. The development of the formulation can be understood by first describing the variable set over which the energy function is optimized. First, consider the set $i = 1,\ldots,n$, which defines the number of residue positions along the backbone. At each position i there can be a set of mutations represented by $j\{i\} = 1,\ldots,m_i$, where, for the general case, $m_i = 20 \ \forall \ i$. The equivalent sets $k \equiv i$ and $l \equiv j$ are defined, and $k > i$ is required to represent all unique pairwise interactions. With this in mind, the binary variables y_i^j and y_k^l can be introduced to indicate the possible mutations at a given position. That is, the y_i^j variable will indicate which type of amino acid is active at a position in the sequence by taking the value of 1 for that specification. Then, the formulation, for which the goal is to minimize the energy according to the parameters that multiply the binary variables, can be expressed as:

$$\min_{y_i^j, y_k^l} \sum_{i=1}^{n} \sum_{j=1}^{m_i} \sum_{k=i+1}^{n} \sum_{l=1}^{m_k} E_{ik}^{jl}(x_i, x_k) y_i^j y_k^l$$

subject to

$$\sum_{j=1}^{m_i} y_i^j = 1 \qquad \forall i$$

$$y_i^j, y_k^l = 0-1 \qquad \forall i, j, k, l$$

The parameters $E_{ik}^{jl}(x_i, x_k)$ depend on the distance between the alpha-carbons at the two backbone positions (x_i, x_k) as well as the type of amino acids at those positions. The composition constraints require that there is exactly one type of amino acid at each position. For the general case, the binary variables appear as bilinear combinations in the objective function. Fortunately, this objective can be reformulated as a strictly linear (integer linear programming) problem [56]:

$$\min_{y_i^j, y_k^l} \sum_{i=1}^{n} \sum_{j=1}^{m_i} \sum_{k=i+1}^{n} \sum_{l=1}^{m_k} E_{ik}^{jl}(x_i, x_k) w_{ik}^{jl}$$

subject to

$$\sum_{j=1}^{m_i} y_i^j = 1 \qquad \forall i$$

$$y_i^j + y_k^l - 1 \le w_{ik}^{jl} \le y_i^j \qquad \forall i, j, k, l$$

$$0 \le w_{ik}^{jl} \le y_k^l \qquad \forall i, j, k, l$$

$$y_i^j, y_k^l = 0-1 \qquad \forall i, j, k, l$$

This reformulation relies on the transformation of the bilinear combinations to a new set of linear variables, w_{ik}^{jl}, while the addition of the four sets of constraints serves to reproduce the characteristics of the original formulation. For example, for a given *i, j, k, l* combination, the four constraints require w_{ik}^{jl} to be 0 when either y_i^j or y_k^l is equal (or when both are equal to 0). If both y_i^j and y_k^l are equal to 1, then w_{ik}^{jl} is also enforced to be 1.

The solution of the integer linear programming problem (ILP) can be accomplished rigorously using branch and bound techniques [56,57] making convergence to the global minimum energy sequence consistent and reliable. Furthermore, the performance of the branch and bound algorithm is significantly enhanced through the introduction of reformulation linearization techniques (RLT). The basic strategy is to multiply appropriate constraints by bounded nonnegative factors (such as the reformulated variables) and introduce the products of the original variables by new variables in order to derive higher dimensional lower bounding linear programming (LP) relaxations for the original problem [58]. These LP relaxations are solved during the course of the overall branch and bound algorithm, and thus speed convergence to the global minimum. The following set of constraints illustrates the application of the RLT approach to the original composition constraint. First, the equations are reformulated by forming the product of the equation with some binary variables or their complement. For example, by multiplying by the set of variables y_k^l, the following additional set of constraints $\forall\ j, k, l$ is produced:

$$y_k^l \sum_{j=1}^{m_i} y_i^j = y_k^l \qquad \forall i,k,l \tag{14}$$

This equation can now be linearized using the same variable substitution as introduced for the objective. The set of RLT constraints then become:

$$\sum_{j=1}^{m_i} w_{ik}^{jl} = y_k^l \qquad \forall i,k,l \tag{15}$$

Finally, for such an ILP problem it is straightforward to identify a rank-ordered list of the low-lying energy sequences through the introduction of integer cuts [56], and repetitive solution of the ILP problem. By using the enhancements outlined above, in combination with the commercial (LP) solver CPLEX [57], a globally optimal (ILP) solution is generated.

Fold Specificity

Once a set of low-lying energy sequences have been identified via the sequence selection procedure, the fold stability and specificity validation stage is used to identify the most optimal sequences according to a rigorous quantification of conformational probabilities. The approach is based on the development of conformational ensembles for the selected sequences under two sets of conditions. In the first circumstance the structure is constrained to vary, with some imposed fluctuations, around the template structure. In the second condition, a free folding calculation is performed for which only a limited number of restraints are likely to be incorporated (in the case of compstatin and its analogs, only the disulfide bridge constraint is enforced) and with the underlying template structure not being enforced. In terms of practical considerations, the distance constraints introduced for the template-constrained simulation can be based on the structural boundaries defined by the NMR ensemble (in the case of compstatin and its analogs a deviation of 1.5 Å is allowed for each nonconsecutive C_α–C_α distance from the known NMR structures), or simply by allowing some deviation from a subset of distances provided by the structural template, and hence they allow for a flexible template on the backbone.

The formulations for the folding calculations are reminiscent of structure prediction problems in protein folding [59]. In particular, a novel constrained global optimization problem first introduced for structure prediction using NMR data [60], and later employed in a generic framework for the structure prediction of proteins [61], is used. The global minimization of a detailed atomistic energy force field E_{ff} is performed over the set of independent dihedral angles, ϕ, which can be used to describe any possible configuration of the system. The bounds on these variables are enforced by simple box constraints. Finally, a set of distance constraints, E_l^{dis}, $l = 1,\ldots, N$, which are nonconvex in the internal coordinate system, can be used to constrain the system. The formulation is represented by the following set of equations:

$$\min_{\phi} E_{\text{ff}}$$

subject to

$$\begin{aligned} E_j^{\text{dis}}(\phi) \le E_j^{\text{ref}} \quad & j = 1,\ldots,N \\ \phi_i^L \le \phi_i \le \phi_i^U \quad & i = 1,\ldots,N_\phi \end{aligned}$$

Here, $i = 1,\ldots, N_\phi$ corresponds to the set of dihedral angles, ϕ_i, with ϕ_i^L and ϕ_i^U representing lower and upper bounds on these dihedral angles. In general, the lower and upper bounds for these variables are set to $-\pi$ and π. E_j^{ref} are reference parameters for the distance constraints, which assume the form of a typical square-well potential for both upper and lower distance violations. The set of constraints is completely general, and can represent the full combination of distance constraints or smaller subsets of the defined restraints. The force field energy function E_{ff} can take on a number of forms, although the current work employs the ECEPP/3 model [62].

The folding formulation represents a general nonconvex constrained global optimization problem, a class of problems for which several methods have been developed. In this work, the formulations are solved via the αBB deterministic global optimization approach, a branch and bound method applicable to the identification of the global minimum of nonlinear optimization problems with twice-differentiable functions [59,60,63–67]. In the αBB approach, a converging sequence of upper and lower bounds is generated. The upper bounds on the global minimum are obtained by local minimizations of the original nonconvex problem. The lower bounds belong to the set of solutions of the convex lower bounding problems that are constructed by augmenting the objective and constraint functions through the addition of separable quadratic terms.

In addition to identifying the global minimum energy conformation, the global optimization algorithm provides the means for identifying a consistent ensemble of low-energy conformations [66,68,69]. Such ensembles are useful in deriving quantitative comparisons between the free folding and template-constrained simulations. In this way, the complications inherent to the specification of an appropriate reference state are avoided because a relative probability is calculated for each sequence studied during this stage of the approach.

The relative probability for template stability, p_{temp}, can be found by summing the statistical weights for those conformers from the free folding simulation that resemble the template structure (denoted as set temp), and dividing this sum by the summation of statistical weights for all conformers from the free folding simulation (denoted as set total):

$$p_{\text{temp}} = \frac{\sum_{i \in \text{temp}} \exp[-\beta E_i]}{\sum_{i \in \text{total}} \exp[-\beta E_i]}$$

where $\exp[-\beta E_i]$ is the statistical weight for conformer i.

COMPUTATIONAL AND EXPERIMENTAL FINDINGS

Compstatin

The target chosen to test the novel protein design framework proposed by Klepeis et al. [24,25] is compstatin. Compstatin is a 13-residue cyclic peptide that has the ability to inhibit the cleavage of C3 to C3a and C3b. The effect of targeting the C3 cleavage is triple and results to hindrance in (i) the generation of the proinflammatory peptide C3a, (ii) the generation of opsonin C3b (or its fragment C3d), and (iii) further complement activation of the common pathway (beyond C3) with end result the generation of the membrane attack complex (MAC). A C3-binding complement inhibitor was identified as a 27-residue peptide using a phage-displayed random peptide library [70]. This peptide was truncated to an equally active 13-residue peptide named compstatin with sequence *I[CVVQDWGHHRC]T-*NH_2, where the brackets denote cyclization through a disulfide bridge formed by Cys^2-Cys^{12} [70,71]. Acetylation of the N-terminus of compstatin (Ac-compstatin) resulted in a 3-fold increase in activity [72–74].

Compstatin blocked the cleavage of C3 to the proinflammatory peptide C3a and the opsonin C3b in both hemolytic assays, and in human normal serum [70,72]; prevented heparin/protamine-induced complement activation in baboons in a situation resembling heart surgery [75]; inhibited complement activation during the contact of blood with biomaterial in a model of extracorporeal circulation [76]; increased the lifetime of survival of porcine kidneys perfused with human blood in a hyperacute rejection xenotransplantation model [77]; blocked the *E. coli*-induced oxidative burst of granulocytes and monocytes [78]; and inhibited complement activation by cell lines SH-SY5Y, U-937, THP-1, and ECV304 [79]. Compstatin was stable in biotranformation studies in vitro in human blood, normal human plasma, and serum, with increased stability upon N-terminal acetylation [72]. Compstatin showed little or low toxicity, and no adverse effects when these were measured [75–77]. Finally, compstatin showed species specificity and is active only with human and primate C3 [80].

In Silico Sequence Selection

The first stage of the design approach involves the selection of sequences compatible with the backbone template through the solution of the ILP problem. The formulation relies only on the alpha-carbon coordinates of the backbone residues, which were taken from the NMR-average solution structure of compstatin [71].

A full computational design study from compstatin would result in a combinatorial search of $20^{13} \approx 8 \times 10^{16}$ sequences. However, in light of the results of the experimental studies of the rationally designed peptides, a directed, rather than full, set of computational design studies was performed. First, since the disulfide bridge was found to be essential for aiding in the formation of the hydrophobic cluster and prohibiting the termini from drifting apart, both residues Cys^2 and Cys^{12} were maintained. In addition, because the structure of the type-I β turn was not found to be a sufficient condition for activity, the turn residues were fixed to be those of the parent compstatin sequence, namely, Gln^5-Asp^6-Trp^7-Gly^8. In fact, when stronger type I β sequences were constructed, which was supported by NMR data indicating that these sequences provided higher β-turn populations than compstatin, these sequences resulted in lower or no activity [73]. Therefore, the further stabilization of the turn residues, which would likely be a consequence of the computational peptide design procedure, may not enhance compstatin activity. This is especially true for Trp^7, which was found to be a likely candidate for direct interaction with C3. For similar reasons, Val^3 was maintained throughout the computational experiments.

After designing the compstatin system to be consistent with those features found to be essential for compstatin activity, six residue positions were selected for optimization. Of these six residues, positions 1, 4, and 13 have been shown to be structurally involved in the formation of a hydrophobic cluster involving residues at positions 1, 2, 3, 4, 12, and 13, a necessary but not sufficient component for compstatin binding and activity. The remaining residues, namely those at positions 9, 10, and 11, span the three positions between the turn residues and the C-terminal cystine. For the wild-type sequence these positions are populated by positively charged residues, with a total charge of +2 coming from two histidine residues and one arginine residue.

Based on the structural and functional characteristics of those residues involved in the hydrophobic cluster, a base case was studied with positions 1, 4, and 13 selected only from those residues defined as belonging to the hydrophobic set (A,F,I,L,M,V,Y). In addition, this set included threonine for position 13 to allow for the selection of the wild-type residue at this position. For positions 9, 10, and 11 in the base case, all residues were allowed, excluding cystine and tryptophan.

The sequence selection results exhibit several important and consistent features. First, position 10 is dominated by the selection of a histidine residue, a result that directly reinforces the composition of the wild-type compstatin sequence. In contrast, position 11 is found to have the largest variation in composition, with polar, hydrophobic, and charged residues all being part of the set of optimal low-lying energy sequences. At position 9, a subset of those residues chosen for position 11 is selected. When considering those positions involved in the hydrophobic cluster of compstatin, it is evident that valine provides strong forces at each position. However, the results for position 4 contrast with those at positions 1 and 13 in that tyrosine, rather than valine, is the preferred choice for the lowest as well as a large majority of the low-lying energy sequences.

It should be noted that because the compstatin structure was determined via NMR methods, there exists an ensemble of 21 structures for which alternative templates could be derived. These alternative templates were studied as a means of incorporating backbone flexibility into the sequence selection process, and the results proved to be consistent and in qualitative agreement with those for the average template structure.

In view of the experimental studies on compstatin by Mallik et al. [81], who proposed that tryptophan (W) or fused-ring nonnatural amino acids at position 4 would contribute to high inhibitory activity of the peptide, an additional run was performed with the inclusion of tryptophan (W) in the selection set for position 4. Table 2.1 summarizes the preferred selection at each position according to the composition of the lowest-lying energy sequences. Tryptophan (W) was indeed strongly favored at position 4, if it was included in the selection set. This observation agrees with the experimental finding by Mallik et al. [81]. It should be noted that if tryptophan (W) is allowed to be in the aforementioned hydrophobic set for all six positions, then sequences with tryptophan (W) in position 4, and alanine (A), phenyl (F), or tryptophan (W) in position 9, are predicted among the most promising ones by the proposed novel in silico sequence selection framework (position 1 is I, position 10 is R, and position 13 is T, as in set D of figure 2.1).

Table 2.1 Preferred residue selection for positions 1, 4, 9, 10, 11, and 13 of compstatin, as compared to the wild-type sequence

Position	Wild type	Optimal[a]	Optimal[b]
1	I	A,V	V,A
4	V	Y,V	W,Y,V
9	H	T,F,A	F,T
10	H	H	H,K,S
11	R	T,V,A,F,H	H,F,T
13	T	V,A,F	V,A,F

Only residues with greater than 10% representation among the lowest-lying energy sequences are considered optimal. Provided in decreasing order.
[a]Base case: positions 1 and 4 selected from {A,F,I,L,M,V,Y}; position 13 selected from {A,F,I,L,M,V,Y,T}; positions 9,10, and 11 selected from all residues except C and W.
[b]Base case with position 4 among {A,F,I,L,M,V,Y,W}.

Fold Specificity Calculations For Selected Sequences

Based on the sequence selection results, a handful of optimal sequences were constructed for use in the second stage of the computational design procedure. Figure 2.1 presents the peptides studied which are further classified into sets A, B, C, and D.

For all sequences further characterized via the fold stability calculations, residue 10 was set to histidine, a prediction consistent with the

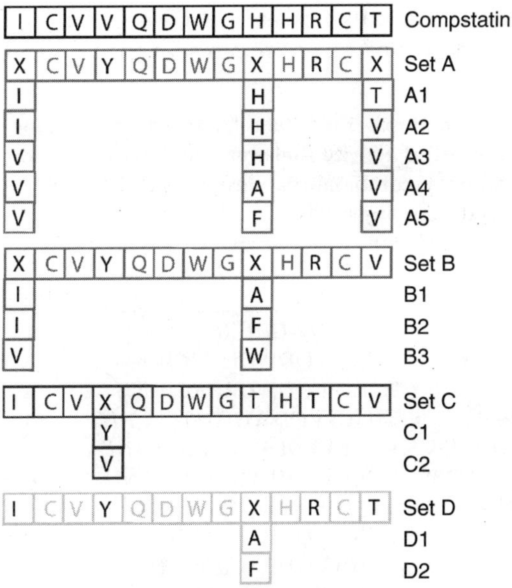

Figure 2.1 Set of sequences tested for fold specificity.

composition of the parent peptide sequence. Moreover, since the variation in the residue composition for position 11 is predicted to be rather broad, position 11 was restricted to be arginine in subsequent sequences (except set C). The first set of sequences was constructed to better analyze the effect of the tyrosine substitution at position 4, with the justification to focus on this substitution being an attempt to assess the unusually dominant selection of tyrosine at position 4. The consistent element of the sequences belonging to set A is the assignment of tyrosine to position 4. To further isolate any substitution with respect to the parent peptide sequence, sequences A1, A2, and A3 assume the parent compstatin composition of histidine at position 9. Moreover, sequence A1 resembles the parent peptide sequence at positions 1 and 13 as well, while sequences A2 and A3 are constructed so as to add the valine substitutions incrementally, first at position 13 for sequence A2 and then at both positions 1 and 13 for sequence A3. Sequences A1 and A3 exhibit substantial increases in fold stability over the parent peptide sequence (table 2.2). These results highlight the significance of the tyrosine substitution at position 4, and may help to further clarify certain features of the proposed binding model for the compstatin–C3 complex [73].

To further explore the combination of position 9 substitutions with the presence of tyrosine at position 4, several additional sequences were constructed. The B1 and B2 constructions represent a reduction in the number of simultaneous mutations from the parent peptide sequence. In effect these two sequences correspond to the individual combinations of sequence A2 with both sequence A4 and sequence A5

Table 2.2 Sequence and experimental relative activity of compstatin analogs with improved activity that were identified by rational design, experimental combinatorial design, and the novel in silico de novo protein design approach

Peptide	Sequence	Relative activity	Reference
Compstatin	I[**CV**V**QDWG**HHR**C**]T-NH_2	1	[70]
Ac-Compstatin	Ac-I[**CV**V**QDWG**HHR**C**]T-NH_2	3	[72]
Ac-H9A	Ac-I[**CV**V**QDWG**AHR**C**]T-NH_2	4	[73]
Ac-I1L/H9W/T13G	Ac-L[**CV**V**QDWG**WHR**C**]G-NH_2	4	[74]
Ac-I1L/V4Y/H9F/T13V	Ac-V[**CV**Y**QDWG**FHR**C**]V-NH_2	6	[24]
Ac-I1L/V4Y/H9A/T13V	Ac-V[**CV**Y**QDWG**AHR**C**]V-NH_2	9	[24]
Ac-V4Y/H9F/T13V	Ac-I[**CV**Y**QDWG**FHR**C**]V-NH_2	11	[24]
Ac-V4Y/H9A/T13V	Ac-I[**CV**Y**QDWG**AHR**C**]V-NH_2	14	[24]
Ac-V4Y/H9A	Ac-I[**CV**Y**QDWG**AHR**C**]T-NH_2	16	[24]
Ac-V4W/H9A	Ac-I[CVWQDWGAHRC]T-NH_2	45	[81]

Boldface is used to indicate that amino acids were fixed. Brackets indicate the disulfide bridge. Relative complement inhibitory activity is derived from IC_{50} measurements such that position 1 is taken from sequence A2, while position 9 matches the substitutions incorporated into sequences A4 and A5. An additional sequence, B3, is formulated as a combination of sequence A3 and the position 9 substitution of histidine to tryptophan as taken from control sequence X2. Each of the three designed sequences demonstrates significant increases in fold stability relative to the original compstatin sequence (table 2.2).

Another set of two additional sequences was identified with the only difference between them being the specification of the residue at position 4. For sequence C1, tyrosine was assigned to position 4, while sequence C2 was selected to have valine at this position. For both sequences, threonine was specified at positions 9 and 11, while positions 1 and 13 were set to isoleucine and valine, respectively. The choice of isoleucine for position 1 helps to reduce the number of simultaneous changes from the parent peptide sequence.

For both sequences C1 and C2 the stability calculations indicate a substantial decrease in stability when compared to the parent peptide sequence. Nevertheless, between sequences C1 and C2 there is strong evidence for the preference of tyrosine at position 4. This prompted closer examination of the residue selections at positions 9 and 11, the two remaining positions not involved in the hydrophobic clustering of compstatin. In particular, the specification of threonine at both positions 9 and 11 results in a negative net charge balance due to the aspartate at position 6, especially because of the replacement of arginine by threonine at position 11. This validates further the placement of arginine at position 11 for the previous set of sequences (table 2.2).

The final set of sequences was designed in accordance with additional reductions in the number of simultaneous mutations relative to the parent peptide sequence. Specifically, sequences D1 and D2 resemble sequences B1 and B2 with threonine instead of valine as the C-terminal residue, a specification matching the composition of the original parent peptide sequence. Both sequences provide significant increases in fold stability. For sequences D1 and D2 the differences with respect to the parent peptide sequence are isolated to the residue before and after the β turn. Both the position 4 tyrosine and position 9 phenylalanine substitutions provide enhancements to the fold stability of the compstatin structure, and represent unforeseen and unpredictable enhancements over the parent peptide sequence (table 2.2).

Experimental Validation

A number of the designed sequences presented above were constructed and tested experimentally for their activity, without performing NMR-based structural analyses. Since the ultimate goal is to enhance the functional activity of compstatin, such achievements must be complemented and verified through experimental studies. Rather than performing massive chemical synthesis of peptide analogs, a few selected analogs were tested against the theoretical prediction. Table 2.2 shows the experimentally measured percent complement inhibition and peptide D1 is currently the most active compstatin analog available. The C2A/C12A analog is inactive [73] and has been used as a negative control for the inhibition measurements. Table 2.2 summarizes the results from the inhibitory activity experiments in comparison to the theoretical fold stability results.

Qualitatively, the predicted increases in fold stability and specificity are in excellent agreement with the results from the experimental studies. This is especially significant, given that the predictions correspond more directly to fold stability enhancements, while the experiments directly test inhibitory function.

The comparison between experimental and computational results indicates that the most active compstatin analogs are sequences D1 and B1, as suggested by the optimization study. The common characteristic of these two sequences is the substitutions at positions 4 and 9, the two positions flanking the β- turn residues, Gln^5-Asp^6-Trp^7-Gly^8. In particular, the combination of tyrosine at position 4 and alanine at position 9 are key residues for increased activity and lead to a 16-fold improvement over the parent peptide compstatin (see table 2.2).

CONCLUSIONS AND FUTURE WORK

A novel computational structure/activity-based methodology for the de novo design of peptides and proteins was presented. The method is completely general in nature, with the main steps of the approach being the availability of NMR-derived structural templates, combinatorial selection of sequences based on optimization of parameterized pairwise residue interaction potentials, and validation of fold stability and specificity using deterministic global optimization. The optimization study led to the identification of many active analogs including a 16-fold more active analog, as validated through immunological activity measurements. Allowing tryptophan in position 4, the in silico sequence prediction framework demonstrates that tryptophan is preferred over tyrosine and tyrosine is preferred over valine. This is in agreement with recent experimental results [81] which showed a 45-fold improvement in the inhibitory activity of the peptide $Ac\text{-}I[CVWQDWGAHRC]T\text{-}NH_2$.

These results are extremely impressive and represent significant enhancements in inhibitory activity over analogs identified by either purely rational or experimental combinatorial design techniques. The work provides direct evidence that an integrated experimental and theoretical approach can make possible the engineering of compounds with enhanced immunological properties. Future work will be focused on algorithmic improvement on the novel de novo protein design framework to enhance computational efficiency [82], trial and incorporation of non-energy-based formulations to increase accuracy of predictions, and application of the framework on more protein targets.

ACKNOWLEDGMENTS

C.A.F. gratefully acknowledges financial support from the National Science Foundation and the National Institutes of Health (R24 GM069736).

REFERENCES

1. Drexler, K. E. Molecular engineering: an approach to the development of general capabilities for molecular manipulation. *Proceedings of the National Academy of Sciences USA*, 78:5275–8, 1981.
2. Pabo, C. Molecular technology. Designing proteins and peptides. *Nature*, 301:200, 1983.
3. Hardin, C., T. V. Pogorelov and Z. Luthey-Schulten. Ab initio protein structure prediction. *Current Opinion in Structural Biology*, 12:176–81, 2002.
4. Ponder, J. W. and F. M. Richards. Tertiary templates for proteins. *Journal of Molecular Biology*, 193:775–91, 1987.
5. Hellinga, H. W. and F. M. Richards. Construction of new ligand binding sites in proteins of known structure. I. Computer aided modeling of sites with predefined geometry. *Journal of Molecular Biology*, 222:763–85, 1991.
6. Desjarlais, J. R. and T. M. Handel. De novo design of the hydrophobic cores of proteins. *Protein Science*, 4:2006–18, 1995.
7. Dahiyat, B. I. and S. L. Mayo. Protein design automation. *Protein Science*, 5:895–903, 1996.
8. Dahiyat, B. I., D. B. Gordon and S. L. Mayo. Automated design of the surface positions of protein helices. *Protein Science*, 6:1333–7, 1997.
9. Raha, K., A. M. Wollacott, M. J. Italia and J. R. Desjarlais. Prediction of amino acid sequence from structure. *Protein Science*, 9:1106–19, 2000.
10. Street, A. G. and S. L. Mayo. Pairwise calculation of protein solvent-accessible surface areas. *Folding and Design*, 3:253–8, 1998.
11. Nohaile, M. J., Z. S. Hendsch, B. Tidor and R. T. Sauer. Altering dimerization specificity by changes in surface electrostatics. *Proceedings of the National Academy of Sciences USA*, 98:3109–14, 2001.
12. Koehl, P. and M. Levitt. De novo protein design. I. In search of stability and specificity. *Journal of Molecular Biology*, 293:1161–81, 1999.
13. Wernisch, L., S. Hery and S. J. Wodak. Automatic protein design with all atom force-fields by exact and heuristic optimization. *Journal of Molecular Biology*, 301:713–36, 2000.
14. Lim, W. A., A. Hodel, R. T. Sauer and F. M. Richards. Crystal structure of a mutant protein with altered but improved hydrophobic core packing. *Proceedings of the National Academy of Sciences USA*, 91:423–7, 1994.

15. Emberly, E. G., R. Mukhopadhyay, C. Tang and N. S. Wingreen. Flexibility of α-helices: results of a statistical analysis of database protein structures. *Journal of Molecular Biology*, 327:229–37, 2003.
16. Emberly, E. G., R. Mukhopadhyay, C. Tang and N. S. Wingreen. Flexibility of β-sheets: principal component analysis of database protein structures. *Proteins: Structure, Function, and Genetics*, 55:91–8, 2004.
17. Su, A. and S. L. Mayo. Coupling backbone flexibility and amino acid sequence selection in protein design. *Protein Science*, 6:1701–7, 1997.
18. Mooers, B. H. M., D. Datta, W. A. Baase, E. S. Zollars, S. L. Mayo and B. W. Matthews. Repacking the core of T4 lysozyme by automated design. *Journal of Molecular Biology*, 332:741–56, 2003.
19. Kuhlman, B. and D. Baker. Native protein sequences are close to optimal for their structures. *Proceedings of the National Academy of Sciences USA*, 97:10383–8, 2000.
20. Desjarlais, J. R. and T. M. Handel. Side chain and backbone flexibility in protein core design. *Journal of Molecular Biology*, 290:305–18, 1999.
21. Kraemer-Pecore, C. M., J. T. Lecomte and J. R. Desjarlais. A de novo redesign of the WW domain. *Protein Science*, 12:2194–205, 2003.
22. Farinas, E. and L. Regan. The de novo design of a rubredoxin-like Fe site. *Protein Science*, 7:1939–46, 1998.
23. Harbury, P. B., J. J. Plecs, B. Tidor, T. Alber and P. S. Kim. High-resolution protein design with backbone freedom. *Science*, 282:1462–7, 1998.
24. Klepeis, J. L., C. A. Floudas, D. Morikis, C. G. Tsokos, E. Argyropoulos, L. Spruce and J. D. Lambris. Integrated computational and experimental approach for lead optimization and design of compstatin variants with improved activity. *Journal of the American Chemical Society*, 125:8422–3, 2003.
25. Klepeis, J. L., C. A. Floudas, D. Morikis, C. G. Tsokos and J. D. Lambris. Design of peptide analogs with improved activity using a novel de novo protein design approach. *Industrial and Engineering Chemistry Research*, 43:3817, 2004.
26. Lee, C. Predicting protein mutant energetics by self-consistent ensemble optimization. *Journal of Molecular Biology*, 236:918–39, 1994.
27. Desmet, J., M. De Maeyer, B. Hazes and I. Lasters. The dead-end elimination theorem and its use in side-chain positioning. *Nature*, 356:539–42, 1992.
28. Koehl, P. and M. Delarue. Application of a self-consistent mean field theory to predict protein side-chains conformation and estimate their conformational entropy. *Journal of Molecular Biology*, 239:249–75, 1994.
29. Koehl, P. and M. Delarue. A self consistent mean field approach to simultaneouos gap closure and side-chain positioning in homology modeling. *Nature Structural Biology*, 2:163–70, 1995.
30. Koehl, P. and M. Delarue. Mean-field minimization methods for biological macromolecules. *Current Opinion in Structural Biology*, 6:222–6, 1996.
31. Pierce, N. A., J. A. Spriet, J. Desmet and S. L. Mayo. Conformational splitting: a more powerful criterion for dead-end elimination. *Journal of Computational Chemistry*, 21:999–1009, 2000.
32. Voigt, C. A., D. B. Gordon and S. L. Mayo. Trading accuracy for speed: a quantitative comparison of search algorithms in protein sequence design. *Journal of Molecular Biology*, 299:789–803, 2000.

33. Malakauskas, S. M. and S. L. Mayo. Design, structure, and stability of a hyperthermophilic protein variant. *Nature Structural Biology*, 5:470–5, 1998.
34. Strop, P. and S. L. Mayo. Rubredoxin variant folds without irons. *Journal of the American Chemical Society*, 121:2341–5, 1999.
35. Shimaoka, M., J. M. Shifman, H. Jing, L. Takagi, S. L. Mayo and T. A. Springer. Computational design of an intergrin I domain stabilized in the open high affinity conformation. *Nature Structural Biology*, 7:674–8, 2000.
36. Bolon, D. N. and S. L. Mayo. Enzyme-like proteins by computational design. *Proceedings of the National Academy of Sciences USA*, 98: 14274–9, 2001.
37. Goldstein, R. F. Efficient rotamer elimination applied to protein side-chains and related spin glasses. *Biophysics Journal*, 66:1335–40, 1994.
38. Gordon, D. B. and S. L. Mayo. Radical performance enhancements for combinatorial optimization algorithms based on the dead-end elimination theorem. *Journal of Computational Chemistry*, 19:1505–14, 1998.
39. Gordon, D. B., G. K. Hom, S. L. Mayo and N. A. Pierce. Exact rotamer optimization for protein design. *Journal of Computational Chemistry*, 24:232–43, 2003.
40. Pierce, N. A. and E. Winfree. Protein design is NP-hard. *Protein Engineering*, 15:779–82, 2002.
41. Dantas, G., B. Kuhlman, D. Callender, M. Wong and D. Baker. A large scale test of computational protein design: folding and stability of nine completely redesigned globular proteins. *Journal of Molecular Biology*, 332:449–60, 2003.
42. Tuffery, P., C. Etchebest, S. Hazout and R. Lavery. A new approach to the rapid determination of protein side chain conformations. *Journal of Biomolecular Structure and Dynamics*, 8:1267–89, 1991.
43. Jin, W., O. Kambara, H. Sasakawa, A. Tamura and S. Takada. De novo design of foldable proteins with smooth folding funnel: automated negative design and experimental verification. *Structure*, 11:581–90, 2003.
44. Zhou, J. and J. G. Saven. Statistical theory of combinatorial libraries of folding proteins: energetic discrimination of a target structure. *Journal of Molecular Biology*, 296:281–94, 2000.
45. Kono, H. and J. G. Saven. Statistical theory for protein combinatorial libraries. Packing interactions, backbone flexibility, and the sequence variability of a main-chain structure. *Journal of Molecular Biology*, 306:607–28, 2001.
46. Saven, J. G. Connecting statistical and optimized potentials in protein folding via a generalized foldability criterion. *Journal of Chemical Physics*, 118:6133–6, 2003.
47. Park, S., X. Yang and J. G. Saven. Advances in computational protein design. *Current Opinion in Structural Biology*, 14:487–94, 2004.
48. Hecht, M. H., A. Das, A. Go, L. H. Bradley and Y. Wei. De novo proteins from designed combinatorial libraries. *Protein Science*, 13:1711–23, 2004.
49. Dahiyat, B. I. and S. L. Mayo. De novo protein design: fully automated sequence selection. *Science*, 278:82–7, 1997.
50. Kuhlman, B., G. Dantae, G. C. Ireton, G. Verani, B. Stoddard and D. Baker. Design of a novel globular protein fold with atomic-level accuracy. *Science*, 302:1364–8, 2003.

51. Park, B. and M. Levitt. Energy functions that discriminate x-ray and near native folds from well-constructed decoys. *Journal of Molecular Biology*, 258:367–92, 1996.
52. Meller, J. and R. Elber. Linear programming optimization and a double statistical filter for protein threading protocols. *Proteins*, 45:241–61, 2001.
53. Tobi, D. and R. Elber. Distance-dependent, pair potential for protein folding: results from linear optimization. *Proteins: Structure, Function, and Genetics*, 41:40–6, 2000.
54. Tobi, D., G. Shafran, N. Linial and R. Elber. On the design and analysis of protein folding potentials. *Proteins*, 40:71–85, 2000.
55. Loose, C., J. L. Klepeis and C. A. Floudas. A new pairwise folding potential based on improved decoy generation and side-chain packing. *Proteins: Structure, Function, and Bioinformatics*, 54:303–14, 2004.
56. Floudas, C. A. *Nonlinear and Mixed-Integer Optimization: Fundamentals and Applications*. Oxford University Press, New York, 1995.
57. CPLEX. *Using the CPLEX Callable Library*. ILOG, Inc., Mountain View, Cal., 1997.
58. Sherali, H. D. and W. P. Adams. *A Reformulation Linearization Technique for Solving Discrete and Continuous Nonconvex Problems*. Kluwer Academic Publishing, Boston, 1999.
59. Klepeis, J. L., H. D. Schafroth, K. M. Westerberg and C. A. Floudas. Deterministic global optimization and ab initio approaches for the structure prediction of polypeptides, dynamics of protein folding and protein-protein interaction. In R. A. Friesner (Ed.), *Advances in Chemical Physics*, Vol. 120 (pp. 254–457). Wiley, New York, 2002.
60. Klepeis, J. L., C. A. Floudas, D. Morikis and J. Lambris. Predicting peptide structures using NMR data and deterministic global optimization. *Journal of Computational Chemistry*, 20:1354–70, 1999.
61. Klepeis, J. L. and C. A. Floudas. Ab initio tertiary structure prediction of proteins. *Journal of Global Optimization*, 25:113–40, 2003.
62. Némethy, G., K. Gibson, K. Palmer, C. Yoon, G. Paterlini, A. Zagari, S. Rumsey and H. Scheraga. Energy parameters in polypeptides. 10. Improved geometrical parameters and nonbonded interactions for use in the ECEPP/3 algorithm, with application to proline-containing peptides. *Journal of Physical Chemistry*, 96:6472–84, 1992.
63. Adjiman, C., I. Androulakis and C. A. Floudas. A global optimization method, αBB, for general twice-differentiable constrained NPLs. I. Theoretical advances. *Computers and Chemical Engineering*, 22:1137, 1998.
64. Adjiman, C., I. Androulakis and C. A. Floudas. A global optimization method, αBB, for general twice-differentiable constrained NPLs. II. Implementtation and computational results. *Computers and Chemical Engineering*, 22:1159, 1998.
65. Adjiman, C., I. Androulakis and C. A. Floudas. Global optimization of mixed-integer nonlinear problems. *AIChE Journal*, 46:1769–97, 2000.
66. Klepeis, J. L. and C. A. Floudas. Free energy calculations for peptides via deterministic global optimization. *Journal of Chemical Physics*, 110:7491–512, 1999.
67. Floudas, C. A. *Deterministic Global Optimization: Theory, Methods and Applications*. Kluwer Academic Publishers, Boston, 2000.

68. Klepeis, J. L., M. T. Pieja and C. A. Floudas. A new class of hybrid global optimization algorithms for peptide structure prediction: integrated hybrids. *Computer Physics Communications*, 151:121–40, 2003.
69. Klepeis, J. L., M. T. Pieja and C. A. Floudas. Hybrid global optimization algorithms for protein structure prediction: alternating hybrids. *Biophysical Journal*, 84:869–82, 2003.
70. Sahu, A., B. K. Kay and J. D. Lambris. Inhibition of human complement by a C3-binding peptide isolated from a phage displayed random peptide library. *Journal of Immunology*, 157:884–91, 1996.
71. Morikis, D., N. Assa-Munt, A. Sahu and J. D. Lambris. Solution structure of compstatin, a potent complement inhibitor. *Protein Science*, 7:619–27, 1998.
72. Sahu, A., A. M. Soulika, D. Morikis, L. Spruce, W. T. Moore and J. D. Lambris. Binding kinetics, structure activity relationship and biotransformation of the complement inhibitor compstatin. *Journal of Immunology*, 165:2491–9, 2000.
73. Morikis, D., M. Roy, A. Sahu, A. Torganis, A. Jennings, G. C. Tsokos and J. D. Lambris. The structural basis of compstatin activity examined by structure-function-based design of peptide analogs and NMR. *Journal of Biological Chemistry*, 277:14942–53, 2002.
74. Soulika, A. M., D. Morikis, M. R. Sarias, M. Roy, L. Spruce, A. Sahu and J. D. Lambris. Studies of structure-activity relations of complement inhibitor compstatin. *Journal of Immunology*, 170:1881–90, 2003.
75. Soulika, A. M., M. M. Khan, T. Hattori, F. W. Bowen, B. A. Richardson, C. E. Hack, A. Sahu, L. H. Edmunds and J. D. Lambris. Inhibition of heparin/protamine complex-induced complement activation by compstatin in baboons. *Clinical Immunology*, 96:212–21, 2000.
76. Nillson, B., R. Larsson, J. Hong, G. Elgue, K. N. Ekdahl, A. Sahu and J. D. Lambris. Compstatin inhibits complement and cellular activation in whole blood in two models of extracorporeal circulation. *Blood*, 92:1661–7, 1998.
77. Fiane, A. E., T. E. Mollnes, V. Videm, T. Hovig, K. Hogasen, O. J. Mellbye, L. Spruce, W. T. Moore, A. Sahu and J. D. Lambris. Compstatin, a peptide inhibitor of C3, prolongs survival of ex-vivo perfused pig xenografts. *Xenotransplantation*, 6:52–65, 1999.
78. Mollnes, T. E., O. L. Brekke, M. Fung, H. Fure, D. Christiansen, G. Bergseth, V. Videm, K. T. Lappegard, J. Kohl and J. D. Lambris. Essential role of the C5a receptor in *E. coli*-induced oxidative burst and phagocytosis revealed by a novel lepirudin-based human whole blood model of inflammation. *Blood*, 100:1869–77, 2002.
79. Klegeris, A., E. A. Singh and P. L. McGeer. Effects of c-reactive protein and pentosan polysulphate on human complement activation. *Immunology*, 106:381–8, 2002.
80. Sahu, A., D. Morikis and J. D. Lambris. Compstatin, a peptide inhibitor of complement, exhibits species-specific binging to complement component C3. *Molecular Immunology*, 39:557–66, 2003.
81. Mallik, B., M. Katragadda, L. A. Spruce, C. Carafides, C. G. Tsokos, D. Morikis and J. D. Lambris. Design and NMR characterization of active

analogues of compstatin containing non-natural amino acids. *Journal of Medicinal Chemistry*, 48:274–86, 2005.

82. Fung, H. K., S. Rao, C. A. Floudas, O. Prokopyev, P. M. Pardalos and F. Rendl. Computational comparison studies of quadratic assignment like formulations for the in silico sequence selection problem in de novo protein design. *Journal of Combinatorial Optimization*, 10:41–60, 2005.

3

Molecular Simulation and Systems Biology

William C. Swope, Jed W. Pitera, & Robert S. Germain

Although we receive a relatively static view of molecular structure from spectroscopic tools such as x-ray crystallography and nuclear magnetic resonance (NMR), the reality is that molecules are in constant motion at biological temperatures. Intermolecular motions, such as the binding or unbinding of an antibody–antigen complex, have an important role in biological processes. In addition, biomolecules are always flexing, bending, and stretching in ways that affect their function. For example, many proteins display *allosteric* behavior in which the binding of a ligand to some site on the protein causes the protein to change its shape. This can result in the active site of an enzyme becoming operational. The folding and unfolding of a protein are more extreme examples of intramolecular motions that have a profound impact on biological function.

One way to understand the motions of biological molecules is by using a computer to simulate those motions explicitly. The computational techniques used to model intra- and intermolecular motions are known as molecular simulations. Molecular simulations are a set of computational methods that allow the modeling of the motions of molecules. Molecular motions are coupled to the environment—other biomolecules, cofactors, counterions, and water. A typical molecular simulation involves simulating the motions of all of the atoms of a protein or nucleic acid along with all of the surrounding water molecules. The simulation is carried out by calculating the forces that atoms exert on each other and using those forces to propagate the motions of those atoms. A single simulation cannot capture all of the possible motions of a molecular system, but will at best achieve a statistical sampling of the important states and motions. Some representative images from a molecular simulation of a protein are shown in figure 3.1.

CONNECTION TO SYSTEMS BIOLOGY

In general, molecular simulations can be used in systems biology in two ways. First, detailed simulations can be used to calculate unknown parameters of more simplified models. For example, it is possible to

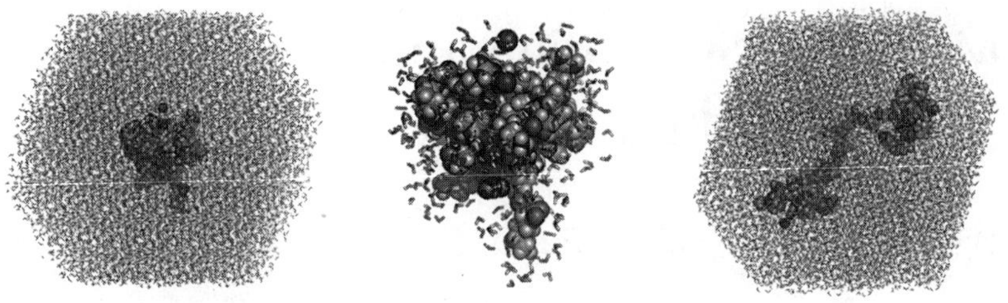

Figure 3.1 Images from a molecular dynamics simulation of the protein 1BBL in explicit solvent. The left-hand panel shows the folded 1BBL protein [82] (40 amino acids, 618 atoms) surrounded by 15,000 water molecules. The central panel is a detail view showing a shell of water molecules within 5 Å of the folded protein. The right-hand image shows a representative extended (unfolded) structure of 1BBL. The protein and two chloride counterions are shown as spheres, while water molecules are shown in a stick representation. This and other molecular graphics in this chapter were generated with the PyMOL program [83] unless otherwise noted.

use molecular simulations to calculate the lateral diffusion coefficient for an integral membrane protein in order to understand how rapidly such proteins move around the cell surface. Molecular simulation methods can also be used to estimate association and dissociation rate constants for protein–protein complexes involved in a signaling cascade. Also, the rates of enzymatic processes relevant for systems biology, such as phosphorylation, can be calculated from coupled quantum mechanical/molecular mechanical (QM/MM) simulations. QM/MM simulations use classical models in combination with a quantum mechanical treatment of a subregion of the system.

The most probable way molecular simulations will be used to provide parameters for simplified models is through the calculation of free energy differences between different states of a system. Significant advances have been made in calculating absolute protein–ligand binding free energies [1] and in estimating biomolecular binding free energies from simulation [2–4]. New techniques [5] promise the possibility of simulating a single model protein–DNA complex and using those data to predict the optimum binding site for the transcription factor in question. Clearly, these free energy calculations can be used to supply missing parameters for models of biological systems, but they also show promise in helping to elucidate the underlying processes that must be modeled, for example, by predicting transcription factor binding sites to help decode gene regulatory networks.

At a more detailed level, molecular simulations can be used to develop intermediate coarse-grained models of biological systems, where collections of tens of atoms are represented by a single "particle."

These coarse-grained particles interact with an effective potential that cannot be derived directly from basic physical principles but can be inferred by observing the results of more detailed simulations. The use of atomistic molecular simulation to develop potentials for coarse-grained models is a standard practice in the simulation of nonbiological polymers [6]. Researchers are beginning to use coarse-grained models for biological simulations of large systems, such as lipid membranes [7].

The second way that molecular simulation methods can be used in systems biology is to add more detail to a simpler model. Simulations of wild-type and mutant forms of a protein can give a detailed, atomistic picture of how the mutation affects the protein's biological function. Simulations of a protein–protein binding event under different conditions (ionic strength, pH) may give insights into how the process will occur in different cellular compartments or environments. Comparative simulations of two homologous proteins can provide insights into differences or similarities of function. In all of these cases, the simulation results should always be verified by quantitative, and ideally predictive, comparison with experimental data. With a judicious choice of methods and their careful application, however, molecular simulation is an important tool for gaining a complete picture of biological processes—the central goal of systems biology.

THE ROLE OF HIGH-PERFORMANCE COMPUTING

Almost from the very beginning of electronic computing, advances in the scale of molecular dynamics simulations have marched in step with advances in computing performance (figure 3.2). At the same time, both the methods and models used in carrying out the calculations have advanced as well. Often, the availability of increased computational capability has meant that more realistic models can be employed. For example, the first molecular dynamics simulations of biomolecules were performed without inclusion of solvent, while now simulations of proteins with explicit representation of solvent molecules as part of the calculation are routine (to the point where most of the computer time goes to simulating water rather than protein) (see table 3.1, page 95).

Advances in high-performance computing capabilities can be characterized by their ability to increase *capability* or *capacity*. Capability refers to the ability to apply computational resources to a single problem. Large shared memory computers and clusters equipped with very high performance interconnects are examples of systems that enable *capability* calculations. There are many problems, particularly in the life sciences, that involve performing many independent calculations, such as large numbers of BLAST queries, or trying to screen a library

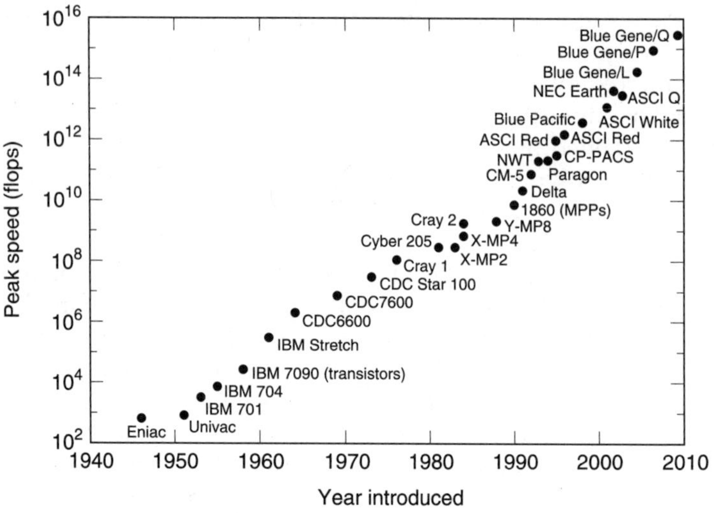

Figure 3.2 Chart showing the approximate progression of peak supercomputing capability (measured in floating point operations/second or instructions/second) as a function of the year in which that capability became available.

of potential drug candidates against a particular receptor. These problems require large amounts of aggregate *capacity*, but for which the individual calculations are of modest size. Grids and loosely coupled networks of distributed workstations are examples of *capacity* systems. Of course, any capability machine can also be used for capacity calculations, but it is often not cost-effective to do so. Simulations of large molecular systems that exhibit phenomena on long time scales are *capability* calculations, but often one wishes to compute statistical properties for which a number of separate trajectories may be used. Also, some techniques require a number of n-body simulations to be run in parallel with only loose coupling between them.

As the time scales accessible via simulation have become longer and experimentally accessible time scales have become shorter, the ability to connect simulation data to experiments in a significant way has started to become a reality. This also allows the models (force fields) used in simulation to be tested more extensively against experiment.

MOLECULAR SIMULATION

Molecular modeling and simulation by computer are techniques based on the fundamental principles of physics. They have been used for many years as a means to gain an understanding of materials and processes at the molecular level and are capable of describing biological systems and processes with great detail. Simply stated, modeling and simulation refer to computations that are based on approximations of the energies of systems of one or more molecules as a function of their conformations and relative orientations. (The term "simulation" usually also includes the idea that we are interested in modeling not only the energies, but also the temporal behavior of the system.)

The techniques available for computing the energies as a function of molecular conformation within a modeling or simulation study range from approaches based on quantum mechanics all the way to rather coarse-grained classical ones. Those based on quantum mechanics are capable of a higher degree of accuracy but are computationally much more expensive. Consequently, they are practical only for relatively small molecular systems. The classical approaches approximate the inherently quantum mechanical molecular interactions with a parameterized function that is very easy to evaluate, and so they may be used to study larger molecular systems like proteins or cell membranes, and processes that involve long time scales such as protein folding or drug–protein binding phenomena. However, the accuracy of these approximations is always an important issue; although they have been shown to work for a range of problems, they must be revalidated for each new type of use. There are also hybrid methods where some regions of the system are treated quantum mechanically and the rest is treated classically.

The right level of approximation needed for modeling a molecular system depends critically on the types of processes and phenomena one is interested in studying. For example, the behavior of electronically excited molecules, and chemical bond breaking and formation, involve the rearrangement of electrons in the molecular system and one must use quantum mechanics to evaluate the energies associated with these kinds of processes. Therefore, if one is interested in how enzymes catalyze the making and breaking of bonds during, say, metabolism, one would need to include the role of electronic structure and a quantum mechanical approach would be required.

If one is not interested in these kinds of phenomena, classical approaches may be adequate. These have their greatest applicability for biological phenomena that do not involve large changes in the electronic state of the molecules of interest. This includes a very wide range of important problems including the structure, function, energetics, and dynamics of biologically functional molecules and systems. For example, to study protein folding, classical models of the process have been shown to be adequate for a variety of proteins.

The mathematical expression that approximates the conformational energies and interaction energies between molecules in a classical molecular simulation is commonly called a potential energy function, or force field. It is a scalar function of all the coordinates of all the atoms in the system, and the force on any particular atom is the gradient of this function with respect to motion of that atom.

Types and Uses of Classical Molecular Simulations

Given a model for the interactions between atoms and molecules, one can perform computations of various types to explore different aspects of the behavior of the molecular system. These aspects usually can be characterized as structural, thermodynamic, or kinetic. The first relates to the determination of the most stable molecular conformation, or structure, of the molecular system. Protein structure refinement falls into this category. This problem is one of optimization: given an energy expression, find the set of atomic coordinates that minimizes the energy. This problem is nontrivial because of the high dimensionality of the search space (a small protein with 2000 atoms has a search space with 6000 dimensions). Furthermore, the energy function has a large number of local minima, and so the global minimum can be very difficult to find. Most energy functions can be analytically differentiated, so one can use gradients and second, or even higher, derivatives of the energy function to facilitate the search.

The Velocity Verlet Algorithm

Start with current atomic positions, velocities and forces:

$$\{r_i(t), v_i(t), F_i(t)\}$$

Using the current atomic coordinates, velocities, and forces, update the coordinates:

$$r_i(t+\Delta t) = r_i(t) + v_i(t)\cdot \Delta t + \frac{F_i(t)}{2m}\cdot \Delta t^2$$

Using the updated positions of all the atoms, update the forces:

$$F_i(t+\Delta t) = -\nabla_i U(r_1(t+\Delta t), r_2(t+\Delta t), \ldots)$$

Using the updated forces, update the velocities:

$$v_i(t+\Delta t) = v_i(t) + \frac{\Delta t}{2m}[F_i(t+\Delta t) + F_i(t)]$$

Finish with new atomic positions, velocities, and forces:

$$\{r_i(t+\Delta t), v_i(t+\Delta t), F_i(t+\Delta t)\}$$

Ligand–receptor docking [8] is another case where molecular modeling is used to determine the lowest energy conformation of a system; in this case, one is interested in discovering the lowest energy conformation for a ligand–receptor complex. Various poses of the ligand and receptor are generated, and their energies evaluated. The calculation is repeated hundreds of thousands of times for each ligand in a compound library, allowing the ligands in the library to be ranked by their binding energy. This ranking can then be used to prioritize experimental binding assays.

A second type of information that can be obtained from models relates to the thermodynamics. To model thermodynamics one uses a technique meant to mimic a particular set of thermodynamic conditions, such as a specification of the temperature and density, or temperature and pressure, of the system. These techniques include molecular dynamics (MD) and several variants, as well as Monte Carlo (MC). Traditional MD involves computing a sequence of coordinates for the molecular system that represent its temporal evolution according to the laws of classical mechanics (see box on page 72). This is done by repeatedly updating the atomic positions and velocities in a way that reflects the positions, velocities, and forces a short time earlier. Of course, this technique can be used to simulate the kinetics of a system,

The Metropolis Monte Carlo Algorithm

Start with a "state," a set of coordinates for all the atoms of the system, and the potential energy of that state:

$S_k=\{r_1,r_2,\ldots\};\ U_k=U(S_k)$

Apply a suitable random process to generate a trial state and evaluate the potential energy of the trial state:

$S^T=\{r_1^T, r_2^T,\ldots\};\ U^T=U(S^T)$

Accept the trial state with a probability of

$P_{accept}=\min(1,e^{-(U^T-U_k)/kT})$

If the trial state is accepted, the next state of the chain is the trial state:

$S_{k+1}=S^T;\ U_{k+1}=U^T$

Otherwise, the original state is repeated in the chain:

$S_{k+1}=S_k;\ U_{k+1}=U_k$

but it also can be used to generate a sample set of the configurations that a system might take subject to conditions of fixed density and energy. If one wants to model conditions where the molecular system is in contact with a heat bath of some specified temperature, or one where it is in a container subject to some external pressure, one can use MC or variations of standard MD where the motion of the molecules is modified in a way to reflect the effect of the heat bath and/or pressure on the system. It must be remembered, however, that such variations in the MD algorithm necessarily affect the dynamics of the molecules under study.

In addition to MD, Monte Carlo techniques are often employed to simulate molecular systems in contact with a heat bath or under conditions of some externally applied pressure. In Metropolis MC (see box on page 73), one produces a large set of configurations for the system by the generation of a Markov chain. With this procedure, one makes various types of random moves of molecules in the system to produce a trial state. The trial state is either accepted or rejected. If the trial state is rejected, the state from which the trial state was obtained is retained. This process is then repeated many times. The acceptance/rejection criterion depends on the temperature as well as on the types of moves that are being performed, and it is carefully designed such that if one averages properties over all of the states produced, one gets a result that is characteristic of the particular thermodynamic condition being studied. With MC techniques the system evolves in a way that attempts to sample new configurations with the appropriate *thermal* weighting; the succession of configurations does not represent any real "time evolution" of the system.

The third type of information that can be derived from simulations relates to the kinetics or dynamical motion of the system. Molecular dynamics is the method required for these types of studies since it describes the behavior of molecules that are subject only to the forces on them from their molecular interactions. Questions that require an understanding of kinetics include "How quickly do proteins fold?" Biological processes occur on a variety of time scales, and the length of the simulation required to study them depends on the particular phenomena under study. In order to have adequate statistical confidence in kinetic results, enough effort must be expended to observe not just one but dozens or hundreds of the events of interest. Because of this, there are still many interesting and important biological processes that are currently beyond the reach of meaningful simulation.

Force Fields

Typically, a force field expresses the energy as a sum of expressions of different types: one type to describe contributions from chemical bond stretching, one for angle bending, one for twisting, or torsional, motion about chemical bonds, and one for nonbonded interactions. Nonbonded interactions involve atoms separated by more than three chemical bonds or atoms belonging to different molecules. These interactions are of three types, electrostatic, short-ranged repulsion, and long-ranged dispersion. The electrostatic interactions are usually approximated by assuming that each atom in the system has a partial positive or negative charge. The short-ranged repulsion and long-ranged dispersion interactions are usually approximated together with a function that has a Lennard-Jones, or 6-12, form. In this functional form, the r^{-12} term describes the repulsion and the r^{-6} term describes the attraction. The terms of a typical intermolecular force field and the corresponding potential energy expressions are illustrated in figure 3.3.

In these expressions, r_{ij} represents the distance between atoms i and j, θ_{ijk} represents the angle formed by atoms i, j, and k, ϕ_{ijkl} represents the torsion angle formed by atoms i, j, k, and l. A force field has a large number of parameters. The bond parameters, for example, include the bond spring constants, k_{ij}, and equilibrium bond lengths, r_0. These, in turn, depend on what are the atom types of atoms i and j, as well as the nature of the chemical bond between them.

Force field expressions can have various amounts of complexity and can include physical effects (not included in the expressions in figure 3.3)

$$U_{Total} = \sum_{Bond\,(ij)} U^{(ij)}_{Bond}(r_{ij}) + \sum_{Angle\,(ijk)} U^{(ijk)}_{Angle}(\theta_{ijk}) + \sum_{Torsion\,(ijkl)} U^{(ijkl)}_{Torsion}(\phi_{ijkl}) + \sum_{LJ\,(ij)} U^{(ij)}_{LJ}(r_{ij}) + \sum_{Coulomb\,(ij)} U^{(ij)}_{Coulomb}(r_{ij})$$

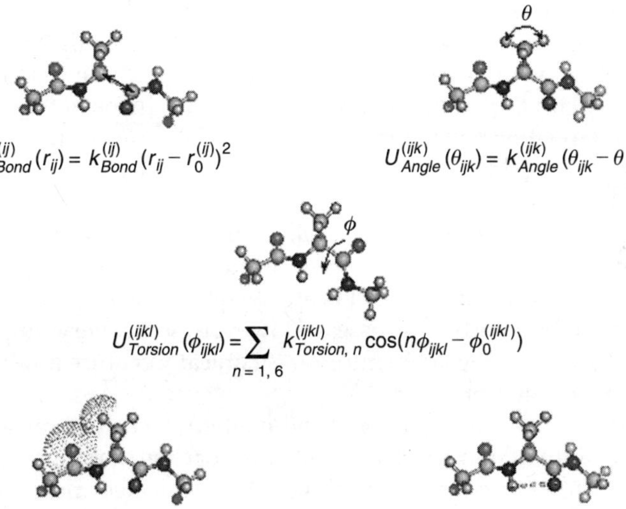

$$U^{(ij)}_{Bond}(r_{ij}) = k^{(ij)}_{Bond}(r_{ij} - r^{(ij)}_0)^2$$

$$U^{(ijk)}_{Angle}(\theta_{ijk}) = k^{(ijk)}_{Angle}(\theta_{ijk} - \theta^{(ijk)}_0)^2$$

$$U^{(ijkl)}_{Torsion}(\phi_{ijkl}) = \sum_{n=1,6} k^{(ijkl)}_{Torsion,\,n}\cos(n\phi_{ijkl} - \phi^{(ijkl)}_0)$$

$$U^{(ij)}_{LJ}(r_{ij}) = 4\varepsilon_{ij}[(\sigma_{ij}/r_{ij})^{12} - (\sigma_{ij}/r_{ij})^6]$$

$$U^{(ij)}_{Coulomb}(r_{ij}) = q_i q_j / r_{ij}$$

Figure 3.3 Intramolecular interactions and corresponding potential energy terms in a typical molecular mechanics force field.

such as electron polarization and charge transfer. A number of force fields have been developed for important and common classes of biomolecules. These have been hand-optimized to give more or less realistic approximations to the intermolecular interactions for a number of important purposes. This effort has engaged hundreds of researchers over a period of close to twenty years. And still, when using a force field or any other approximation to the intermolecular interactions, one must always be careful to verify that the interaction model is adequate for the purpose at hand.

Key Challenges in Molecular Simulation

In spite of the large number of approximations involved, and the availability of very powerful computers, building, performing, and interpreting simulations is not a trivial task. There are several critical challenges that must be met in the field of molecular simulation.

Both MD and MC techniques require a model that specifies the interactions between particles in the system. One of the most significant challenges for the field has to do with the development of accurate representations of the molecular interactions. This is sometimes referred to as force field quality, but it appears also in the context of the various quantum mechanical approaches where it could relate to the degree of accuracy associated with the choice of quantum method and basis set. Quantum mechanical approaches can be limited in terms of the quality of the functional (in the case of density functional theoretical approaches), the size and quality of the basis set, treatment of solvation effects, temperature, and boundary conditions.

For classical modeling, force field approximations to the molecular interactions come in many flavors. Some of these include treatment of electronic polarization effects and some do not. The ones that do not include polarization, sometimes known as *fixed charge* force fields, have been developed and tuned over many years for various materials. For protein simulations, there are a handful of fixed charge force fields, including CHARMm [9], OPLS-AA [10], AMBER [11], and GROMOS [12]. These have reached a certain degree of maturity, and many experts in the field know which force fields are adequate for the study of various types of biological processes. However, some very important phenomena are inherently quantum mechanical in nature and the use of force fields is inappropriate.

Related to force field quality is the method used to represent the solvent in simulations. There are some approaches that involve treating the electrostatic effect of the solvent on a solvated molecule as if the solvent were a continuum dielectric. These are known as *implicit* solvent models. Usually used in conjunction with continuum solvent models is Langevin dynamics, where the dynamical effects of the solvent are emulated by imposing short random impulses on the solvated molecule. If the solvent can be treated in this manner, the number of particles that must be simulated can be significantly reduced. However, these models have their deficiencies and are known to fail in predicting the structure and thermodynamics for certain types of solvated molecules [13].

An alternative to implicit solvent models is the use of an *explicit* molecular model for the solvent. Explicit solvent models treat the solvent as a large number of discrete molecules interacting with the solute. Several accurate force fields exist for the simulation of a handful of solvent types. The most studied solvent is water, and there are several popular fixed charge force fields for this important solvent, including SPC [14], TIP3P, TIP4P [15], SPC/E [16], TIP5P [17], and TIP4P-Ew [18]. Several water force fields that include polarization effects are under active development and testing. Although polarizable models show a lot of promise, at this time they have not been fully proven in terms of their accuracy in the context of large biomolecular simulations on proteins.

After force field quality, the second major challenge for the field of simulation relates to the time scale of the biophysical processes of interest. One can see from figure 3.4 that bond vibrations have periods measured in tens of femtoseconds (fs; 1 femtosecond = 10^{-15} seconds). This fast motion determines the time scale for MD time step sizes. Typically, time step sizes are in the range of 0.5 to 2.0 fs. Simulations with time step sizes of 2.0 fs require the bond vibrations to be frozen,

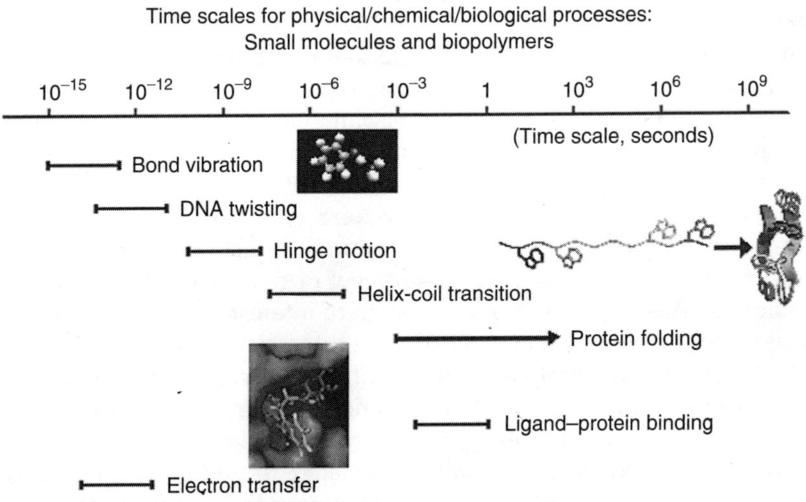

Figure 3.4 A hierarchy of representative time scales for processes in biological systems.

using algorithms such as SHAKE [19] and RATTLE [20]. The figure also shows that many biophysical phenomena of interest, such as protein folding, happen on time scales in the microsecond (10^{-6} seconds) to millisecond (10^{-3} seconds) range. This means that the simulations must be run for 10^9 to 10^{12} time steps to simulate a single folding event. One of the most ambitious simulations performed to date, by Duan and Kollman [21] in 1998, was the study of the folding of a small peptide for a microsecond.

Long time scales cause a challenge for two phases of simulations: equilibration and sampling. First, simulations must be performed for a long enough period of time for the material to adopt conformations that are typical at the temperatures of interest. This is known as the equilibration phase of a simulation, and the length of time needed for equilibration depends on the thermodynamic conditions such as temperature and density, as well as on the nature of the material under study and the degree to which the starting conformation is different from those observed in an equilibrated system. Imagine, for example, that you begin a simulation on a peptide in its folded structure, but at a temperature where the folded structure is unstable. It will take some amount of time for the folded structure to unfold and adopt conformations that are more representative of those at the temperature of interest. Generally, the slower the time scales are in a system, the longer it will take to equilibrate.

The second challenge caused by long time scales is that they imply the need for longer simulations for the molecular system to adopt all the conformations of relevance at the temperature of interest. Suppose, for example, a molecule is known to adopt some conformation A 80% of the time and conformation B 20% of the time. We do not know these percentages a priori, but try to determine them from running a long simulation and looking to see what conformations the molecule adopts. The precision with which we can determine the ratio of occurrences of A to those of B is determined by the number of observations of changes of state during the simulation. If the interconversion rate is fast, we do not need to run very long to get precise measures of the ratio. However, if the interconversion rate is slow, we will need to run much longer to obtain a given level of precision. This argument holds for most properties of the system of interest. For example, if one wishes to determine the average of the solute–solvent interaction energy, the temporal behavior of the solute conformational change as well as of the solvent reorganization will determine the amount of time we need to simulate to measure this observable to any particular degree of precision.

Long time scales can usually be related to energy surfaces with many local minima and large barriers between them. If the barriers are large compared to the thermal energy in the system, they will not be crossed frequently, and all thermally accessible states will not be visited by the system during the simulation. Such energy surfaces are characterized as *rough,* and they present formidable challenges in the field of simulation. Rough energy landscapes result in the occurrence of *rare events*—the infrequent crossings of high free energy regions—that are difficult to characterize.

EXAMPLES OF MOLECULAR SIMULATION APPLIED TO BIOLOGICAL SYSTEMS

Some Scientific Questions Addressed by Molecular Simulation

How does protein folding happen? Many proteins fold *reversibly* into unique shapes. What this means is, first, that there is a unique three-dimensional shape for the protein in the sense that most or all copies of it in an ensemble adopt the same shape. Second, if conditions are altered, for example, by heating or by adding salts of different types, or by adding acid or base so as to change the pH, the protein will unfold into a *denatured* state, where its functional properties are lost. Reversibility means that when the conditions are restored, the protein will refold and its function will be restored. The reason this is so remarkable is that even very small proteins can adopt a very large number of conformations that are locally stable, and yet reversibly folding proteins always refold to one of a very small set of the most stable conformations! The hypothesis is that there is a driving force for folding that is actually coded into the amino acid sequence; this force helps to guide the protein into the right shape. A great deal of research is underway to understand exactly how this happens.

Is there a pathway to protein folding? Are there well-defined structural intermediates and traps observed during protein folding? A key question for those involved in the field of protein folding relates to the pathway or mechanism of folding. Many view protein folding as the progression of the protein from one structure to another until the final folded structure develops. In this view, the mechanism of folding is a specification of the intermediate structures. Of course, there may not be only one, but rather many pathways possible, all of which have different likelihoods of occurring. At the other end of the spectrum of possibilities, there may be an infinite number of pathways, such that any arbitrary unfolded structure takes its own path to the folded state. The answer to this question of the number and nature of the folding pathways will probably be different for different proteins. Simulations can shed light on these aspects of the nature of protein folding. A view of the kinetics of folding needs to acknowledge the fact that in general there are multiple pathways each with different likelihoods of occurrence as well as different times over which folding will be observed.

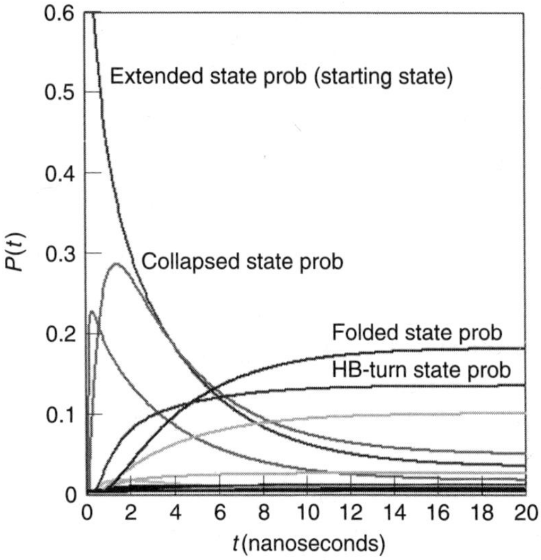

Figure 3.5 The relative concentration versus time of various conformations of an ensemble of peptides. The starting ensemble consisted entirely of fully extended conformations.

(See figure 3.5, which shows the relative populations of various types of protein conformations observed as a function of time for the folding of a small hairpin-shaped peptide, the hairpin motif from protein G.)

How is protein structure affected by temperature? Another issue of concern to people who study protein folding relates to the thermodynamic stability of proteins. As a collection of proteins is heated, they will unfold, but each member of the collection may adopt a different unfolding path. An important aspect of thermally induced folding and unfolding is the determination of the relative concentrations of various intermediate structures that are observed at each temperature. A tabulation of these structures can give some insight into the ways proteins might misfold, which, in turn, can be related to the effect a genetic mutation can have on protein stability and misfolding (figure 3.6).

What is the effect of mutations on both thermodynamics and kinetics? These questions are of more than academic interest. Genes contain the instructions for building proteins from amino acids. Genetic mutations usually cause disease because defective proteins are manufactured. There are many genetic diseases that have become known as *conformational diseases* [22]. These are diseases associated with mutations that cause proteins to misfold or to fold so slowly that they are unable to perform their function before being degraded by natural protein cleanup processes that exist in the cell. In light of this, a very important area that can be studied by simulation is how various types of mutation

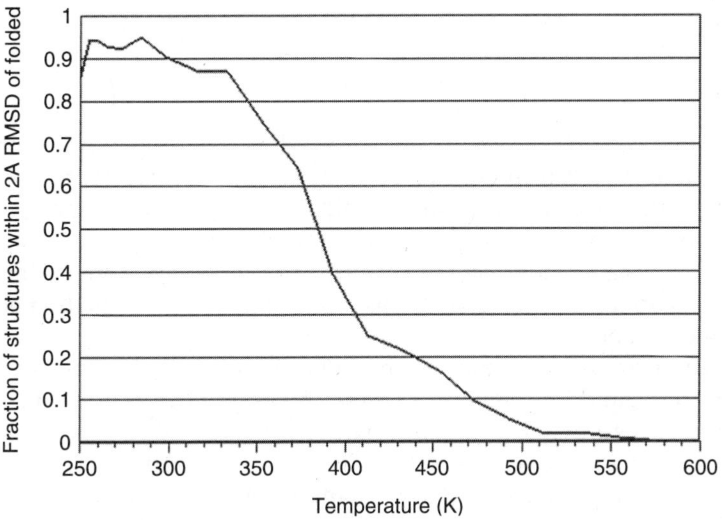

Figure 3.6 Melting curve for a protein generated from molecular dynamics simulations.

affect the structure, thermodynamic stability, and folding kinetics of proteins.

How do lipid membranes reorganize in the presence of a protein? Lipids are molecules with a water-soluble acidic (charged) head group connected to a long hydrophobic chain. In the presence of water, lipid molecules form into stable structures called bilayers (figure 3.7). In a bilayer, two

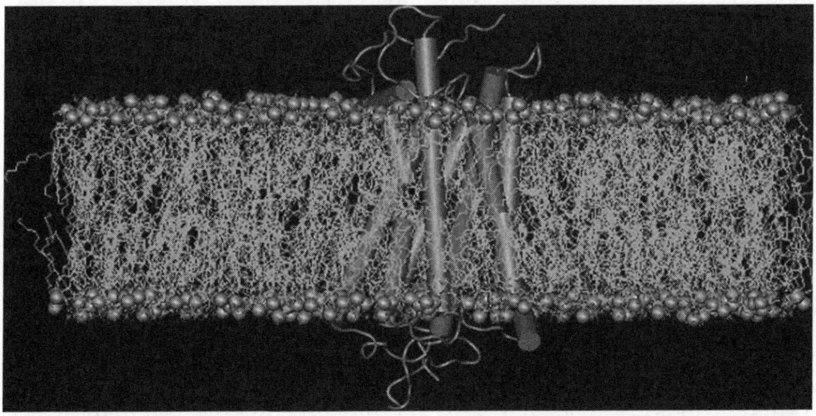

Figure 3.7 Image of the light-sensitive protein rhodopsin embedded in a model lipid bilayer representing a cell membrane. The protein is shown as a secondary structure cartoon, with the lipids drawn as lines (fatty acid tails) and spheres (head groups). This image was created with the program VMD [84].

sheets of acidic head groups form that are in contact with water, and the hydrophobic chains are contained between these layers. Lipid bilayers are a primary constituent of the cell membranes of all cells. Cells have a number of other molecules dissolved within these bilayers, such as proteins that are involved in controlled transport of molecular material into and out of the cell (transport proteins), and in the sensing of material in the cell's environment and the transfer of information from outside to inside the cell (signaling proteins). Membranes and their components are extremely important since most drugs work by affecting the transfer of material or information across the membrane.

Cell membranes contain a number of other molecules, such as cholesterol, and it is believed that the cholesterol has a profound effect on the structure and dynamical nature of the membrane. Cholesterol also surrounds and affects the behavior of membrane-bound proteins. The exact role of cholesterol and its importance to the cell membrane is not fully understood, and one important use of simulation is to study the cholesterol-induced dynamics and organization of material in the membrane [23].

A Short Historical Overview

Arguably, the first biomolecular simulation was a simulation of 216 water molecules for 2.17 picoseconds by Rahman and Stillinger in 1971 [24]. This simulation demonstrated that liquid water contains an extensive, but dynamic, network of water–water hydrogen bonds that fluctuate on a time scale of several picoseconds. Diffusion in water was also observed to proceed by a continuous, cooperative flow of several molecules of water rather than by a single molecule hopping from one site to another in the liquid.

The first detailed dynamical simulation of a protein was carried out in 1977 [25]. The protein in question was bovine pancreatic trypsin inhibitor (BPTI, PDB [26] id 1PTI), and it was simulated for a total of 8.8 ps in vacuum. Despite the short simulation time, it was possible to observe significant, low-frequency anharmonic motions of the protein. The first simulation of a protein in water occurred many years later with the simulation of BPTI in a hydrated protein crystal by van Gunsteren et al. [27].

In addition to dynamical properties, molecular simulations can also be used to study the function of proteins. For example, simulations can be used to determine the relative affinity of a protein for two different ligands. The first such binding free energy calculation was carried out to compare two ligands binding to the protein thermolysin by Bash et al. in 1987 [28].

Beyond ligand binding, proteins also act as enzymes to accelerate chemical reactions. These reaction processes involve the breaking and formation of covalent bonds, and therefore need to be treated quantum mechanically. Although it is not practical to model an entire protein quantum mechanically, hybrid quantum mechanical/molecular mechanical (QM/MM) methods have been developed. QM/MM methods embed a small quantum mechanical region (e.g., the substrate and active site side chains) in a larger context of molecular mechanics atoms (the surrounding protein and solvent). These methods were pioneered for the study of the reaction pathway of triosephosphate isomerase [29]. There is an extensive literature discussing the specific ways enzymes accelerate reactions [30–32], but the general mechanism involves decreasing the activation free energy for the reaction—in effect, stabilizing the transition state relative to the reactants.

Biomolecular simulations are not limited to proteins or other biopolymers. Simulations have contributed a great deal of detail to our understanding of other biological structures such as the lipid bilayers that make up cell membranes. The first simulation of a simplified bilayer was carried out in 1982 [33]. Because of computational resource limitations, the membrane was modeled as a bilayer of united atom aliphatic chains with an empirical restraint potential replacing the water above and below the membrane. The resulting 320 atoms were simulated for a total of 80 ps.

Computer power is starting to reach the point where biomolecular simulations can access the time scales and degree of sampling necessary to provide important insights into the thermodynamics and kinetics of protein folding. The first folding free energy landscape of a protein in explicit water was calculated by Boczko and Brooks in 1995 [34]. By examining the free energy of the protein as a function of a specific coordinate, such as the radius of gyration, it is possible to find potential transition states for folding and get a better picture of the set of conformations that are sampled by the protein at a particular temperature. In order to determine the free energy landscape, Boczko and Brooks combined data from a large number of separate simulations, each of which restrained the protein to a particular range of values for selected order parameters like the radius of gyration and fraction of native contacts. This allowed them to build up the free energy landscape for folding without simulating the 10^9–10^{15} time steps necessary to simulate a folding event.

The direct, long time scale simulation of protein folding is a formidable technical challenge. The longest single atomistic trajectory of a protein in water to date was carried out by Duan and Kollman in 1998 [21]. They simulated the 36-residue villin headpiece for 1 μs (5×10^8 time steps) starting from a thermally unfolded state. During the simulation, early events in protein folding such as helix formation and hydrophobic collapse were observed. In addition, the simulation clearly shows the protein collapsing and expanding as it searches for lower free energy structures. It is important to realize that folding is a stochastic, rather than a directed, process—a partially folded protein need not proceed directly to the folded state, but might unfold partially or completely before attaining a stable structure. In addition, folded proteins undergo continuous thermal fluctuations that result in periodic local denaturation and global unfolding events. These events are difficult to study experimentally, but are clearly observable in molecular simulations.

More recently, molecular dynamics simulations have been used to study peptide folding in a predictive fashion. Simulations have been used to predict the relative rates of folding for different $\beta\beta\alpha$ mutants [35] as well as to predict the structures of novel peptide sequences [36]. The continuous advances in computer power suggest that similar successes are on the horizon for systems as large as a typical protein (100–300 amino acids).

The Beta Hairpin Peptide from Protein G: A Case Study for the Interaction Between Experiment, Simulation, and Theory

The 16 C-terminal amino acids of the B1 domain of protein G form a beta hairpin structure—two strands connected by a turn. (See figure 3.8,

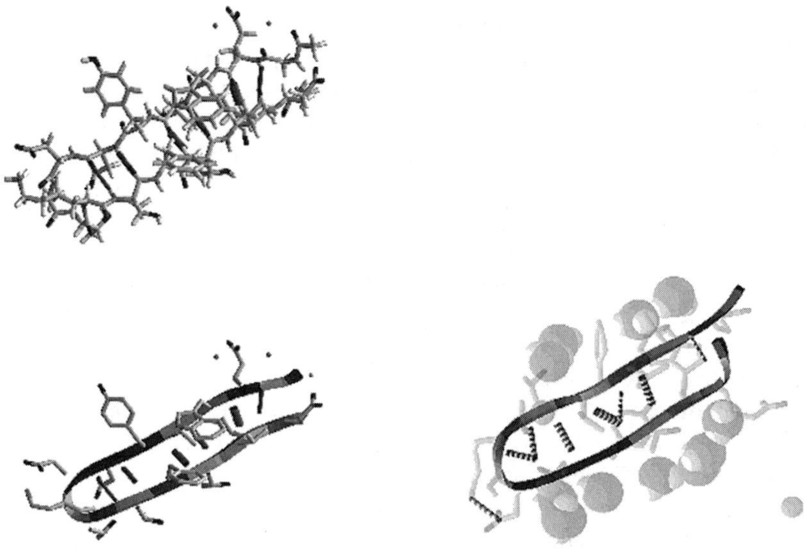

Figure 3.8 Three different views of the 16 amino acid C-terminal beta hairpin from the B1 domain of protein G. Counterclockwise, from the upper left, the images show a stick representation, a mixed stick and ribbon representation, and a ribbon representation with transparent side chains and nearby water molecules. The striped cylinders show native and nonnative hydrogen bonds between the two strands.

showing three different representations of this peptide.) This structure is stably folded on its own, and has emerged as one of the standard model systems used to study protein folding mechanisms. A range of experimental and computational approaches have been applied to elucidate the thermodynamics and kinetics of folding in this system.

Experimental work on this beta hairpin was initiated by Blanco et al. in 1994 [37]. They were the first to synthesize the 16-residue peptide and demonstrate that it was folded in a beta hairpin structure by NMR spectroscopy. The experimental study of the kinetics of folding in this system was pioneered by Munoz et al., who determined the folding time to be 6 μs using a laser temperature-jump experiment [38]. They also suggested that folding proceeded by a "zipping" mechanism, where the turn forms first, and then the beta-sheet hydrogen bonds form outward from the turn. In this model, formation of native hydrophobic contacts happens relatively late in the folding process.

This hydrogen-bond-centric view of folding was challenged by early simulations of the folding thermodynamics of this system. Dinner et al. studied the beta hairpin in implicit solvent using the CHARMm force field [39]. Their simulations showed a folding mechanism that was initiated by hydrophobic collapse. Following the formation of contacts between the hydrophobic side chains, native-like hydrogen bonds formed between those core residues and propagated outward toward the turn and the termini.

Pande and Rokhsar simulated the unfolding and refolding of the beta hairpin in water using an explicit solvent model [40]. Their studies showed an unfolding process that proceeded by an initial loss of hydrogen bonding, followed by dissolution of the hydrophobic core. This suggested a folding process that contained four distinct metastable states: the unfolded state, a more compact state with a partially solvated hydrophobic core, a state with a well-packed hydrophobic core but no native hydrogen bonds, and the folded state.

Detailed thermodynamic measurements of the folding transition were carried out by Honda et al. [41] Their data show no evidence for a significant folding intermediate and instead indicate that there is a single, cooperative folding transition. In contrast, many of the simulation studies have identified a partially folded state with native-like hydrophobic contacts as a significant intermediate. This conflict remains to be resolved; it is possible that one or more intermediates exists, but at a population which is so small that folding is effectively two-state.

Garcia and Sanbonmatsu were the first to apply a thermodynamic sampling technique called replica exchange molecular dynamics to the beta hairpin [42]. Replica exchange molecular dynamics is a method to enhance sampling in molecular simulations by running multiple coupled simulations of the same system at different temperatures.

In this case, Garcia and Sanbonmatsu simulated the beta hairpin in explicit solvent using the 1994 version of the AMBER force field. They observed an energy landscape with little or no barrier to folding, but also found nonnative trapped states where the peptide adopted an alpha-helical conformation. A subsequent study by Zhou et al. also used replica exchange in explicit solvent but made use of a more recent OPLS force field [43]. No helical intermediates were observed in that case. Further thermodynamic studies have concentrated on more detailed force field or solvent model comparisons (especially comparing implicit and explicit solvent results), though the current consensus is that stable nonnative helical structures are likely to be an artifact of the force fields used in early studies.

Attempts have also been made to simulate the kinetics of beta hairpin folding. One attempt made use of a distributed computing effort, where tens of thousands of independent short simulations of the beta hairpin in implicit solvent were used. The rate at which individual simulations reach the folded state could be observed, and correlated to the experimentally observed folding time constant of 6 μs. The kinetics have also been studied using a simulation technique called transition path sampling, which allows accurate estimations of the rates for rare transitions between two well-defined end states. Bolhuis modeled the folding process by two transitions—one from the unfolded manifold to an intermediate with native hydrophobic contacts, and one from that intermediate to the native state—and calculated a simulated folding time constant of 5 μs [44]. Further work has concentrated on the application of Markov chain-type models to this system, to produce a more complete and detailed picture of the folding kinetics [45,46]. A key insight of later studies has been that the distinction between folding mechanisms where hydrogen bonds form first and those where hydrophobic interactions form first is largely semantic. The same simulation can support either model depending on the exact definitions used for the identification of hydrogen bonds and hydrophobic contacts.

These extensive simulation efforts have prompted further experimental studies to determine the mechanism of folding in this system. Dyer and coworkers studied the kinetics of a series of insertion mutants where the hairpin loop was moved progressively further from the hydrophobic core and established that longer loops have slower folding kinetics [47]. This suggests that searching for the correct intrastrand contacts may be a diffusive, entropy-dominated process. The stability of the wild-type hairpin has also been revisited by by a comparison of earlier circular dichroism experiments with measurements of chemical shifts as a function of temperature [48]. This analysis indicated that the wild-type hairpin may be approximately 30% folded in water at room temperature.

RECENT ADVANCES IN MOLECULAR SIMULATION

Methodological Developments

There are always two major areas of methodological development in molecular simulation: improvements in the representation of the system (force field improvements) and improvements in how efficiently a simulation can visit all of the important conformations of a system (sampling improvements).

In terms of force field improvements, the next generation of biomolecular force fields moves beyond fixed charge models of electrostatic interactions to ones that include polarizability. In a polarizable force field, the electrostatic interactions between atoms are no longer described by a simple charge–charge interaction but rather by one that can include both fixed and polarizable charge–charge, charge–dipole, dipole–dipole, and higher multipolar interactions. Recent polarizable force fields for biomolecules include variants of the AMBER [49], CHARMm [50,51], and OPLS-AA [52] potentials as well as the newly developed AMOEBA [53,54] model. It remains to be seen whether the inclusion of polarizability will significantly improve the predictive abilities of these models. Another notable improvement in force fields is a movement away from a Fourier series representation of the dihedral angle term toward a spline-based model which offers more flexibility in fitting to the true quantum mechanical potential [55].

The most significant recent development in sampling techniques has been the wide adoption of enhanced sampling methods, including multicanonical sampling and replica exchange simulation. Multicanonical sampling corresponds to simulating the system with an effective potential that generates a random walk in energy space, allowing a broad sampling of molecular conformations [56]. The drawback of multicanonical sampling is that the appropriate effective potential needs to be determined by trial and error. Replica exchange simulation allows a similar broad sampling of energies and conformations, but achieves it by carrying out a series of coupled simulations at different temperatures. Each simulation has one copy (or replica) of the simulated system, and periodically the replicas at adjacent temperatures are exchanged with a Metropolis Monte Carlo-like acceptance rule [57]. A schematic of the replica exchange method is shown in figure 3.9. A common weakness of all enhanced sampling methods is that their computational cost increases significantly with system size—a method that works for alanine dipeptide in water may be impractical for simulating a solvated protein.

Another recent development relevant to systems biology has been the widespread use of approximate methods that estimate free energies by postprocessing simulation data. In these approaches, an

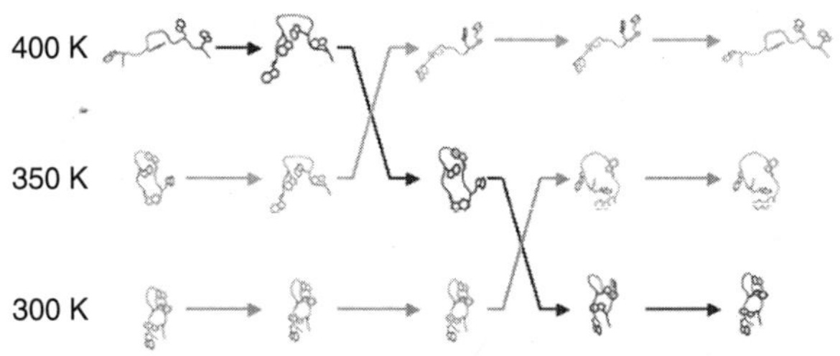

Figure 3.9 Schematic representation of the replica exchange simulation method. Independent replicas of the system are simulated at different temperatures and periodically exchange between temperatures based on Metropolis Monte Carlo-like acceptance criteria.

initial molecular dynamics simulation is carried out in explicit solvent. The coordinates of the biomolecule or biomolecular complex (excluding water and counterions) are saved at regular intervals. Afterward, the energy of each coordinate is recalculated using a molecular mechanics force field plus an implicit solvent model. These energies are averaged and used as an estimate of the free energy of the system. Such methods are often referred to as ES/IS (explicit solvent/implicit solvent), MM-PBSA (Molecular Mechanics–Poisson-Boltzman Surface Area), or MM-GBSA (Molecular Mechanics–Generalized Born Surface Area), depending on the implicit solvent model used [58,59]. They have been used to approximate binding free energies of protein–ligand [60], protein–protein [61], and protein–nucleic acid [3] complexes, as well as to compare the relative free energies of different biomolecular conformations [62,63]. All of these methods are empirical, with little connection to statistical mechanics. As a result, it is difficult or impossible to correctly account for entropic contributions to the free energies calculated by these approaches, though this is an active area of research [64]. For biomolecular association, this entropic term is often significant, making it the largest source of error in such calculations.

New Applications

Several recent examples of molecular modeling and simulation work are provided below. This is not an exhaustive list by any means, but is only meant to provide a flavor for the diversity of approaches that can be applied to model biological systems.

ELECTROSTATICS OF MICROTUBULES

One example of the modeling of a very large cellular structure is in the work of Baker et al. [65]. This work characterized the electrostatic

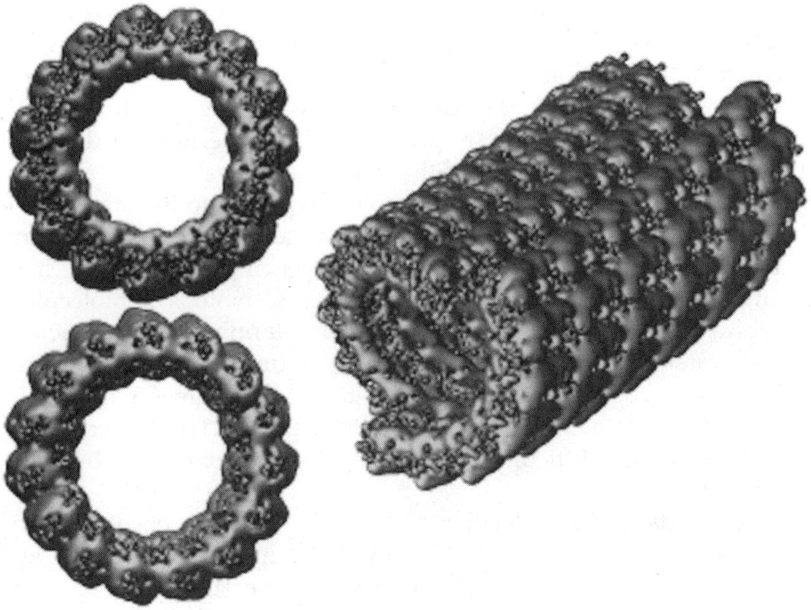

Figure 3.10 Electrostatic isopotential surfaces of a microtubule calculated by Baker et al. [65]. To view this figure in color, see the companion web site for *Systems Biology*, http://www.oup.com/us/sysbio.

field around a microtubule (figure 3.10). Microtubules are hollow cylindrical structures in the cells of eukaryotes that are part of their rigid cytoskeleton. These structures provide a number of different roles in cells besides providing structure; they are involved in material transport within the cell as well as cell motility and division. They are approximately 25 nm in diameter and, depending on their particular function, can have lengths from nanometers to millimeters. The system studied by Baker et al. consisted of over 600,000 atoms and the study employed a finite element approach for solving the Poisson–Boltzmann electrostatic equations for the field.

HIGH-PRECISION SYSTEMATIC CHARACTERIZATION AND COMPARISON OF FORCE FIELDS

Work is continuing on the development of improved force fields for modeling water. Most biological processes take place in an aqueous environment and, in fact, the structure of most biological molecules is determined to a large extent by the nature of their interaction with water. Therefore, the quality of the force field for modeling water–water interactions is of critical importance to the study of most biological molecular systems. Recently, Horn et al. [18] developed a fixed charge water model that is hoped to be superior for simulations that employ Ewald techniques for the treatment of long-ranged electrostatic interactions in systems studied with periodic boundary conditions. This model was developed by finding the force field parameters that give the best agreement with experimental data for the temperature dependence of the density and heat of vaporization of pure water. These observables are very sensitive to the strength of the water–water intermolecular hydrogen bond. Clearly it is important to have an accurate representation of hydrogen bonding, since it is so important to the structure and stability of biological molecules. Notably, even though it was not used to determine the parameters of the water–water interaction, the temperature dependence of the self-diffusion coefficient predicted by the model agrees remarkably well with experimental measurements (figure 3.11). Other water models have had difficulty reproducing this very important kinetic observable.

Force fields for biomolecular simulations have also been carefully assessed with respect to their ability to predict hydration free energies. Hydration free energies are a rather sensitive measure of the strength of interaction between a solute and water. These free energies can be accurately measured by a variety of experimental techniques. The various amino acid side chains exhibit very different hydration free energies and these differences are a key factor in determining the folding rates and stabilities of proteins. In a folded protein, hydrophilic amino acid side chains are mostly in contact with water, whereas hydrophobic amino acid side chains are mostly in contact with other hydrophobic groups.

Recent work by Shirts et al. [66] demonstrated the ability of classical models to reproduce experimentally determined hydration free energies. The computation of hydration free energies by simulation is a computationally demanding task. Simulations that are long enough to produce adequately precise measurements of the hydration free energy for the comparison and assessment of force field accuracy required the Folding@Home [67] distributed cluster. This novel computational system makes use of tens of thousands (now over 100,000) of contributed computers that run a specially developed screen saver that is capable of performing simulations on small peptides and biomolecules. The hydration free energy results (figure 3.12) are illuminating in that, although they are not accurate in an *absolute* sense, most force fields do very well at reproducing the *relative* solvation free energies of the various amino acid side-chain analogs. This is remarkable because this observable was not used to develop the force fields in the first place. Although improvements are possible, the force fields are probably adequate for many purposes related to structure prediction and, perhaps, also ligand binding.

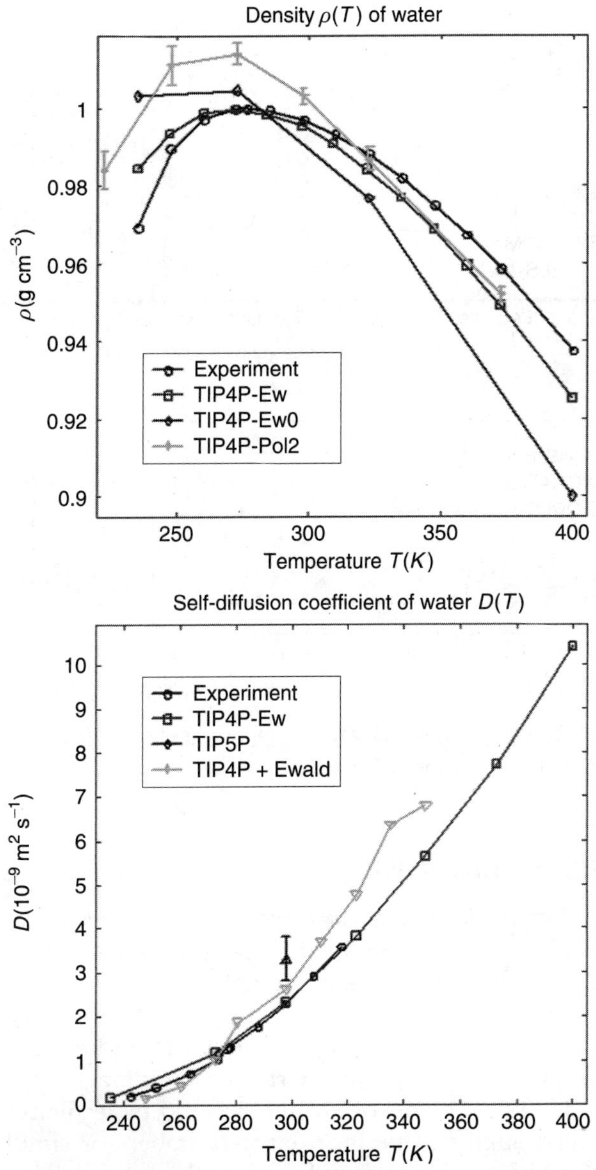

Figure 3.11 Density and self-diffusion coefficient of several water models as a function of temperature.

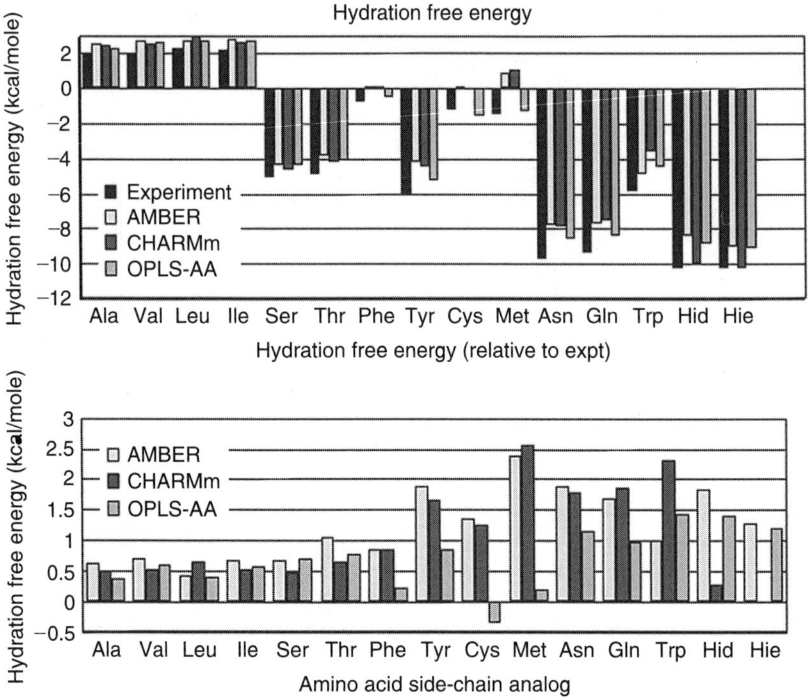

Figure 3.12 Hydration free energies of amino acid side chains calculated from molecular simulations using different force fields.

CONCLUSIONS: THE FUTURE

At the end of their 1987 book, *Dynamics of Proteins and Nucleic Acids* [68], McCammon and Harvey outlined the key challenges to simulation and modeling. It is very interesting to note that although considerable progress has been made since that time, the challenge areas are still much the same. Future efforts will continue to address the issues of improved representations of the intermolecular interactions, improved ability to address problems of greater size and molecular complexity, and improved ability to perform larger numbers of simulations on systems at a given level of complexity. In the future, these improvements will almost certainly be realized from efforts and advances in the same areas of (1) more, faster, and less expensive computer hardware; (2) hardware architectural features, such as deep pipelining to support vector operations and high-speed interprocessor communication to support parallelism; (3) simulation and modeling software designed to take advantage of these modern computer hardware architectures; and (4) improved theoretical approaches to modeling and simulation.

Exploiting Advances in Computer Architecture

As has been the case since the invention of the integrated circuit, the landscape of computational capability is changing at an amazing rate. Computers have become faster and less expensive, but so has the ability to build clusters of personal computers and workstations with high-speed interconnects. Special on-chip hardware that supports a high degree of pipelining, vector operations, caching, and multiple 64-bit floating point arithmetic units has become almost standard. Most production computer systems consist of nodes with multiple processors having these capabilities. Advances in and standardization of software interfaces for the development of parallel software have enabled the ability to exploit these kinds of computer configurations. Many of the standard modeling and simulation software packages have been modified to exploit these parallel configurations.

The Internet has produced a new computing paradigm that we are still learning how to exploit. With grid computing one will be able to transparently distribute tasks and data to computational capability across the Internet. One form of this exists in the highly successful Folding@Home project at Stanford, where 100,000 people have contributed otherwise idle computer time to protein folding simulations performed by a screen saver [67]. A similar project operated by Oxford University has been established with over 2.6 million contributed computers that are performing a search for anticancer drug candidates through the use of ligand-protein docking [69].

In addition to parallel clusters and grid computing approaches, there is always interest in supercomputing efforts as well. The most notable current example of this is the BlueGene project at IBM [70]. This system consists of cabinets of 1024 nodes of two processors per node. An important aspect of this system is the existence of multiple high-performance interconnection networks between the processors that allow scaling to over 64,000 nodes. These include both a nearest neighbor type of topology and a tree-type topology.

Advances in Theoretical Methods

Among the most important theoretical advances for the future will be the development of better interaction models. These will take several forms. It may be the case that further improvements can be made using the functional forms in the current generation of fixed charge force fields. With the computational capacity now available, effort could be spent performing a thorough assessment of these force fields for their ability to predict a variety of biologically important properties. Second, after quantifying inadequacies exhibited by these force fields, one needs to consider what additional physical phenomena need to be included, such as electronic polarization. Considerable effort will then most certainly involve the development and validation of a systematic methodology to produce the parameters needed for a polarizable force field. Very few such force fields exist today, and generally they have not been very well characterized with respect to their ability to improve the accuracy of biomolecular modeling and simulation studies. For important classes of biologically important molecules, such as proteins, fixed charge force fields have been improved over a period of decades. Therefore, in spite of their greater computational cost, in some respects the current generation of polarizable force fields may even be worse than the current generation of fixed charge models.

There will always be interest in the simulation and modeling of very large biological structures, many of which are too large for an atomistic simulation. Therefore, the future will involve considerable effort in the development of theoretical approaches to support coarse-grained and multiscale modeling. In coarse-grained models, groups of atoms are treated as a single unit and in the simulation these units interact with an effective potential that can be obtained in principle from atomistic modeling. In multiscale models, some regions of a molecular system are treated atomistically (or even quantum mechanically) and these regions are placed in contact with others treated with much less detail, such as with a continuum or finite element model of the material. Currently several research groups are exploring these areas and a great deal of theoretical work has yet to be done. However, this work is very important, since there will always be technologically and medically important biological phenomena that happen on time and length scales that defy a detailed atomistic description.

Coarse graining and multiscale approaches allow the modeling of larger molecular systems. However, there are also small as well as large molecular systems that exhibit extremely long time scale phenomena. Theoretical advances to be developed in the near future will certainly include methods to treat the kinetic modeling of rare events. Some of these will certainly include development and application of transition path sampling [71] and Markov modeling [72–75].

Benefits and Trends

Improvements in the areas mentioned above will permit qualitative improvements in our understanding of biology. As stated by McCammon and Harvey, with each improvement we can perform a larger number of simulations at a given level of complexity as well as simulate more complicated systems. The ability to model larger, more complex molecular systems and for longer time scales will help us understand entirely new types of biophysical processes and phenomena that cannot be addressed today.

Greater computational capacity allows the use of more sophisticated models for the molecular interactions (table 3.1). Since McCammon and Harvey's 1987 work, great strides have been taken even in the use of fixed charge force fields: united atom simulations, the norm in the 1980s, are now much less common; explicit representation of solvent, rare in biomolecular simulations of the 1980s, has now become the standard; Ewald electrostatics, rare in the 1980s, are now used for the more accurate modeling of electrostatic interactions. These kinds of trends in the quality of modeling will certainly continue into the future.

Finally, with greater computing capability and capacity comes also improved sampling. The future should bring much greater and much needed statistical precision associated with the properties measured from simulations and this will greatly increase their utility and drive further improvements.

Table 3.1 This table gives some indication of how the scale of simulation has increased over time and particularly how the increase in the scale of simulations tracks the increase in supercomputer capability shown in figure 3.2.

Year	Investigators	System	Computational platform	Time scale
1955	Fermi et al. [76]	64-particle chain with nonlinear interactions	MANIAC I	50,000 cycles
1959	Alder and Wainwright [77]	108 hard spheres	IBM 704	3000 collisions
1964	Rahman [78]	Lennard-Jones liquid (argon) 864 particles	CDC 3600	10 ps
1974	Stillinger and Rahman [79]	Water 216 molecules	IBM 360/195	22 ps
1977	McCammon et al. [25]	BPTI (in vacuum)		8.8 ps
1983	Van Gunsteren et al. [27]	BPTI (in hydrated crystal)	Cyber 170/760	20 ps
1994	York et al. [80]	BPTI (in hydrated crystal)	Cray Y-MP	1 ns
1998	Sheinerman and Brooks [81]	Segment B1 of protein G	CrayT3D,T3E, C-90	7.6 ns
1998	Duan and Kollman [21]	Villin headpiece (36 residues) in 3000 water molecules	Cray T3D,T3E	1 μs

The Growing Connection Between Modeling and Experiment

An important new trend that will become very strong in the future is the blending of experimental and modeling approaches. Over the last two decades, as the capabilities and accuracy of modeling and simulation have improved, experimental measurement techniques have also improved. Experiments using atomic force microscopy and scanning tunneling microscopy are now capable of probing and resolving atomic dimensions. Time-resolved spectroscopic methods on proteins and peptides are capable of resolving phenomena that occur on microsecond time scales. In fact, we are close to entering an era of nanotechnology where experimental and modeling technologies can address similar time and length scales. This is certainly no threat to either field, since experimental and modeling techniques give very different kinds of information. In fact, the best benefit is obtained for science if they are used together. Models are more believable when they make predictions that are supported by experiment. And when they are believable they provide additional useful information that cannot be obtained from experiment. Therefore, in the future one should expect a growing synergy between experiment and modeling, with modeling being a key tool in the interpretation of experiments.

Grand Challenge Problems

There is no end of important phenomena that can be addressed with the realization of some of the improvements mentioned above. Protein structure prediction, and the understanding of the mechanisms, pathways, and kinetics of protein folding and other forms of biological self-assembly, are still outstanding problems with high technological and medical relevance.

There are also large-scale simulation challenges, such as RNA folding and the structure and function of large cellular structures that need to be understood at the molecular level. These include the ribosome, the molecular factory used by the cell to build proteins, and the proteosome, the molecular factory used by the cell to degrade them. Large-scale simulations could advance our understanding of the role and function of chaperones, proteins that aid in the folding of other proteins. Other frontiers include the simulation of protein–ligand and protein–protein interactions, and the detailed modeling of the functioning of membrane-bound proteins.

ACKNOWLEDGMENTS

The authors would like to acknowledge Frank Suits, Mike Pitman, and Alexander Balaeff for help with the figures in this chapter.

REFERENCES

1. Hermans, J. and L. Wang. Inclusion of loss of translational and rotational freedom in theoretical estimates of free energies of binding. Application to a complex of benzene and mutant T4 lysozyme. *Journal of the American Chemical Society*, 119(11):2707–14, 1997.
2. Kombo, D. C., B. Jayaram, K. J. McConnell, et al. Calculation of the affinity of the lambda repressor-operator complex based on free energy component analysis. *Molecular Simulation*, 28(1-2):187–211, 2002.
3. Jayaram, B., K. McConnell, S. B. Dixit, et al. Free-energy component analysis of 40 protein-DNA complexes: a consensus view on the thermodynamics of binding at the molecular level. *Journal of Computational Chemistry*, 23:1–14, 2002.
4. Misra, V. K., J. L. Hecht, A. S. Yang, et al. Electrostatic contributions to the binding free energy of the lambda cl repressor to DNA. *Biophysical Journal*, 75:2262–73, 1998.
5. Paillard, G., C. Deremble and R. Lavery. Looking into DNA recognition: zinc finger binding specificity. *Nucleic Acids Research*, 32(22):6673–82, 2004.
6. Muller-Plathe, F. Coarse-graining in polymer simulation: from the atomistic to the mesoscopic scale and back. *Chemphyschem*, 3:754–69, 2002.
7. Lopez, C. F., P. B. Moore, J. C. Shelley, et al. Computer simulation studies of biomembranes using a coarse grain model. *Computer Physics Communications*, 147(1-2):1–6, 2002.
8. Kuntz, I. D., E. C. Meng, and B. K. Shoichet. Structure-based molecular design. *Accounts of Chemical Research*, 27:117–23, 1994.
9. MacKerell, A. D. J., B. Brooks, C. L. Brooks, et al. CHARMm: the energy function and its parameterization with an overview of the program. In P. V. R. Schleyer (Ed.), *The Encyclopedia of Computational Chemistry* (pp. 271–7). John Wiley, New York, 1998.
10. Kaminski, G., R. A. Friesner, J. Tirado-Rives, et al. Evaluation and improvement of the OPLS-AA force field for proteins via comparison with accurate quantum chemical calculations on peptides. *Journal of Physical Chemistry B*, 105:6474–87, 2001.
11. Cornell, W. D., P. Cieplak, C. I. Bayly, et al. A second generation force field for the simulation of proteins and nucleic acids. *Journal of the American Chemical Society*, 117:5179–97, 1995.
12. van Gunsteren, W. F., X. Daura and A. E. Mark. The GROMOS force field. In P. V. R. Schleyer (Ed.), *The Encyclopedia of Computational Chemistry* (pp. 1211–16). John Wiley, New York, 1998.
13. Zhou, R. and B. J. Berne. Can a continuum solvent model reproduce the free energy landscape of a beta-hairpin folding in water? *Proceedings of the National Academy of Sciences USA*, 99(20):12777–82, 2002.
14. Berendsen, H. J. C., J. P. M. Postma, W. F. van Gunsteren, et al. Interaction models for water in relation to protein hydration. In B. Pullman (Ed.), *Intermolecular Forces* (pp. 331–42). Reidel, Dordrecht, 1981.
15. Jorgensen, W. L., J. Chandrasekhar, J.D. Madura, et al. Comparison of simple potential functions for simulating liquid water. *Journal of Chemical Physics*, 79:926–35, 1983.

16. Berendsen, H. J. C., J. R. Grigera and T. P. Straatsma. The missing term in effective pair potentials. *Journal of Physical Chemistry*, 91(24):6269–71, 1987.
17. Mahoney, M. W. and W. L. Jorgensen. A five-site model for liquid water and the reproduction of the density anomaly by rigid, nonpolarizable potential functions. *Journal of Chemical Physics*, 112(20):8910–22, 2000.
18. Horn, H. W., W. C. Swope, J.W. Pitera, et al. Development of an improved four-site water model for biomolecular simulations: TIP4P-Ew. *Journal of Chemical Physics*, 120(20):9665–78, 2004.
19. Ryckaert, J. P., G. Ciccotti, and H. J. C. Berendsen. Numerical integration of the Cartesian equations of motion of a system with constraints: molecular dynamics of *n*-alkanes. *Journal of Computational Physics*, 23:327–41, 1977.
20. Andersen, H. C. RATTLE: a velocity version of the SHAKE algorithm for molecular dynamics. *Journal of Computational Physics*, 52:24–34, 1983.
21. Duan, Y. and P. A. Kollman. Pathways to a protein folding intermediate observed in a 1-microsecond simulation in aqueous solution. *Science*, 282: 740, 1998.
22. Dobson, C. M. Protein folding and its links with human disease. In *Biochemical Society Symposia Volume 68: From Protein Folding to New Enzymes* (pp. 1–26). Portland Press, London, 2001.
23. Pitman, M. C., F. Suits, A. D. Mackerell, et al. Molecular-level organization of saturated and polyunsaturated fatty acids in a phosphatidylcholine bilayer containing cholesterol. *Biochemistry*, 43(49):15318–28, 2004.
24. Rahman, A. and F. H. Stillinger. Molecular dynamics study of liquid water. *Journal of Chemical Physics*, 55:3336–59,1971.
25. McCammon, J. A., B. R. Gelin and M. Karplus. Dynamics of folded proteins. *Nature*, 267(5612):585–90, 1977.
26. Berman, H. M., J. Westbrook, Z. Feng, et al. The Protein Data Bank. *Nucleic Acids Research*, 28:235–42, 2000.
27. van Gunsteren, W. F., H. J. C. Berendsen, J. Hermans, et al. Computer-simulation of the dynamics of hydrated protein crystals and its comparison with x-ray data. *Proceedings of the National Academy of Sciences USA*, 80(14):4315–19, 1983.
28. Bash, P. A., U. C. Singh, F. K. Brown, et al. Calculation of the relative change in binding free energy of a protein-inhibitor complex. *Science*, 235(4788): 574–6, 1987.
29. Bash, P. A., M. J. Field, R. C. Davenport, et al. Computer simulation and analysis of the reaction pathway of triosephosphate isomerase. *Biochemistry*, 30(24):5826–32, 1991.
30. Warshel, A. Energetics of enzyme catalysis. *Proceedings of the National Academy of Sciences USA*, 75(11):5250–4, 1978.
31. Kollman, P. A., B. Kuhn, O. Donini, et al. Elucidating the nature of enzyme catalysis utilizing a new twist on an old methodology: quantum mechanical-free energy calculations on chemical reactions in enzymes and in aqueous solution. *Accounts of Chemical Research*, 34:72–9, 2001.
32. Garcia-Viloca, M., J. Gao, M. Karplus, et al. How enzymes work: analysis by modern rate theory and computer simulations. *Science*, 303(5655): 186–95, 2004.
33. Vanderploeg, P. and H. J. C. Berendsen. Molecular-dynamics simulation of a bilayer-membrane. *Journal of Chemical Physics*, 76:3271–6, 1982.

34. Boczko, E. M. and C. L. Brooks. First-principles calculation of the folding free-energy of a 3-helix bundle protein. *Science*, 269(5222):393–6, 1995.
35. Snow, C. D., H. Nguyen, V. S. Pande, et al. Absolute comparison of simulated and experimental protein-folding dynamics. *Nature*, 420(6911): 102–6, 2002.
36. Simmerling, C., B. Strockbine and A.E. Roitberg. All-atom structure prediction and folding simulations of a stable protein. *Journal of the American Chemical Society*, 124(38):11258–9, 2002.
37. Blanco, F. J., G. Rivas and L. Serrano. A short linear peptide that folds into a native stable beta-hairpin in aqueous-solution. *Nature Structural Biology*, 1:584–90, 1994.
38. Munoz, V., P. A. Thompson, J. Hofrichter, et al. Folding dynamics and mechanism of beta-hairpin formation. *Nature*, 390:196–8, 1997.
39. Dinner, A. R., T. Lazaridis and M. Karplus. Understanding beta-hairpin formation. *Proceedings of the National Academy of Sciences USA*, 96(16): 9068–73, 1999.
40. Pande, V. S. and D. S. Rokhsar. Molecular dynamics simulations of unfolding and refolding of a beta-hairpin fragment of protein G. *Proceedings of the National Academy of Sciences USA*, 96(16):9062–7, 1999.
41. Honda, S., N. Kobayashi and E. Munekata. Thermodynamics of a beta-hairpin structure: evidence for cooperative formation of folding nucleus. *Journal of Molecular Biology*, 295:269–78, 2000.
42. Garcia, A. E. and K. Y. Sanbonmatsu. Exploring the energy landscape of a beta hairpin in explicit solvent. *Proteins: Structure, Function, and Genetics*, 42:345–54, 2001.
43. Zhou, R. H., B. J. Berne and R. Germain. The free energy landscape for beta hairpin folding in explicit water. *Proceedings of the National Academy of Sciences USA*, 98(26):14931–6, 2001.
44. Bolhuis, P. G. Transition-path sampling of beta-hairpin folding. *Proceedings of the National Academy of Sciences USA*, 100(21):12129–34, 2003.
45. Swope, W. C., J. W. Pitera, F. Suits, et al. Describing protein folding kinetics by molecular dynamics simulations. 2. Example applications to alanine dipeptide and beta-hairpin peptide. *Journal of Physical Chemistry B*, 108(21):6582–94, 2004.
46. Evans, D. A. and D. J. Wales. Folding of the GB1 hairpin peptide from discrete path sampling. *Journal of Chemical Physics*, 121:1080–90, 2004.
47. Dyer, R. B., S. J. Maness, E.S. Peterson, et al. The mechanism of beta-hairpin formation. *Biochemistry*, 43(36):11560–6, 2004.
48. Fesinmeyer, R. M., F. M. Hudson, and N. H. Andersen. Enhanced hairpin stability through loop design: the case of the protein G B1 domain hairpin. *Journal of the American Chemical Society*, 126(23):7238–43, 2004.
49. Cieplak, P., J. Caldwell and P. Kollman. Molecular mechanical models for organic and biological systems going beyond the atom centered two body additive approximation: aqueous solution free energies of methanol and *N*-methyl acetamide, nucleic acid base, and amide hydrogen bonding and chloroform/water partition coefficients of the nucleic acid bases. *Journal of Computational Chemistry*, 22(10):1048–57, 2001.
50. Patel, S., A. D. Mackerell and C. L. Brooks. CHARMm fluctuating charge force field for proteins. II. Protein/solvent properties from molecular

dynamics simulations using a nonadditive electrostatic model. *Journal of Computational Chemistry*, 25(12):1504–14, 2004.

51. Patel, S. and C. L. Brooks. CHARMm fluctuating charge force field for proteins. I. Parameterization and application to bulk organic liquid simulations. *Journal of Computational Chemistry*, 25:1–15, 2004.
52. Kaminski, G. A., H. A. Stern, B. J. Berne, et al. Development of a polarizable force field for proteins via ab initio quantum chemistry: first generation model and gas phase tests. *Journal of Computational Chemistry*, 23(16):1515–31, 2002.
53. Grossfield, A., P. Y. Ren and J.W. Ponder. Ion solvation thermodynamics from simulation with a polarizable force field. *Journal of the American Chemical Society*, 125(50):15671–82, 2003.
54. Ren, P. Y. and J. W. Ponder. Temperature and pressure dependence of the AMOEBA water model. *Journal of Physical Chemistry B*, 108(35): 13427–37, 2004.
55. Feig, M., A. D. MacKerell and C. L. Brooks. Force field influence on the observation of pi-helical protein structures in molecular dynamics simulations. *Journal of Physical Chemistry B*, 107(12):2831–6, 2003.
56. Berg, B. A. and T. Neuhaus. Multicanonical algorithms for 1st order phase-transitions. *Physics Letters B*, 267:249–53, 1991.
57. Hansmann, U. H. E. Parallel tempering algorithm for conformational studies of biological molecules. *Chemical Physics Letters*, 281(1-3):140–50, 1997.
58. Vorobjev, Y. N. and J. Hermans. ES/IS: estimation of conformational free energy by combining dynamics simulations with explicit solvent with an implicit solvent continuum model. *Biophysical Chemistry*, 78(1-2):195–205, 1999.
59. Kollman, P. A., I. Massova, C. Reyes, et al. Calculating structures and free energies of complex molecules: combining molecular mechanics and continuum models. *Accounts of Chemical Research*, 33(12):889–97, 2000.
60. Hou, T. J., S. L. Guo, and X. J. Xu. Predictions of binding of a diverse set of ligands to gelatinase-A by a combination of molecular dynamics and continuum solvent models. *Journal of Physical Chemistry B*, 106(21): 5527–35, 2002.
61. Gohlke, H., C. Kiel and D. A. Case. Insights into protein-protein binding by binding free energy calculation and free energy decomposition for the Ras-Raf and Ras-RaIGDS complexes. *Journal of Molecular Biology*, 330:891–913, 2003.
62. Vorobjev, Y. N., J. C. Almagro and J. Hermans. Discrimination between native and intentionally misfolded conformations of proteins: ES/IS, a new method for calculating conformational free energy that uses both dynamics simulations with an explicit solvent and an implicit solvent continuum model. *Protein: Structure, Function, and Genetics*, 32:399–413, 1998.
63. Lee, M. R., D. Baker and P. A. Kollman. 2.1 and 1.8 angstrom average C-alpha RMSD structure predictions on two small proteins, HP-36 and S15. *Journal of the American Chemical Society*, 123:1040–6, 2001.
64. Swanson, J. M. J., R. H. Henchman and J. A. McCammon. Revisiting free energy calculations: a theoretical connection to MM/PBSA and direct

calculation of the association free energy. *Biophysical Journal*, 86:67–74, 2004.

65. Baker, N. A., D. Sept, M. J. Holst, et al. The adaptive multilevel finite element solution of the Poisson-Boltzmann equation on massively parallel computers. *IBM Journal of Research and Development*, 45(3-4): 427–38, 2001.
66. Shirts, M. R., J. W. Pitera, W. C. Swope, et al. Extremely precise free energy calculations of amino acid side chain analogs: comparison of common molecular mechanics force fields for proteins. *Journal of Chemical Physics*, 119(11):5740–61, 2003.
67. Folding@Home, http://folding.stanford.edu.
68. McCammon, J. A. and S. C. Harvey. *Dynamics of Proteins and Nucleic Acids*. Cambridge University Press, Cambridge, 1987.
69. University of Oxford, http://www.chem.ox.ac.uk/curecancer.html.
70. Allen, F., G. Almasi, W. Andreoni, et al. Blue Gene: a vision for protein science using a petaflop supercomputer. *IBM Systems Journal*, 40:310–27, 2001.
71. Bolhuis, P. G., C. Dellago, P. L. Geissler, et al. Transition path sampling: throwing ropes over mountains in the dark. *Journal of Physics: Condensed Matter*, 12:A147, 2000.
72. de Groot, B. L., X. Daura, A. E. Mark, et al. Essential dynamics of reversible peptide folding: memory-free conformational dynamics governed by internal hydrogen bonds. *Journal of Molecular Biology*, 309:299–313, 2001.
73. Apaydin, M. S., D. L. Brutlag, C. Guestrin, et al. Stochastic roadmap simulation: an efficient representation and algorithm for analyzing molecular motion. *Journal of Computational Biology*, 10(3-4):257–81, 2003.
74. Swope, W. C., J. W. Pitera and F. Suits. Describing protein folding kinetics by molecular dynamics simulations. 1. Theory. *Journal of Physical Chemistry B*, 108(21):6571–81, 2004.
75. Singhal, N., C. D. Snow and V. S. Pande. Using path sampling to build better Markovian state models: predicting the folding rate and mechanism of a tryptophan zipper beta hairpin. *Journal of Chemical Physics*, 121:415–25, 2004.
76. Fermi, E., J. Pasta and S. Ulam, Studies of nonlinear problems. In *Collected Works of Enrico Fermi* (pp. 978–88). University of Chicago Press, Chicago, 1965.
77. Alder, B. J. and T. E. Wainwright. Studies in molecular dynamics. 1. General Method. *Journal of Chemical Physics*, 31:459–66, 1959.
78. Rahman, A. Correlations in motion of atoms in liquid argon. *Physical Review A: General Physics*, 136(2A):A405, 1964.
79. Stillinger, F. H. and A. Rahman. Improved simulation of liquid water by molecular-dynamics. *Journal of Chemical Physics*, 60:1545–57, 1974.
80. York, D. M., A. Wlodawer, L. G. Pedersen, et al. Atomic-level accuracy in simulations of large protein crystals. *Proceedings of the National Academy of Sciences USA*, 91(18):8715–18, 1994.
81. Sheinerman, F. B. and C. L. Brooks III. Molecular picture of folding of a small alpha/beta protein. *Proceedings of the National Academy of Sciences USA*, 95:1562–7, 1998.

82. Robien, M. A., G. M. Clore, J. G. Omichinski, et al. 3-Dimensional solution Structure of the E3-binding domain of the dihydrolipoamide succinyl transferase core from the 2-oxoglutarate dehydrogenase multienzyme complex of *Escherichia coli*. *Biochemistry*, 31(13):3463–71, 1992.
83. Delano, W. PyMOL. http://pymol.sourceforge.net. 2001.
84. Humphrey, W., A. Dalke and K. Schulten. VMD: visual molecular dynamics. *Journal of Molecular Graphics*, 14:33–8, 1996.

4

Global Gene Expression Assays: Quantitative Noise Analysis

G. A. Held, Gustavo Stolovitzky, & Yuhai Tu

The last decade has witnessed a shift in molecular biology from methods that probe hypotheses a few molecules at a time toward whole genome measurements. Global gene expression assays have enabled the monitoring of the transcription levels of tens of thousands of genes simultaneously [1–3]. In the near future, it will be possible to profile all of the nonrepetitive sequences in the genomes of higher organisms [4], including *H. sapiens*, with only a few DNA gene chips. This will allow one to obtain a global view of the transcriptomes corresponding to different cell phenotypes. Such capability will greatly accelerate and perhaps fundamentally change biomedical research and development in many areas, ranging from developing advanced diagnostics to unraveling complex biological pathways and networks, to eventually facilitating individual-based medicine [1,2].

Interestingly, the power of global gene expression assays brings along its own drawbacks—useful information is typically measured amidst high levels of noise. In general, the changes in the measured transcript values between different experiments are due to both biological variations (corresponding to real differences between different cell types and tissues) and experimental noise. To correctly interpret this data, it is crucial to understand the sources of the experimental noise. As will be demonstrated, systematic study of the noise in such data enables one to assign a meaningful statistical significance to observed transcriptional changes. In addition, understanding the sources of noise provides a useful guide in attempting to improve the technologies. Finally, it provides an effective and systematic means of comparing the reliability of the various global gene expression assays available today.

The most mature global gene expression technology is arguably the microarray. In all of its implementations (cDNA arrays [2], oligonucleotide arrays [1,5], etc.), this transcription profiling method exhibits significant technology-dependent noise. Models that characterize this noise through the study of replicate measurements [6] have been developed and provide a measure of security against false discoveries.

An alternative gene expression profiling method employs the sequencing of short sequence tags derived from the ends of messenger RNA.

This methodology encompasses the techniques of SAGE [7,8] and Massively Parallel Signature Sequencing (MPSS) [3,9]. MPSS represents a powerful alternative to microarray technologies. It provides a sensitive measure of gene expression without requiring a priori knowledge of transcribed sequences. The MPSS process is complex; from the extraction of the total RNA to the quantification of transcripts, there are a number of steps that contribute to the total noise. One significant difference between MPSS and microarray technologies is that MPSS produces data in digital format, whereas microarray data is analog [10]. (While the spot intensities obtained from microarray technologies are digital, these values do not correspond to a specific number of target molecules bound to a particular spot; that is, the data is a digital representation of a signal that is analog relative to the intensity resolution of the optical scanner.)

In this chapter, techniques for identifying and quantifying sources of noise through the analysis of replicate experiments [6,11] are discussed. In utilizing these techniques, one systematically studies the variations between equivalent experiments which, in the absence of experimental variability and noise, would yield identical results. An analysis is presented here of replicate experiments carried out with two systems: Affymetrix GeneChip microarrays [12] and MPSS technology [13]. In presenting these analyses, the intent is both to illustrate the general technique and to observe the inherent differences between the analysis of analog and digital gene expression data. Finally, it is shown how noise analysis provides an effective method for quantifying the relative sensitivity of different experimental techniques.

QUANTITATIVE NOISE ANALYSIS OF DNA MICROARRAY EXPERIMENTS

The data and results presented in this section are largely derived from the quantitative study of noise in Affymetrix GeneChips presented in ref. [12], where more details on sample preparation and data processing can be found. The data used in this section can be downloaded from http://www.research.ibm.com/FunGen/.

Materials and Methods

In this section an analysis of the experimental noise introduced during the sequential processing steps in high-density oligonucleotide-based microarray (Affymetrix) assays is presented. Elucidating the sources of noise may be of help in identifying those experimental steps that need to be modified to improve the signal to noise ratio. The results presented here show that it is the hybridization (including the subsequent readout) step, as opposed to the sample preparation step, where most of the noise originates. Based on these results, a data analysis method that takes into consideration the quantitative characterization of the noise, and thus provides a tool for evaluating the statistical significance of gene expression changes from different microarray experiments, is proposed.

The measurement noise is studied through replicate experiments in which gene expression levels of a cell line are measured multiple times. Two sources of experimental noise can be identified, starting from the extracted mRNA, to the final readout of the gene expression levels: the prehybridization target sample preparation steps on the one hand, and the hybridization and subsequent readout processes (including staining and scanning) on the other. For simplicity, these two sources of noise will be referred to as sample preparation noise and hybridization noise. In order to separate the noise sources due to these two factors, an analysis will be presented of previously reported [12] multiple replicate experiments where, at different stages of the experiment, the sample is divided equally into multiple aliquots, and the subsequent steps of the experiment are carried out independently. In this analysis of Affymetrix GeneChips, mRNA from cells of a human Burkitt's lymphoma cell line (Ramos) is used for the replicate experiments. Total RNA is extracted from the Ramos cells. The purified RNA sample is subsequently separated equally into several subgroups. Each subgroup independently goes through the target preparation steps, composed of the reverse transcription (RT) step and the in vitro transcription (IVT) step. At the end of the target sample preparation, each of the subgroups is again split into several samples, each of which is independently hybridized to different Affymetrix U95A GeneChip arrays. The experimental design is shown schematically in figure 4.1. To have sound

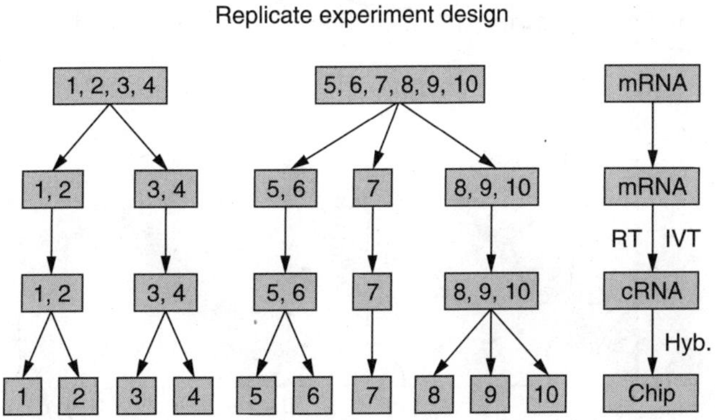

Figure 4.1 Illustration of the DNA microarray replicate experiments setup. Two different mRNA samples are used, each being probed multiple times (replicates) with varying degrees of differences in measurement steps in order to separate the preparation error that occurred during the reverse transcription (RT) and in vitro transcription (IVT) processes and the final hybridization error.

statistics and to ensure that the experimental statistics are independent of the starting mRNA, the above replicate experiments were repeated with total RNA taken from two different cultures of the Ramos cells, as represented in figure 4.1, where experiments 1–4 and experiments 5–10 start from the different RNAs.

Sample preparation starting from 5 μg total RNA, hybridization, staining, and scanning were performed according to the Affymetrix protocol [14]. The analysis uses the (average difference-based) expression values obtained by Affymetrix Microarray Suite (MAS) 5.0 with all the default parameters and TGT set to be 250.

Results and Analysis

From the experiments described above, one obtains a gene expression value matrix $E_{i,j}$, where $i = 1, 2, \ldots, I$ labels all the individual genes being probed and $j = 1, 2, \ldots, 10$ represents all the experiments shown in figure 4.1. For the U95A chip used, $I \sim 12{,}600$. Because of the large variation in measured gene expression values, the analysis is performed using the logarithm of the expression level: $\theta_{i,j} = \log_{10} E_{i,j}$.

For a pair of experiments j_1 and j_2, the overall differences in gene expression can be visualized by plotting θ_{i,j_1} versus θ_{i,j_2} for all of the genes on the microarray. In figure 4.2, two pairs of experiments, (1,3) and (1,10), are shown. Each point corresponds to a gene i and, ideally, these points should lie along the diagonal. Deviations from the diagonal are due to noise. Although figures 4.2a and 4.2b appear similar, the reasons for the deviation of the expression values from the diagonal line are different. Experiments 1 and 3 measure mRNA levels of exactly the same sample, so the observed expression differences between these experiment are caused by measurement error alone. On the other hand, samples 1 and 10 are from different cultures of the

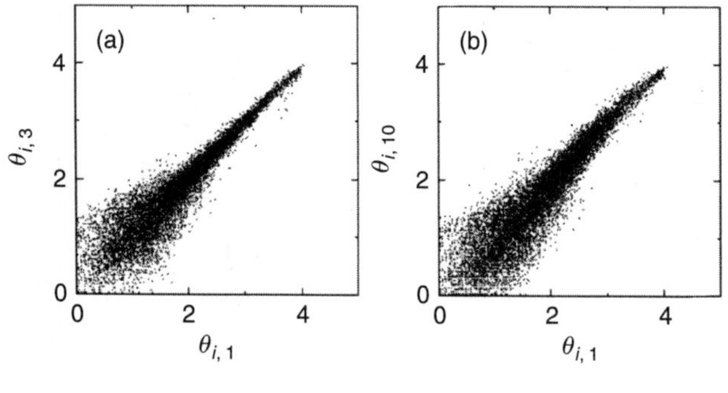

Figure 4.2 The scatter plots of gene expression value pairs θ_{i,j_1} versus θ_{i,j_2} for all genes i in the array and for: (a) experiment pair (1,3), where the deviation from the diagonal axis is caused purely by experimental error; (b) experiment pair (1,10), where true differences exist between the two transcriptomes.

cell line, so the measured expression value differences as shown in figure 4.2b contain the combined effect of the genuine gene expression differences between the two cultures together with differences caused by measurement error. Therefore, to correctly assess the statistical relevance of the measured gene expression differences between two experiments, such as 1 and 10, it is crucial to characterize the fluctuations caused purely by experimental measurement, such as the noise shown in figure 4.2a.

Although experimental noise is known to be a feature of microarray experiments, only recently has it been studied systematically by replicate experiments [6,11]. In particular, for oligonucleotide microarrays, Novak et al. [11] characterized the dispersion between two experiments by the standard deviation of their corresponding gene expression levels. Using this measure of dispersion, they studied the different effects of experimental, physiological, and sampling variability, which provide important guidance for microarray experiment design. Following this methodology, this section focuses on understanding how different experimental steps contribute to the total noise and what the possible mechanism for the noise could be. The distribution of the noise is studied in detail; this is used in devising a statistical method to determine differentially expressed genes.

In order to separate the different noise sources, all the replicate experiment pairs are placed into two groups. Group G_1 consists of all the pairs that differ only in the hybridization step:

$$G_1=\{(1,2),(3,4),(5,6),(8,9),(9,10),(8,10)\}$$

Group G_2 consists of all the replicate experiment pairs that are carried out separately right after the extraction of the mRNA:

$$G_2=\{(1,3),(1,4),(2,3),(2,4),(5,7),(5,8),(5,9),(5,10),(6,7),(6,8),(6,9),\\ (6,10),(7,8),(7,9),(7,10)\}$$

While gene expression differences between pairs of experiments in G_2 represent the full experimental noise, G_1 has been constructed to extract the noise due to hybridization alone. For reference, all the nonreplicate experiment pairs are placed into a third group $G_3=\{(i,j), 1\le i\le 4, 5\le j\le 10\}$.

NOISE DISTRIBUTION

As is evident from figure 4.2, the noise depends strongly on the expression level. Therefore, an expression-dependent distribution function is needed to characterize the variability between replicates. Given two measured gene expression values, θ_1 and θ_2, for the same gene from two replicate experiments, the estimated value of the true expression level, $\bar{\theta}$, and the size of the measurement error, $\delta\theta$, can be defined as $\bar{\theta} \equiv (\theta_1 + \theta_2)/2$ and $\delta\theta \equiv (\theta_1 - \theta_2)/\sqrt{2}$. $\bar{\theta}$ is discretized with a relatively small bin size so as to maintain a good resolution while having sufficient data points per bin. The results are insensitive to the exact choice of the bin size. For a given $\bar{\theta}$, the average of $\delta\theta$ between two experiments should be zero: $<\delta\theta \mid \bar{\theta}> = 0$. Any significantly nonzero value of $<\delta\theta \mid \bar{\theta}>$ is caused by systematic experimental errors whose source is beyond the scope of the current discussion. This error typically appears as a departure from the diagonal of scatter plots such as those of figure 4.2. A hint of it can be seen at the higher values of figure 4.2b. Even though this was not a significant problem for the data sets presented here, such errors have been compensated for whenever they occurred by subtracting any nonzero $<\delta\theta \mid \bar{\theta}>$ from $\delta\theta$ for each of the replicate experiment pairs in all of the subsequent analysis.

Within each group G_k (k = 1,2), the distribution of $\delta\theta$ for a given $\bar{\theta}$ can be obtained from each pair of replicate experiments; these distributions are found to be highly consistent with each other. Better statistics are obtained by using the gene expression values from all the pairs of replicate experiments in G_k to construct the noise distribution: $P_k(\delta\theta \mid \theta_0) = Prob_k(\delta\theta \mid \bar{\theta} = \theta_0)$. In figure 4.3a, the noise distribution functions for different values of θ_0 are shown. One may use the second-order moment to quantify the strength of the noise and its dependence on the value of the expected expression level θ_0:

$$\sigma_k^2(\theta_0) = \int_{-\infty}^{\infty} \delta\theta^2 P_k(\delta\theta \mid \theta_0) d\delta\theta \tag{1}$$

In figure 4.3c, the dependence of σ_2 on θ_0 is shown. For reference, σ_3, the difference in gene expression between pairs of experiments in G_3, is calculated in the same way as were $\sigma_{1,2}$, and is plotted in figure 4.3c as well. It is interesting that σ_3 is consistently larger than σ_2 for $\theta_0 \geq 1$, indicating the existence of signal beyond noise even for the small differences between the same cell line from different cultures.

For a given θ_0 one can define the rescaled noise $\delta\theta' \equiv \delta\theta / \sigma_k(\theta_0)$ and obtain the distribution function for $\delta\theta'$: $Q_k(\delta\theta' \mid \theta_0)$. In doing so, one finds that except for very small values of θ_0, the $Q_k(\delta\theta' \mid \theta_0)$ collapse onto a single curve $\Phi(\delta\theta')$ independent of θ_0 and k, as shown in figure 4.3b (for $k = 2$ only). Equivalently, this means that the distribution for $\delta\theta'$ can be well approximated by:

$$P_k(\delta\theta \mid \theta_0) \approx \frac{1}{\sigma_k(\theta_0)} \Phi\left(\frac{\delta\theta}{\sigma_k(\theta_0)}\right) \tag{2}$$

for $\theta_0 \geq 1$, which includes more than 90% of the data. The rescaled distribution function is found to have an exponentially decaying tail, in

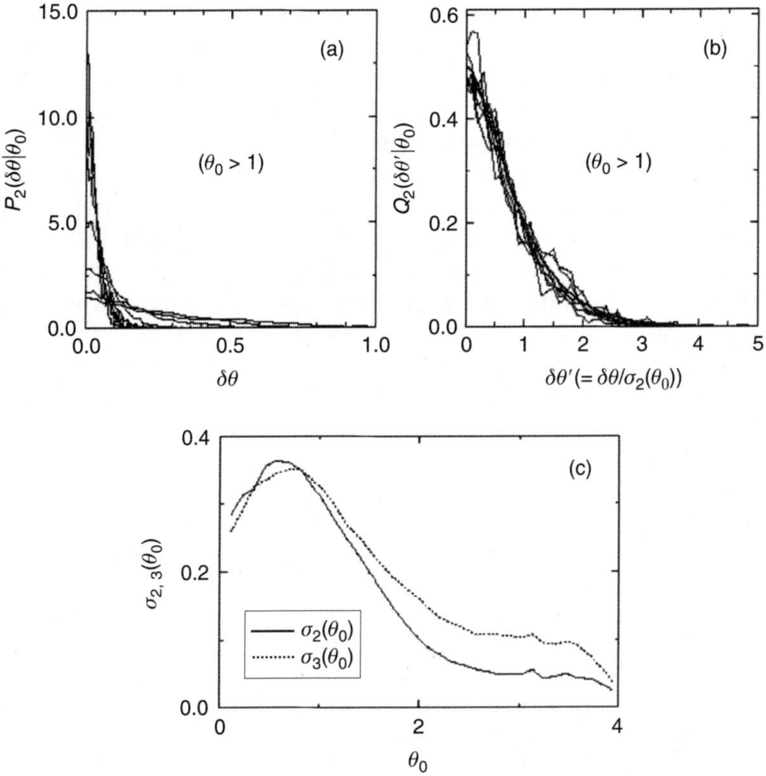

Figure 4.3 The noise distribution functions at different values of mean expression values: $\theta_0 = 0.9, 1.3, 1.7, 2.2, 2.6, 3, 3.5, 3.9$ (a) before and (b) after rescaling by the standard deviation $\sigma_2(\theta_0)$, which is shown in (c). Only the positive region of $\delta\theta > 0$ is shown in (a) and (b) for symmetry reasons. The rescaled distribution functions collapse onto a single curve well fitted by $\Phi(x) = 0.5\exp(-x^2/0.5 + 0.6|x|)$ plotted as the thick line shown in (b).

contrast with a Gaussian distribution. In fact, $\Phi(x)$ can be approximated very well by an empirical function $\Phi(x) \approx 0.5\exp(-x^2/0.5+0.6|x|)$ as shown in figure 4.3b (thick solid line).

From eq. (2), one observes that all of the expression-dependent information in the noise is given by the variance $\sigma_k^2(\theta_0)$ for $\theta_0 \geq 1$. The following subsection focuses on analyzing the dependence of the noise strength $\sigma_k^2(\theta_0)$ on the expression value.

NOISE DUE TO SAMPLE PREPARATION AND HYBRIDIZATION

To dissect the origins of noise, the total measurement noise is divided into two parts: the first is sample preparation noise $\delta\theta_{prep}$ caused by the prehybridization steps such as reverse transcription and IVT; the second is the hybridization noise $\delta\theta_{hyb.}$ For replicate pairs in group G_1 and G_2, the noise can be expressed respectively as: $\delta\theta_1 = \delta\theta_{hyb}$, $\delta\theta_2 = \delta\theta_{prep} + \delta\theta_{hyb}$. Assuming the two sources of noise are independent of each other, their variances can be obtained by $\sigma^2_{hyb} = \langle\delta\theta^2_{hyb}\rangle = \sigma^2_1$, $\sigma^2_{prep} = \langle\delta\theta^2_{prep}\rangle = \sigma^2_2 - \sigma^2_1$, where $\sigma^2_{1,2}$ can be computed from eq. (1).

In figure 4.4 is shown $\sigma_1(\theta_0)$ (dotted line) and $\sigma_2(\theta_0)$ (solid line) versus the expected value of the expression level θ_0. Although the difference between σ_2 and σ_1 is small in comparison with σ_2, $\sigma_1(\theta_0)$ is consistently smaller than $\sigma_2(\theta_0)$ for all the values of $\theta_0 \geq 1$. This should be so because the difference between σ_2 and σ_1 accounts for the sample preparation noise: this difference, albeit small, is real.

The dependence of σ_{prep} versus θ_0 is plotted in the inset of figure 4.4. One finds that the dependence of σ^2_{prep} on the expression level θ_0 can be well approximated by:

$$\sigma^2_{prep} \approx 7.2\times10^{-3} + 0.045\times10^{-\theta_0} \tag{3}$$

The constant first term dominates the sample preparation noise for expression values $\theta_0 \geq 2$.

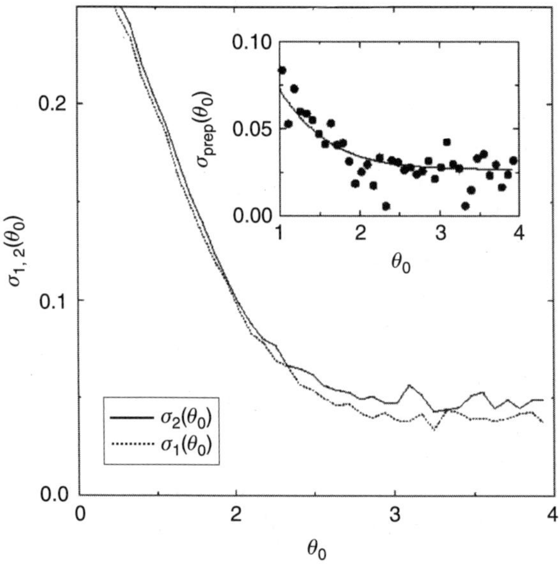

Figure 4.4 The dependence of the noise strength $\sigma_{1,2}$ on the expected values of the gene expression for replicates in groups G_1 and G_2. (inset) The variance of the sample preparation noise $\sigma_{prep} = (\sigma^2_2 - \sigma^2_1)^{1/2}$ is shown; σ_{prep} has very weak dependence on the expression value for the large expression levels $\theta_0 \geq 2$, and can be fitted by $(7.2\times10^{-3} + 0.045\times10^{-\theta_0})^{0.5}$ for $\theta_0 \geq 1$.

To understand the possible mechanisms for such noise behavior as shown in eq. (3), it is convenient to translate the above noise strength in θ (= $\log_{10}(E)$) to the noise strength in intensity $E: \sigma_E^2(E_0) \equiv <\delta E^2> \approx (\ln 10/\sqrt{2})^2 E_0^2 <\delta\theta^2>$, where $E_0 = 10^{\theta_0}$ and $\delta E = E - E_0$. By using the numerical fit for σ^2_{prep}, the variance of the sample preparation noise δE_{prep}, $\sigma^2_{E,\text{prep}}$, can written as

$$\sigma^2_{E,\text{prep}}(E_0) \equiv \left\langle \delta E^2_{\text{prep}} \right\rangle \approx 1.9 \times 10^{-3} E_0^2 + 0.12 E_0 \qquad (4)$$

The two terms in eq. (4) represent two independent sources of noise, the discussion of which follows.

For the first term, δE_{prep} is proportional to the gene expression E_0 itself. To understand this term, it is important to realize that during sample preparation the mRNA is first reverse transcribed into cDNA, and cRNA is subsequently generated from cDNA by IVT. The number of RNA molecules is amplified during the IVT, that is, $N_{\text{cRNA}} = A \times N_{\text{mRNA}}$, where A is the amplification rate and N_{mRNA} and N_{cRNA} are the numbers of mRNA and cRNA molecules, respectively. A varies between one sample preparation process and another due to fluctuations in the reaction conditions, including fluctuation due to handling of the sample ("human factors"). The fluctuation of A between different sample preparation processes, denoted as δA, leads to a fluctuation in N_{cRNA} of the form $\delta A \times N_{\text{mRNA}}$. Since N_{mRNA} is proportional to E_0, the first term in eq. (4) can thus be explained by the fluctuation in A. Furthermore, σ_A, the standard deviation of A, can be estimated: $\sigma_A \equiv <\delta A^2>^{1/2} \approx (1.9 \times 10^{-3})^{1/2} \bar{A}$, where $\bar{A}$ is the mean amplification rate. Assuming a typical value of $\bar{A}$ around 100 [5], one has $\sigma_A \sim 4.4$.

For the second term in eq. (4), δE_{prep} is proportional to the square root of E_0, indicative of Poisson-like noise. Such noise in the sample preparation may arise naturally from the probabilistic nature of the amplification process (IVT).

The accuracy of the sample preparation process inevitably depends on "human factors," whose influence is difficult to estimate. The result here can be best viewed as an upper limit for the noise caused by the intrinsic chemical processes involved in the sample preparation.

Most of the total measurement error comes from the hybridization noise, which depends strongly on the expression level (see figure 4.4). For expression level $\theta_0 \geq 1$, the hybridization noise $\delta\theta_{\text{hyb}}$ decreases rapidly with increasing expression level as shown in figure 4.5, where $\log_{10}(\sigma^2_{\text{hyb}})$ is plotted versus θ_0. Empirically, σ^2_{hyb} can be fitted by:

$$\sigma^2_{\text{hyb}}(\theta_0) \approx \beta \times 10^{-\gamma\theta_0}$$

with $\beta = 1.6 \pm 0.2$ and $\gamma = 1.1 \pm 0.1$ for the region $1.4 \leq \theta_0 \leq 2.7$, before saturating to a constant 1.1×10^{-3}.

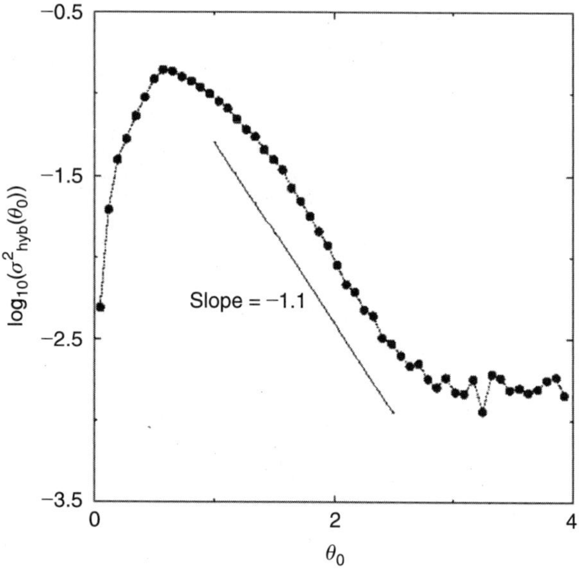

Figure 4.5 Logarithm of the hybridization noise versus the expression level for DNA microarray data.

It is noteworthy that calculation of the hybridization noise strength σ^2_{hyb} for nine pairs of replicate experiments [15], which were performed with a different type of Affymetrix GeneChip array (HuGeneFL), with a different type of cell (human fibroblast cells), and in a different laboratory, yielded an exponentially decaying component of the hybridization noise the same as that observed above [12]. That is, the noise as characterized in the present analysis seems to show a degree of universality, although more work will be needed to confirm this behavior.

USE-FOLD: A METHOD FOR UNIFORM SIGNIFICANCE OF EXPRESSON FOLD CHANGE

The results presented in the previous subsections can be used to design a method for determining the statistical relevance of gene expression changes. The idea is simply that the fold change experienced by a gene under different biological conditions has to be larger than the fold change expected from the noise. The noise distribution function discussed above can be used to evaluate the significance of the difference between a pair of gene expressions (θ_1, θ_2) for the same gene but from different experiments. One begins with the null hypothesis that the θ_1 and θ_2 arise from the same distribution (G_2) and that the difference between them is due to noise. One then defines a gene expression dependent p-value as

$$p(\theta_1,\theta_2|\theta_0) \equiv \int_{|\delta\theta|\geq\Delta\theta_0} P_2(\delta\theta|\theta_0)\,d\delta\theta \tag{5}$$

where $\Delta\theta_0 = |\theta_1 - \theta_2|/\sqrt{2}$ and $\theta_0 = (\theta_1 + \theta_2)/2$. For $\theta_0 \geq 1$, one can use eq. (2), and the p-value can be expressed simply as a function of the signal to noise ratio $R \equiv \Delta\theta_0/\sigma_2(\theta_0)$: $p(\theta_1,\theta_2|\theta_0) \equiv 2\int_R^\infty \Phi(x)dx$.

In figure 4.6, the contour lines for $p(\theta_1,\theta_2|\theta_0) = 0.05$ are shown together with two lines corresponding to a uniform 2-fold expression value change $[|\theta_1 - \theta_2| = \log_{10} 2]$. This clearly shows that, given a fixed confidence level (p-value = 0.05), a requirement of a uniform 2-fold expression change is at once too stringent for high expression levels and inadequate for the low expression levels ($\theta_0 \leq 2$). In fact, given the strong expression level dependence of the noise, no significance criterion based solely on the expression fold change is appropriate. Instead, in order to guarantee a fixed level of statistical relevance p_0, one should enforce a uniform (i.e., expression level independent) lower bound on the signal to noise ratio $R \geq R_0(p_0)$.

QUANTITATIVE NOISE ANALYSIS OF MPSS EXPERIMENTS

The data and results presented in this section are largely derived from the quantitative study of noise in MPSS experiments presented in ref. [13],

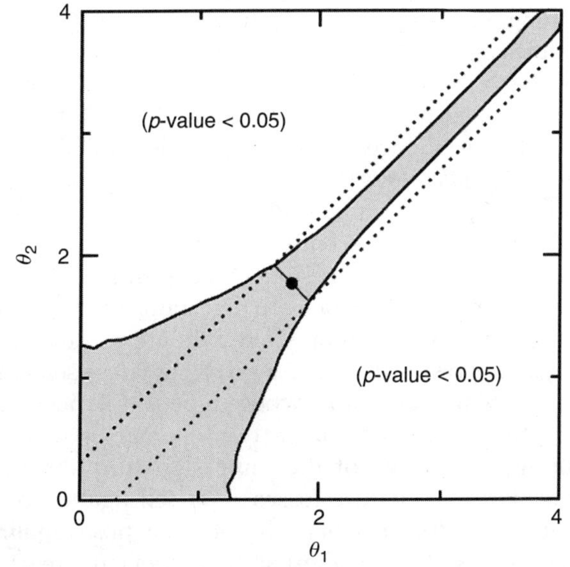

Figure 4.6 The contour line of p-value equal to 0.05. Any pair of expression values (θ_1,θ_2) outside the shaded area represent differentially expressed genes beyond experimental noise with a p-value of 0.05 or smaller. The two dotted lines represent 2-fold expression changes.

where more details on sample preparation and data processing can be found.

Materials and Methods

MASSIVELY PARALLEL SIGNATURE SEQUENCING (MPSS)

A review of the principal stages of the MPSS protocol follows; more detailed discussions are available elsewhere [3,9,10].

1. *cDNA signature/tag conjugate library construction.* Poly(A)$^+$ mRNA is extracted from the tissue of interest from which cDNA is synthesized. The 20 bases adjacent to the 3′-most *Dpn*II site (GATC) of each cDNA are captured using a type IIs restriction enzyme. The GATC and its contiguous 13-mer form a 17-mer sequence referred to as a "signature." These signatures are then PCR amplified and cloned into a collection of 8^8 distinct vectors (the "tags"), thus adding a unique identification tag to each signature.
2. *Microbead loading.* Multiple pools of about 640,000 signature/tag conjugates are amplified (i.e., ~4% of tag complexity) and, after the tags are rendered single-stranded, these conjugates are hybridized with microbeads, each of which has bound to it ~10^4 copies of one of the 8^8 possible antitags. The 4% of these microbeads with antitags corresponding to the tags of the conjugates (and thus each hybridized or "loaded" with ~10^4 copies of the signature/tag conjugate to its antitag) is isolated using a fluorescence-activated cell sorter. Approximately 1.5×10^6 loaded microbeads are assembled in a flow cell and the signature sequence on each bead is determined by MPSS.
3. *Massively Parallel Signature Sequencing.* The signatures are sequenced by the parallel identification of four bases by hybridization to fluorescently labeled encoders, followed by removal of that set of four bases and exposure of the next four bases by a type IIs endonuclease digestion. The process is then repeated. The imaged fluorescent data are processed to yield the number of beads that have a given signature sequence. Two types of initiating adaptors, whose type IIs restriction sites are offset by two bases, are ligated to two separate sets of microbeads containing a replicate of the same signature library. This is done to reduce signature losses from self-ligation of ends of signatures produced when digestion exposes palindromic overhangs. These two alternative sequencing reactions are referred to as "2-stepper" (TS) and "4-stepper" (FS) sequencing.
4. *Matching the signature to the genome.* Each of the sequenced 17-mer signatures (which typically matches only one position in a complex genome [7]) is associated with a proximal gene.

Depending on the position of the signature relative to its associated gene, the signature is given a category [13] indicative of the quality of the association.

RESULT OF AN MPSS RUN AND NOMENCLATURE

The net result of an MPSS run is a list of 17-mer signatures and the count of beads having that signature. MPSS sequencing is typically done in replicate. For a given biological sample, loaded beads are taken in fixed aliquots, and independently sequenced k times with the TS and with the FS protocol ($k = 2$–4). These are called MPSS or sequencing replicates. All of these sequencing replicates correspond to the same biological sample.

From the several replicate measurements, a transcripts per million (tpm) measure for each signature is computed. First, for each signature i, either the TS or FS data is chosen by selecting the stepper that counted the most beads for that signature *across all available experiments*. Since a signature may resist sequencing by one or the other stepper protocol, the stepper with the largest count is most likely to be better suited for measuring that signature. Once the stepper is chosen for each signature, the values of the k independent sequencing replicates are combined to give an *aggregate* tpm value $\tau_i \equiv ((v_{i1} + \cdots + v_{ik})/(N_1 + \cdots + N_k)) \times 10^6$, where the v_i's and the N's are the bead counts for the given signature i and the total number of sequenced beads in each MPSS run, respectively. If, for a given signature, $v_{ij} = 0$, then the MPSS replicate j is excluded from both the numerator and the denominator. The justification for this will be given later in the discussion of the statistics of zero measurements. In addition to the aggregate tpm, one may also define the tpm value obtained from a single replicate measurement as $\tau_{ij} \equiv (v_{ij}/N_j) \times 10^6$. As is the case with DNA microarray data, experimentally observed tpm values can span several orders of magnitude and, thus, it is again useful to define $\theta_{ij} \equiv \log_{10} \tau_{ij}$ and $\theta_i \equiv \log_{10} \tau_i$.

DATA SETS USED IN THIS STUDY

Human breast cancer cells. Estrogen receptor-negative BT-20 cell lines [16] were grown. Two distinct poly(A)$^+$ mRNA samples (A and B) were collected from plated cells and used to generate two signature/tag libraries. One of these two libraries was split in two parts and was used to generate two sets (A1 and A2) of loaded microbeads. The other library was used to generate one set of loaded microbeads (B). After loading, each set of beads was independently processed in multiple MPSS runs (see figure 4.7).

Macrophage samples and data. Plastic adherent monocytes were isolated from peripheral blood mononuclear cells collected from buffy

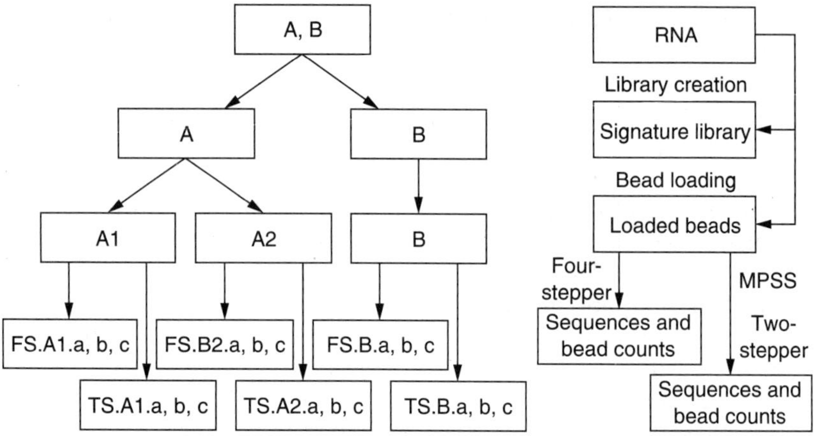

Figure 4.7 Illustration of the MPSS human breast cancer cell replicate experiments setup.

coats from five healthy humans, and cultured for about 10 days in RPMI 1640, supplemented with 20% FBS, L-glutamine, 20 mM Hepes, penicillin, streptomycin, and 50 ng/ml M-CSF, to generate monocyte-derived macrophages. Macrophages were stimulated with 100 ng/ml LPS (*S. minnesota* R595 ultrapure lipopolysaccharide, List Laboratories) and sampled at times 0 h (i.e., before stimulation), 2 h, 4 h, 8 h, and 24 h. For each of these time points, total RNA was isolated with the Trizol reagent (Invitrogen), the total RNA from the individual donors was pooled, and poly(A)$^+$ RNA was isolated with a MicroPoly(A)Pure™ kit (Ambion). Culture supernatants were tested to confirm appropriate induction of cytokines (TNF, IL-6, and IL-12) and an aliquot of total RNA was tested by real-time PCR to ensure appropriate induction of selected genes. The poly(A)$^+$ RNA was processed through the signature library generation and assayed using MPSS. Duplicate samples at times 0 h and 4 h were generated using independent cultures of macrophages and independent pools of RNA for the purpose of replicate noise modeling.

Results and Analysis

ANALYSIS OF NOISE INHERENT IN MPSS

Following the methods of the above DNA microarray analysis, one may seek to separate the sources of measurement noise in MPSS by carrying out multiple replicate experiments where, at different stages of the MPSS process, the sample is divided into multiple aliquots and subsequent steps of the experiments are carried out independently. The experimental design (shown schematically in figure 4.7) allows one to separate the measurement variances resulting from signature library generation, bead loading, and sequencing. Total RNA from estrogen receptor-negative breast cancer cell lines was divided into two aliquots (A and B in figure 4.7). Each of these aliquots was processed independently, generating separate signature/tag libraries. The signature library A was subdivided into two equal parts and each part, along with the signature library B, was independently loaded onto beads, giving rise to loaded bead groups denoted by A1, A2, and B. Finally, each of these loaded bead groups was processed with three MPSS four-stepper (FS) runs (FS a, b, and c) and three MPSS two-stepper (TS) runs (TS a, b, and c). Assuming that each stage of the process is independent of the others, these data sets enable one to estimate the noise introduced at each stage of the process.

In figures 4.8a and 4.8b, two replicate MPSS runs are compared by plotting $(\theta_{ij},\theta_{ij'})$ for replicate MPSS runs j and j'. Recall that the spread

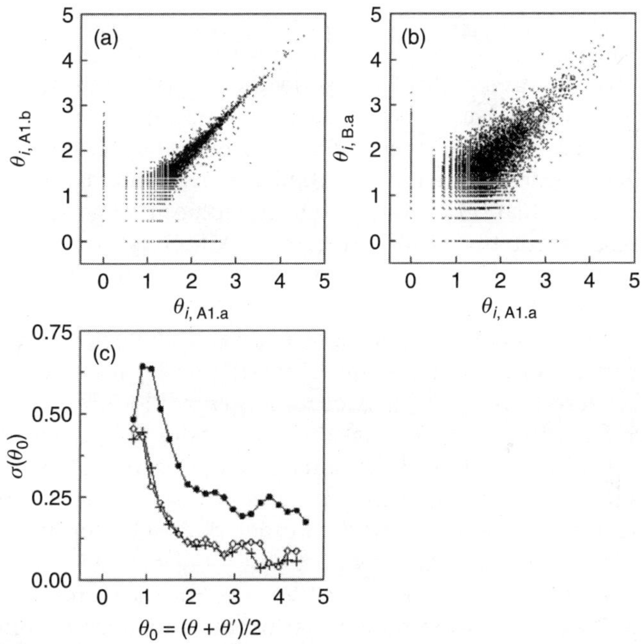

Figure 4.8 (a) Scatter plot of signature log-tpm pairs $(\theta_{i,A1.a},\theta_{i,A1.b})$, where replicates A1.a and A1.b are separate MPSS measurements on samples taken from the same set of loaded beads (see figure 4.7). (b) Scatter plot of signature $\log_{10}$-tpm pairs $(\theta_{i,A1.a},\theta_{i,B.a})$, where replicates A1.a and B.a are taken from distinct mRNA samples collected from plated cells (see figure 4.7). In both (a) and (b) noise appears as deviations of the data points from the diagonal. Note that the noise level is higher in (b) than (a). (c) Standard deviation of measurement noise σ as a function of signal level θ for pairwise comparisons between replicates A1.a and A1.b (+), A1.a and A2.a (◊), and A1.a and B.a (•) (see text).

around the diagonal is a measure of noise. In both figure 4.8a and 4.8b, the x-axes are the $\log_{10}$-tpm value of the signatures in experiment FS.A1.a (see figure 4.7), while the y-axes correspond, respectively, to the $\log_{10}$-tpm of the signatures in experiments FS.A1.b (figure 4.8a) and FS.B.a (figure 4.8b). The spread due to the combined variance introduced by library creation, bead loading, and sequencing (figure 4.8b) is much larger than that due to sequencing alone (figure 4.8a). Adopting the definitions for $\delta\theta$, $\bar{\theta}$, and $P_k(\delta\theta\,|\,\theta)$ from the microarray analysis above, one may compute the function $\sigma(\theta_0)$ using eq. (1). Plots of $\sigma(\theta_0)$ derived from the data of figures 4.8a and 4.8b show that σ decreases with the expression intensity θ_0, with the overall σ (filled circles, figure 4.8c) being about twice as large as the σ arising from the bead loading and sequencing (plus signs, figure 4.8c). Note that the noise from the combination of bead loading and sequencing is almost indistinguishable from that of sequencing alone (diamonds, figure 4.8c), demonstrating that noise stemming from bead loading is negligible.

THE STATISTICS OF THE ZERO COUNTS

Thus far, the analysis presented has dealt only with signatures whose bead counts are at least unity in each of the replicate experiments under consideration. However, one also observes numerous signatures with a finite bead count for one replicate experiment and zero for the other. This situation is unique to technologies that generate a truly digital measure of transcript expression (MPSS as opposed to, say, DNA microarrays). Signatures with a finite bead count for one replicate experiment and zero for the other appear in figures 4.8a and 4.8b as the sets of points forming linear structures at the left and bottom of these figures. (As these plots are in log-log scale, the value of zero counts, i.e. zero tpm, has been arbitrarily given a log-tpm value of 0.) It is clear from these figures that the statistics of the signatures with low but positive counts in both runs are significantly different from the statistics of the signatures measured as zero in one of the replicates. Similar result has been observed in other data sets, for example, in MPSS experiments on *A. thaliana* [17].

To investigate the significance of zero count measurements in an MPSS data set, expression data taken on macrophages 8 hours after LPS stimulation is studied. (This data was chosen because four TS and four FS MPSS runs were taken on this sample.) First, the signatures with exactly four, three, and two nonzero bead counts within the four replicates are identified (see ref. [13] for the method of determining whether the TS or FS data was used for a given signature). Next, one computes the function $\sigma(\theta_0)$ (in computing $\sigma(\theta_0)$, two of the nonzero replicate measurements are chosen at random when dealing with signatures with more than two nonzero values) separately for the data sets in which zero, one, or two out of the four sequencing

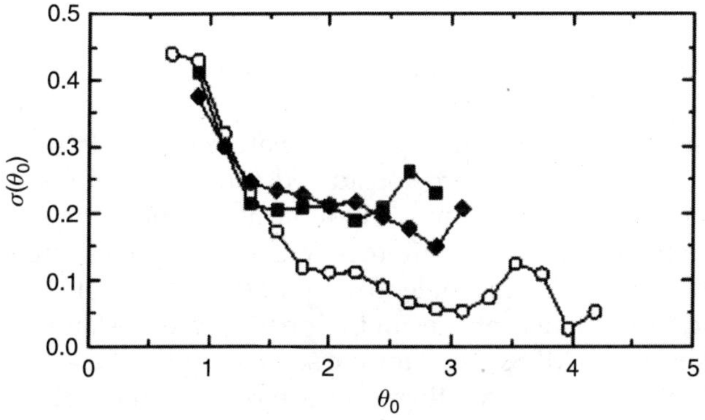

Figure 4.9 Standard deviation of measurement noise σ as a function of signal level θ (see text) for MPSS measurements on LPS-activated macrophages at 8 h after activation. Four replicate MPSS runs were taken and the noise level calculated separately for signatures with no zero measurements (○), one (■), and two (♦) zero measurements (see text). Note that those signatures with no zero measurements exhibit significantly lower noise at higher expression levels.

replicates yielded a zero count. The results are shown in figure 4.9. It can be seen that the abscissa for the nonzero statistics reaches values more than one order of magnitude larger than those for the one- or two-zero statistics. Further, the observed noise strength is considerably smaller for the nonzero statistics than for the other two, which are of similar magnitude.

If a measurement of zero tpm for a given signature represented simply the absence of that signature in the sample being studied, then the three curves in figure 4.9 would be identical. However, it is clear from figure 4.9 that data on signatures for which some of the sequencing replicates are zero exhibits significantly higher variability. This suggests that while some of the observed zeros truly represent the absence of a given signature in the data, other zeros represent the nonmeasurement of a finite signal, or alternatively, some nonzero measurements result from spurious signatures not actually present in the sample. While the physical mechanism of these spurious measurements is not clearly understood, it is noteworthy that signatures that yield zero counts in some replicates and nonzero counts in others are most often (but not always) signatures with less reliable association to a gene in the genome [13]. In any case, it is clear from figure 4.9 that the absence of a signature in one of the sequencing replicates indicates the need for statistical modeling different from that used when the signature is present in all replicate measurements.

EXPANDED DISCUSSION OF ZERO COUNT STATISTICS

This section contains an expanded discussion of the analysis of signatures that yield zero counts in one or more (but not all) of the MPSS sequencing runs constituting a given signature library. In this analysis, only basal measurements are considered (unperturbed macrophages, i.e., $t = 0$ measurements), and of these measurements consideration is further restricted to TS measurements. Exemplary measurements are shown in table 4.1. In this table, B1:1 indicates MPSS run 1 for biological replicate 1 of macrophages in their basal state (B stands for basal), and B1:2 indicates MPSS run 2 for the same sample. Likewise, B2:1 and B2:2 indicate the same for the second biological replicate. One conspicuous feature of the data is the appearance of many signatures for which there is one zero in one of the MPSS runs (e.g., B1:1, signature 2 in table 4.1) and a nonzero in the companion MPSS run (B1:2). Continuity arguments may suggest that if the count of a signature is zero in one run, then it should be small in the other. However, one can readily see that there are signatures that were not measured in three MPSS runs, but were measured in 476 beads in one of the runs (B1:2, signature 6). Likewise, continuity suggests that the statistical behavior observed for sequences for which a zero count is measured in one of the runs must be similar enough to the statistical behavior of the signatures observed at low but positive counts such as 1, 2, and so on. Figures 4.10a and 4.10b (for which the axes are in bead counts, rather than in tpm) show that for comparisons between B1:1 and B1:2, and for comparisons between B2:1 and B2:2, respectively, the behavior at zero is quite different from the behavior at 1 (in the log-log plot, the log of 0 was replaced by the log of 0.5 for clarity). For example, there

Table 4.1 Exemplary signatures and measurements for basal macrophage gene expression

	B1:1	B1:2	B2:1	B2:2	Nomenclature
Signature 1	8	17	15	25	Nonzeros in B1, nonzeros in B2
Signature 2	0	4	5	11	One zero in B1, nonzeros in B2
Signature 3	3	8	2	0	Nonzeros in B1, one zero in B2
Signature 4	22	42	0	0	Nonzeros in B1, two zeros in B2
Signature 5	0	0	15	7	Two zeros in B1, nonzeros in B2
Signature 6	0	476	0	0	One zero in B1, two zeros in B2
Signature 7	0	0	625	0	Two Zeros in B1, one zero in B2
Signature 8	0	23	0	115	One zero in B1, one zero in B2
Signature 9	0	0	0	0	Two zeros in B1, two zeros in B2

The signatures show different possible combinations of zero and nonzero measurements (B1:1, B1:2, B2:1, and B2:2) for two MPSS replicate measurements taken for each of two biological replicates (B1 and B2).

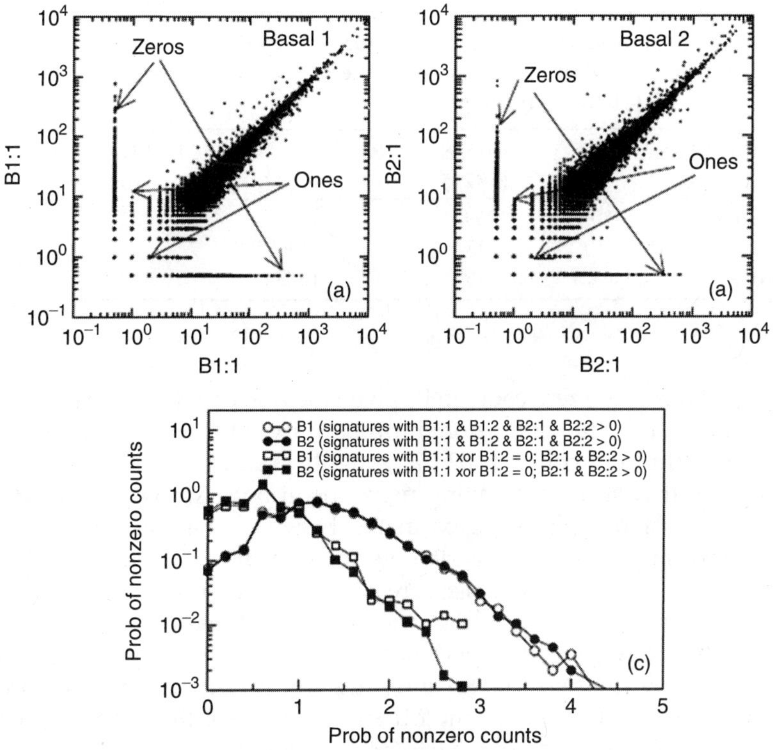

Figure 4.10 (a) Scatter plots of signature bead count in pairs B1:1 versus B1:2 and (b) B2:1 versus B2:2. These plots show that the statistical behavior of the zero counts is quite special, exhibiting a discontinuous behavior with respect to the counts of unity. The axes are in units of bead counts. (c) Probability distribution functions of the counts in B1 and B2 for the ensembles of signatures shown in the legend. Only the nonzero counts were considered for the probability calculation.

are 4487 paired signatures from B1:1 and B1:2 for which exactly one of the pair of measurements is zero, whereas there are 277 pairs for which exactly one of the pair of measurements is 1. Furthermore, the distributions of the points paired with zero and one are quite different, as shown in figures 4.10a and 4.10b. For example, for the B1:1 count of zero, the ordinate reaches values of the order of 10^3, whereas for an abscissa value of 1 or 2, the ordinate reaches only values of 10.

Accepting that the zeros have a distinct behavior, it is worth studying them in further detail. One begins by counting the various combinations of zero and nonzero measurements for those signatures that exhibit at least one nonzero measurement in the four basal MPSS runs. This yields table 4.2, which is posed in terms of conditional

Table 4.2 Counts of the various combinations of zero and nonzero measurements for those signatures that exhibit at least one nonzero measurement in the four basal MPSS runs (B1:1, B1:2, B2:1, and B2:2)

Basal 1 (B1) and Basal 2 (B2)	Nonzeros in B1 (6875)	One zero in B1 (4487)	Two zeros in B1 (8209)
Nonzeros in B2 (10,090)	5239	1558	3293
One zero in B2 (6839)	817	1106	4916
Two zeros in B2 (2642)	819	1823	0

probabilities of the zero count in B2 given the zero count in B1 in table 4.3. The table for the conditional probability of the zero count in B1 given the zero count in B2 has roughly the same characteristics as table 4.3. It is clear that having nonzeros in both MPSS runs in B1 is predictive of having nonzeros in B2. However, having one zero or two zeros in B1 does not clearly determine what to expect in B2.

The data leads to the conjecture (which can be refined by addressing sequence-specific characteristics) that signatures with count numbers below some threshold have a higher probability of not being measured by the two MPSS runs (B1:1 and B1:2 in the B1 case) and, thus, their count in one of the MPSS runs is zero. In these instances, the count of zero does not have to be considered as a real count (i.e., as a possible measurement value given the actual count), but rather as an artifact to be interpreted as a failure in measuring a signature whose value is not necessarily small.

It is straightforward to test a necessary condition that follows from the above conjecture. Consider the signatures whose counts in B1 had exactly one zero in one of the two MPSS runs, but whose count in B2 had two nonzeros (such as, for example, signature 2 in table 4.1). Measure the distribution of counts for those signatures in B2 on the one hand, and in B1 (only the nonzero measurement) on the other. If the measurement of zero in one of the two MPSS runs in B1 were due to the fact that these signatures just did not go through the whole process, then one should expect that the distribution of the counts of

Table 4.3 Data from table 4.2 posed in terms of conditional probabilities of the zero count in B2 given the zero count in B1

Prob(B2 | B1)	Nonzeros in B1	One zero in B1	Two zeros in B1
Nonzeros in B2	76.2%	34.7%	40.1%
One zero in B2	11.9%	24.6%	59.9%
Two zeros in B2	11.9%	40.6%	

these signatures in any one of the MPSS runs in B2 (solid squares in figure 4.10c) and in the nonzero MPSS runs in B1 (open squares in figure 4.10c) coincide. This is exactly what figure 4.10c shows. The overlap of these curves shows that one should not take the zero of one of the MPSS runs as a real measure, but rather as the absence of a measurement. For the sake of completeness, the probability distribution functions of the counts in B1 and B2 of signatures with nonzero counts in all the MPSS runs (open and solid circles) are plotted in figure 4.10c, superimposed on top of the probability distribution functions corresponding to signatures with exactly one zero in one of the B1 MPSS runs. One can see that the dynamic range of counts whose measurements consisted of two nonzeros is considerably larger than that of measurements of signatures that had at least one MPSS run with a count of zero.

The consideration of whether a zero is to be considered as a true measurement or the absence of a measurement is important in deciding what to do with the signatures for which there is one MPSS run claiming some value and the other claiming zero. In particular, it can be argued that if one of the MPSS runs yields zero, then it does not have to be averaged when computing the aggregate tpm for all the runs. To see this, compare, signature by signature, the averages of signatures with two nonzeros in B2 and one nonzero in B1. If the zero in the B1 MPSS run were to be considered a real measurement, then after averaging the zero in B1, one would expect that there will be roughly as many signatures with higher tpm in B1 than in B2 as signatures with higher tpm in B2 than in B1. Following this methodology, of the 1558 signatures with one nonzero in B1 and two nonzeros in B2, the average in B2 is larger than the average in B1 in 1363 of the 1558 cases. This can be explained simply by saying that the averages in B1 have been artificially lowered by adding a zero to signatures that actually were present at a higher level (the zero being an artifact). The global mean (which should be zero) of the differences between the logs of the averages in B1 and B2 was 0.4, which should be compared to the standard error of 0.01, giving a z-score of 40. If instead one ignores the zero in one of the MPSS runs of B1, of the 1558 signatures with one nonzero in B1 and two nonzeros in B2, the average in B2 is larger than the nonzero measurement in B1 in 900 cases (down 463 compared to the averaged B1, and closer to the ideal ~780 of half and half), and the global mean (which should be zero) of the differences between the logs of the averages in B1 and B2 was 0.07, which should be compared to the standard error of 0.01, giving a z-score of 7. This indicates that the trends obtained by ignoring the zero measurements are closer to the ideal unbiased comparison between B1 and B2.

The conclusion is simply that when a zero is measured in one of the two MPSS runs and a nonzero in the other, then the zero has to be ignored (i.e., one takes the maximum of the two runs). If the two runs in B1 measure zero and one of the runs in B2 is nonzero, then the nonzero signature counts are distributed as in figure 4.10c (open and solid squares). This latter distribution can be used to determine the null hypothesis when, for a given signature, the two MPSS runs comprising one of the replicates both yield values of zero.

THREE NULL HYPOTHESES ARE REQUIRED FOR BINARY COMPARISONS

To determine the biological significance of changes in tpm value observed for different signatures in the LPS-activated macrophage data, it is first necessary to formulate null hypotheses using biological replicates. This is accomplished using the biological replicate data from ref. [13] (where each biological replicate data set is composed of two sequencing replicates). Each signature that was measured at least once in a pair of biological replicates yields two aggregate (i.e., determined from two or more sequencing replicates) tpm values, τ_1 and τ_2. Three possibilities can arise in these measurements: (1) none of the counts (v_i's) used to compute τ_1 and τ_2 were zero; (2) at least one of the counts was zero, but neither τ_1 nor τ_2 are zero; (3) either $\tau_1 = 0$ and $\tau_2 > 0$ or $\tau_2 = 0$ and $\tau_1 > 0$. As shown above, the statistics characterizing the expression of signatures when measurements of zero counts are observed are fundamentally different from those resultant when no zeros are observed. Thus, it is necessary to formulate three distinct, conditional hypotheses—one for each of the three conditions above. That is, given two samples and their respective MPSS measurements, one inspects the pattern of zeros obtained in the different sequencing replicas and uses the appropriate null hypothesis on a signature by signature basis.

The formulation of the null hypothesis is begun for signatures with no zero count measurements (case 1) by plotting, in figure 4.11a, all $(\theta_{i,j}, \theta_{i,j'})$ where j and j' are biological replicates taken at $t = 0$ (where each θ is the log of an aggregate tpm count) for all signatures i that have nonzero tpm values in all replicate MPSS runs. Also plotted are equivalent points for which j and j' are biological replicates taken at $t = 4$ h. The standard deviation between measurements as a function of expression level $\sigma(\theta_0)$ is calculated following the methods discussed above. A plot of $\sigma(\theta_0)$ derived from this data (along with a fit of the calculated values of $\sigma(\theta_0)$ to an exponential) is shown in figure 4.11b. For a given θ_0, one can, as before, define a distribution for the rescaled noise $\delta\theta' \equiv \delta\theta/\sigma(\theta_0)$, and obtain the conditional distribution function $P_k(\delta\theta' | \theta_0)$. This distribution is plotted for several ranges of θ_0 in figure 4.11c. These ranges of values for θ_0 correspond to the regions delimited by the dashed lines in figure 4.11a, and the symbols drawn in each region correspond to the symbols in figure 4.11c. Notice that, once normalized by its standard deviation, the distribution of

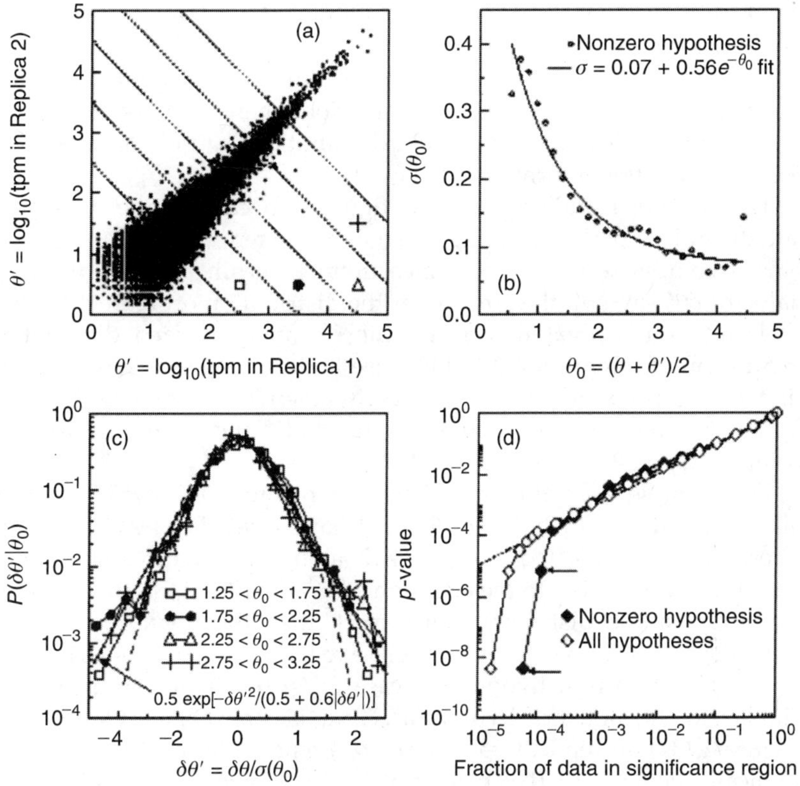

Figure 4.11 (a) Scatter plot of signature $\log_{10}$-tpm pairs $(\theta_{i,j}, \theta_{i,j'})$, where replicates j and j' are biological replicates taken at $t = 0$ (i.e., each θ is the log of an aggregate tpm derived from two biological replicates) for all signatures i that have nonzero tpm counts in all four sequence replicates. Also plotted are equivalent points for which j and j' are biological replicates taken at $t = 4$ h. (b) Standard deviation of measurement noise σ as a function of signal level θ for data shown in (a). Solid line is best fit of calculated values of σ to an exponential decay function. (c) Conditional probability density function $P(\delta\theta' \mid \theta)$ as a function of the rescaled noise $\delta\theta' \equiv \delta\theta/\sigma$ for ranges of signal level $1.25 < \theta < 1.75$ (□), $1.75 < \theta < 2.25$ (•), $2.25 < \theta < 2.75$ (Δ), and $2.75 < \theta < 3.25$ (+). Note that following normalization these distributions are independent of signal level. Fits to the data are discussed in the text. (d) p-value as a function of the fraction of significant data for measurements to which the nonzero null hypothesis is applicable (♦) as well as for all measurements (◊).

the spread away from the diagonal (measured by $\delta\theta$) is independent of the expression strength, as all the distributions coalesce approximately to a single curve $\Phi(\delta\theta')$. The tails of this curve decrease more slowly than a Gaussian (dashed line in figure 4.11c), and are well described by the function $0.5\exp(-x^2/0.5+0.6|x|)$ (thick solid line in figure 4.11c).

It is significant that the noise distributions of both the DNA microarray and MPSS data can be rescaled to expression level independent functional forms, as this demonstrates the general applicability of this methodology of noise analysis to global gene expression assays. It is noteworthy that both of these distributions can be described by the same functional form.

The noise distribution plotted in figure 4.11c can be used to formulate the null hypothesis testing whether a difference in expression in a binary comparison is beyond measurement error. Given a positive value of $\delta\theta'$, say $\delta\theta'_0$, the area under the distribution of figure 4.10c for $|\delta\theta'| > \delta\theta'_0$ is the *p*-value corresponding to a normalized differential expression of magnitude $\delta\theta'_0$. Likewise, for a chosen *p*-value, one can find a corresponding $\delta\theta'_0$. Put differently, given the function $\Phi(\delta\theta')$, one may define the function $P(\delta\theta \mid \theta_0)$ using eq. (2) and from this function compute a *p*-value from eq. (5).

For example, a *p*-value of 0.05 corresponds to a $\delta\theta'_0$ of 2.13. That is, all points with $|\delta\theta| > 2.13\sigma(\theta_0)$ will have a *p*-value less than 0.05. These points are plotted in figure 4.12, along with the two delimiting curves corresponding to the equation $|\delta\theta'| = 2.13\sigma(\theta_0)$. If the parameterization of the distribution is correct, then the fraction of points outside of those curves should be close to 0.05. Indeed, it is 0.04.

The one-zero null hypothesis (case 2) is formulated in a manner similar to the nonzero hypothesis. That is, one begins by plotting (figure 4.13a) all replicate points $(\theta_{i,j}, \theta_{i,j'})$ where j and j' are biological replicates taken at $t = 0$ as well as at $t = 4$ h. However, in this instance only those signatures i for which at least one of the pair of biological replicates is composed of one zero and one nonzero sequencing replicate are considered. The variation between replicates in this data is significantly greater than that observed for the nonzero data.

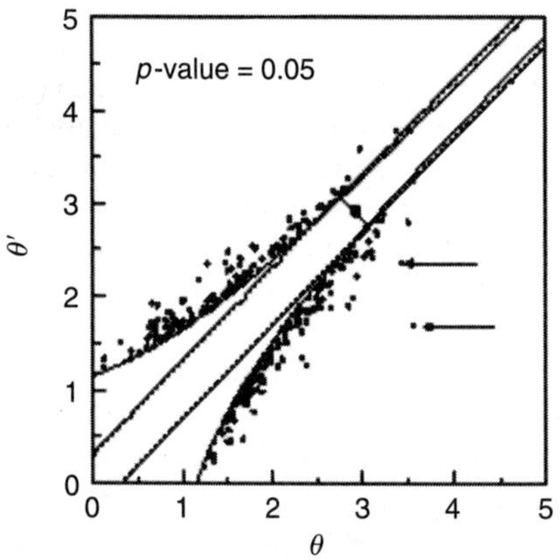

Figure 4.12 Illustration of data in region of significance for figure 4.11(a) and *p*-value equal to 0.05. The two straight lines represent 2-fold expression changes.

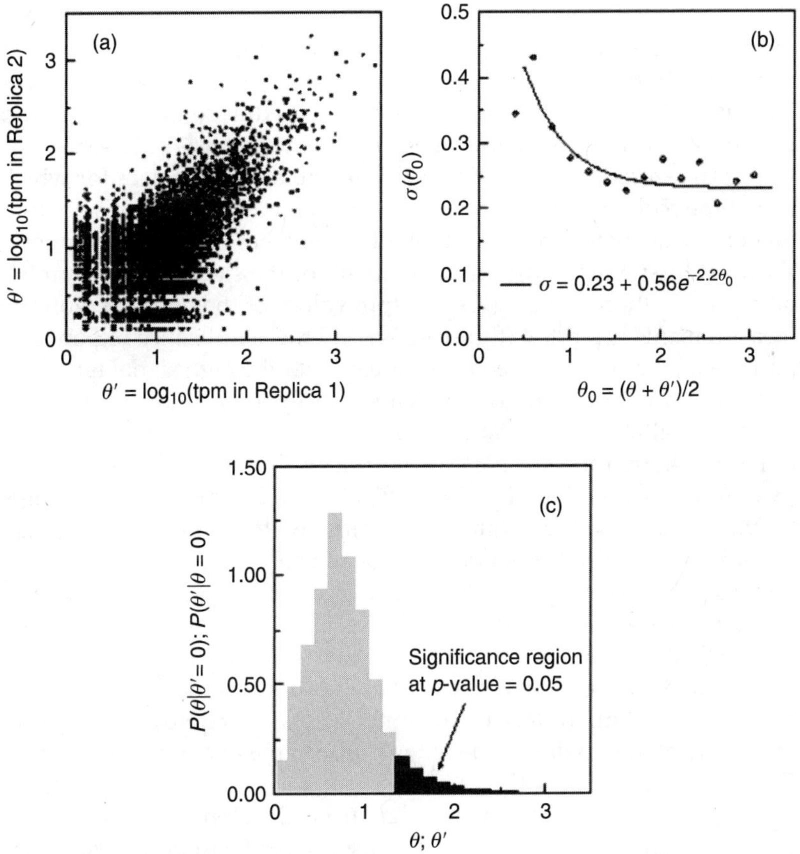

Figure 4.13 (a) One-zero null hypothesis data: scatter plot of signature $\log_{10}$-tpm pairs $(\theta_{i,j}, \theta_{i,j'})$, where replicates j and j' are biological replicates taken at $t = 0$ (i.e., each θ is the log of an aggregate tpm derived from two MPSS replicates) for all signatures i that have one MPSS replicate measurement of zero tpm counts in either j or j' (or both). Also plotted are equivalent points for which j and j' are biological replicates taken at $t = 4$h. (b) Standard deviation of measurement noise σ as a function of signal level θ for data shown in (a). (c) Two-zero null hypothesis data: distribution of nonzero signature tpm values taken from biological replicates taken at $t = 0$. Data taken from biological replicate j where the aggregate tpm measurement in j' was zero and from j' where the aggregate tpm measurement in j was zero. Also plotted are equivalent points for which j and j' are biological replicates taken at $t = 4$ h. The dark shaded region is the significant data for p equal to 0.05.

However, upon calculating $\sigma(\theta_0)$ (figure 4.13b), one finds that the conditional distribution function $P(\delta\theta \mid \theta_0)$ as a function of the rescaled noise distribution $\delta\theta' \equiv \delta\theta/\sigma(\theta_0)$ is roughly independent of θ_0 for this data as well. Thus, one can follow the procedure outlined above to determine p-values using this one-zero null hypothesis data set.

The two-zero hypothesis (case 3) is formulated from data for which one of the biological replicates shows zero counts for a given signature in both sequencing replicates, while the other biological replicate shows at least one nonzero measurement for the signature. The probability distribution of the aggregate tpm values of the nonzero replicate measurements is plotted (figure 4.13c) and the significance region for a particular p-value is defined as the area under the high signal tail of the distribution whose ratio to the area under the entire curve equals the desired p-value.

Figure 4.11d shows a plot of the fraction of points left out of the delimiting curves given by $|\delta\theta| = |\delta\theta'|(\theta_0)$ as a function of the p-value ($|\delta\theta'|$ depends on the p-value). The curve with solid diamonds corresponds to the subset of signatures for which the nonzero hypothesis applies. The open diamonds consider all the measured signatures and the corresponding null hypotheses. Both the nonzero hypothesis and the all-hypothesis curves show that the fraction of points left out of the delimiting curves is very well estimated by the p-value calculation over four orders of magnitude. The precipitous drop-off of the curves at the small p-value range is due to the two outliers indicated with arrows in figure 4.12.

The above p-value formalism has been developed in the context of binary comparisons (typically case/control studies). Using the same formalism it is also possible to estimate error bars for the log-tpm for a given signature. To do this, simply notice that if the log-tpm value of a signature yields a value θ, then $[\theta - 2.13\sigma(\theta), \theta + 2.13\sigma(\theta)]$ is an estimate of the 95% confidence interval. In other words, the probability that a subsequent measurement of that signature falls outside that interval is smaller than 0.05. This confidence interval interpretation is especially useful when data from only a single MPSS run of a given condition is extant, and error bars need to be assigned to these measurements. The calculation of a 95% confidence interval requires an estimate of $\sigma(\theta_0)$. When replicate measurements are not available, $\sigma(\theta_0)$ can be estimated from studies such as the one presented in this chapter. Computational tools to analyze MPSS data for confidence intervals, as well as p-values in case/control measurements and time traces, can be obtained at the website www.research.ibm.com/FunGen.

Finally, note that the method to determine the statistical significance for binary MPSS measurements is essentially the same as the USE-fold method introduced in ref. [12], and presented earlier in this chapter in the context of DNA microarray analysis.

TIME TRACES AND MULTIPLE COMPARISONS: THE MACROPHAGE DATA

Changes in expression level as a function of time are particularly important in understanding the response of cells to a perturbation. Suppose that the aggregate tpm of a signature is measured at n time points $t_0 = 0$ (i.e., before perturbation), $t_1, t_2, \ldots, t_{n-1}$, yielding a series of log-tpms $(\theta_{t_0}, \theta_{t_0}, \ldots, \theta_{t_{n-1}})$. If the perturbation significantly affects the expression of that signature, then one expects a small p-value for at least one of the $n \times (n-1)/2$ pairwise comparisons between temporal data points. It is necessary to consider all pairwise comparisons, and not just those between consecutive measurements, because there are numerous instances where consecutive comparisons are not beyond the level of significance, but those between nonadjacent time points are [13]. A significance index (SI) for the time series of a given signature is defined as the minimum p-value obtained from all possible pairwise comparisons within the series. An SI is considered significant if it is smaller than some chosen threshold p_0. Note that the most significant p-value does not necessarily correspond to the largest fold change, as the significance of a fold change depends on the expression level.

In ref. [13], measurements were reported for the gene expression of macrophages prior to LPS stimulation and at times $t = 2, 4, 8$, and 24 h following it. For each observed signature, the SI of its time series was computed. Often, multiple signatures were found to correspond to a single gene (in the NCBI database). In such cases, the signature with the lowest SI values was associated with the gene. Following this protocol, 12,657 signatures, of which 2356 (20%) underwent statistically significant changes in expression level with SI < 0.05, were identified. Significant signatures corresponded to well-characterized genes [13] in greater proportion than nonsignificant signatures, an indication that the SI allows the identification of meaningful signals from massive amounts of signature data.

COMPARISON OF NOISE BETWEEN DNA MICROARRAY AND MPSS EXPERIMENTS

Conversion of Affymetrix Expression Values into Transcripts per Million

Statistical variability between biologically equivalent replicate experiments provides one useful means of comparing different global gene expression assay technologies. However, before comparing the level of noise observed from DNA microarray and MPSS replicate experiments, it is first necessary to convert the observed signals into equivalent units, given that the noise depends on the expression level in both technologies. MPSS bead counts are converted into transcripts per million (tpm) values following the methods described above. The comparison of MPSS and microarray noise measurements is thus begun by considering the conversion of Affymetrix GeneChip intensities into units of tpm.

The intensity of the fluorescence measured at a given probe of a microarray (i.e., a region containing an oligonucleotide with a given sequence) increases with the concentration of complementary mRNA contained in the target sample [5]. However, the relationship between target concentration and observed fluorescence intensity follows the Langmuir isotherm function:

$$I(c) = \frac{n_p c}{n_e + c} \tag{6}$$

where I and c are the observed intensity and target concetration, respectively, n_p is the saturation value of the intensity, and n_e is a probe-dependent constant [18,19]. It follows from eq. (6) that the observed intensity is a linear function of concentration at low concentrations ($c \ll n_e$) and saturates at high concentrations ($c \gg n_e$).

The Affymetrix MAS 5.0 software calculates the expression level of a particular gene from a set of individual probes. The algorithm followed [14,20] results in the predicted gene expression "signal levels." These signal levels exhibit a dependence on target concentration that follows the same form as the intensity in eq. (6). This may be demonstrated by computing the signal levels for a series of "spike-in" target genes—genes that have been added to a solution of mRNA at known concentration. Specifically, consider the signal levels calculated for a group of 12 genes that have been spiked into hybridization mixtures at known concentrations between 0 and 1024 pM and subsequently hybridized to Affymetrix U95A GeneChips. This data was taken at Affymetrix and is publicly available (http://www.affymetrix.com/support/datasets.affx). In figure 4.14 is plotted the median expression levels (as calculated using MAS 5.0) for all of the spike-in genes reported at each known (i.e., "spike-in") concentration as a function of that known concentration. The dashed line is a fit of the data to eq. (6) with logarithmic weighting (so as to give approximately equal weight to low- and-high concentration data). It is clear that this functional form describes the data well.

The comparison between Affymetrix and MPSS data is considerably simplified if one considers only the data within the concentration range for which both systems yield expression signal levels that scale linearly with target concentration. In doing so, the need to nonlinearly rescale θ for all of the $\sigma(\theta)$ plots computed from the Affymetrix data is avoided. Specifically, one may fit the data in figure 4.14 with spike-in concentration values between 1 pM and 100 pM to a straight line constrained to pass through the origin. This is shown as the solid line

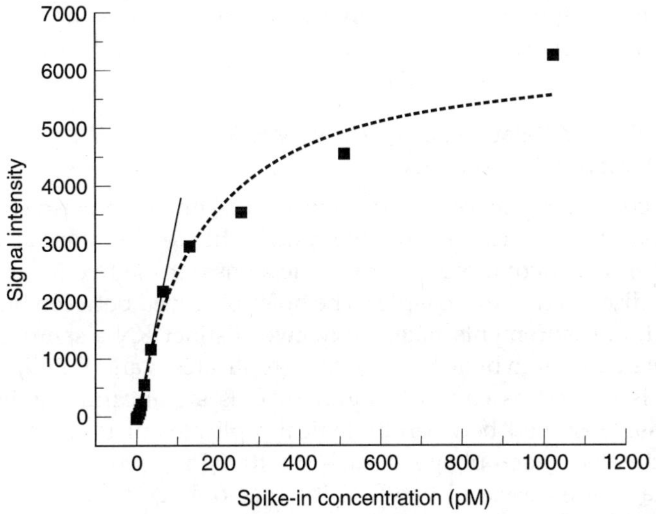

Figure 4.14 Median signal intensity plotted as a function of known "spike-in" concentration for publicly available Affymetrix spike-in data (see text). Data points were calculated using Affymetrix Microarray Suite v. 5.0. Squares are median values resultant from fitting data from 12 reliable genes (i.e., all genes except 407_at and 36889_at) at each concentration between 0 and 1024 pM. Dashed line is fit of data to the Langmuir isotherm function [eq. (6)] and solid line is linear fit through the origin of the data between 1 pM and 64 pM.

in figure 4.14. The upper limit (100 pM) represents the approximate upper limit for which a linear correlation between measured signal level and known (spike-in) concentration is observed. Data below 1 pM was excluded because it contains a significant nonlinear component, presumably due to nonspecific binding. Note that including this low-concentration data does not significantly affect the result. The black line in figure 4.14 is of the form $I = 35c$, where I is the observed intensity and c is the spike-in concentration in pM. Finally, it is necessary to convert the spike-in concentrations to tpm. To do this, it is necessary to know the ratio of the number of spike-in transcripts to the number of transcripts in the background mRNA. This ratio is not available for the specific spike-in data set under consideration. One method of dealing with this is to make the assumption that this ratio is comparable to that reported in earlier spike-in experiments [5]; that is, assume that a spike-in concentration of 1 pM corresponds to a transcript frequency of 1:150,000, or 6.7 tpm. From this assumption follows the approximation that, for spike-in values below approximately 100 pM, each 5.25 units of observed intensity (in the MAS

5.0 units of figure 4.14) corresponds to 1 tpm. This is the conversion factor used below in comparing the relative noise levels of MPSS and Affymetrix expression data.

Comparison of Relative Sensitivity of MPSS and Affymetrix Experiments

When comparing noise levels between dissimilar experiments, it is important to consider the possible contribution of "human factors" on the observed fluctuations. For example, consider the two MPSS experiments discussed in this chapter. The noise observed between biological replicate measurements taken from two distinct RNA samples taken from plated human breast cancer cells (replicates A and B in figure 4.7), which is plotted as circles in figure 4.8c, is significantly higher than the noise observed between biological replicates of RNA taken from LPS-activated macrophage tissues (plotted in figure 4.11b). In fact, the total noise observed between the macrophage tissue replicates is comparable to that observed between replicate bead-loading measurements in the breast cancer replicate experiments of figure 4.7. This strongly suggests that human factors in the tissue preparation and RNA extraction introduced significantly greater variability into the data sets shown in figure 4.8 than they did into the data shown in figure 4.11.

In the following analysis, it is chosen to compare the replicate noise observed in the MPSS experiments on RNA taken from the LPS-activated macrophages with the replicate GeneChip measurements in which RNA was also extracted from different cultures, that is, experimental group G_3 of the Ramos cell cultures. It is crucial to note that human preparation factors involved in the RNA extraction from these different cell lines necessarily introduce different degrees of variability. For this reason, the following discussion simply demonstrates the methodology for comparing replicate noise levels between differing experiments. It is not intended to provide a definitive quantitative comparison of the noise intrinsic to the MPSS and Affymetrix technologies. Such a comparison would require replicate measurements using both technologies to be carried out on the same sample of target RNA as well as a more precise calibration of the expression scales. And, even in this case, the observed results would necessarily include human factors present in the sample preparation steps of each of the technologies. Even with these caveats, it is still interesting to have a qualitative comparison of the repeatability of these quite different gene expression technologies.

In figure 4.15 are replotted the values of $\sigma(\theta)$ plotted in figure 4.11b. Also plotted are the values of $\sigma(\theta)$ from the Affymetrix data of figure 4.3c, where θ, $\delta\theta$, and $\sigma(\theta)$ have been rescaled so that the new plot of $\sigma(\theta_0)$ versus θ_0 is correctly expressed in units of $\log_{10}$(tpm). Note that

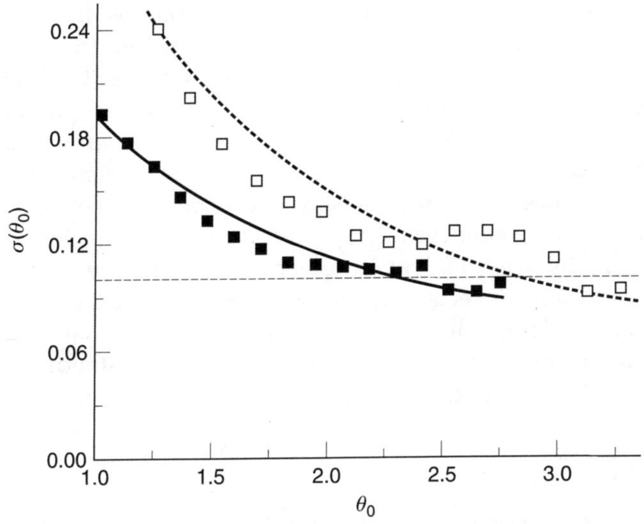

Figure 4.15 Plot of $\sigma(\theta)$ for MPSS macrophage data (□) and GeneChip Ramos cell line data (■). The MPSS data is replotted from figure 4.11b. GeneChip data are taken from $\sigma_3(\theta_0)$ (figure 4.3c) with all data rescaled to units of tpm (see text). Dashed and solid lines are best fits of the macrophage and Ramos data to exponential decaying functions, respectively. Dotted horizontal line is at a value of $\log_{10} 2/\sqrt{2}(2.13)$, which corresponds to the value of σ at which 2-fold changes in expression level are significant at the 95% confidence level.

only data below approximately 700 tpm are shown, as only these data may be linearly rescaled from figure 4.3c. Observe that between 1 and 700 tpm the noise level from the GeneChip replicate experiments is slightly lower than that observed for the nonzero null hypothesis replicate MPSS data. From the data of figure 4.15, one can estimate for each data set the tpm level at which a 2-fold change in expression is significant with 95% confidence (i.e., p-value = 0.05). As the p-values follow from eqs. (2) and (5), where the function $\Phi(x)$ was empirically shown to be the same for both experiments, the value of θ_0 for which $\sigma(\theta_0) = \log_{10} 2/\sqrt{2}(2.13)$ (shown as a dotted line in figure 4.15) is the value for which 2-fold changes are significant with 95% confidence. For the MPSS macrophage experiment this value of θ_0 corresponds to 725 tpm, while for the Affymetrix Ramos cell line experiment it corresponds to 225 tpm.

While the above methodology allows for a comparison of the relative noise levels across experiments done using different global gene expression assay technologies, several other technology-dependent factors should be noted. First, at higher concentrations, the Affymetrix signal strength begins to saturate while in principle the MPSS tpm values should not. Thus, at these higher concentrations, the MPSS data will present a more accurate measure of fold changes. In addition, the MPSS data is obtained without any previous assumptions as to which signatures (and thus which genes) may be present. In contrast, Affymetrix GeneChips only measure the expression levels of those genes for which probe sequences are explicitly incorporated onto the chip. Finally, the MPSS measurements should not be susceptible to cross-hybridization effects that can affect Affymetrix data for experiments that are less well controlled than the spike-in data addressed above. That is, replicate noise measurements notwithstanding, each of these technologies has attributes not present in the other.

SUMMARY

A review has been presented of the means by which sets of biologically equivalent replicate gene expression assays, which bifurcate at different stages of sample processing and data collection, can be used to quantify the sources of noise introduced into these assays. This technique has been utilized to carry out detailed analyses of the noise present in two popular and quite distinct global gene expression assays: Affymetrix GeneChips and Massively Parallel Signature Sequencing (MPSS). In the course of these analyses, the differences between digital and analog expression assays have been illustrated. A general procedure by which similar analyses may be carried out for additional global gene expression assay technologies has also been presented. Finally, it has been shown that the quantitative noise measurements obtained from replicate experiments allows for a comparison of the relative sensitivities of experiments carried out using different technologies.

ACKNOWLEDGMENTS

The results presented in this chapter were largely derived from refs. [12] and [13]. The contributions of the following collaborators in these works are gratefully acknowledged: U. Klein, A. Kundaje, K. Duggar, C. Haudenschild, D. Zhou. T. Vasicek, K. Smith, A. Aderem, and J. Roach.

REFERENCES

1. Lockhart, D. J. and E. A. Winzeler. Genomics, gene expression and DNA arrays. *Nature*, 405(6788):827–36, 2000.
2. Brown, P. O. and D. Botstein. Exploring the new world of the genome with DNA microarrays. *Nature Genetics*, 21(Suppl 1):33–7, 1999.

3. Brenner, S., M. Johnson, J. Bridgham, et al. Gene expression analysis by massively parallel signature sequencing (MPSS) on microbead arrays. *Nature Biotechnology*, 18(6):630–4, 2000.
4. Cawley, S., S. Bekiranov, H. H. Ng, et al. Unbiased mapping of transcription factor binding sites along human chromosomes 21 and 22 points to widespread regulation of noncoding RNAs. *Cell*, 116(4): 499–509, 2004.
5. Lockhart, D. J., H. L. Dong, M. C. Byrne, et al. Expression monitoring by hybridization to high-density oligonucleotide arrays. *Nature Biotechnology*, 14(13):1675–80, 1996.
6. Lee, M. L. T., F. C. Kuo, G. A. Whitmore, et al. Importance of replication in microarray gene expression studies: statistical methods and evidence from repetitive cDNA hybridizations. *Proceedings of the National Academy of Sciences USA*, 97(18):9834–9, 2000.
7. Saha, S., A. B. Sparks, C. Rago, et al. Using the transcriptome to annotate the genome. *Nature Biotechnology*, 20(5):508–12, 2002.
8. Velculescu, V. E., L. Zhang, B. Vogelstein, et al. Serial analysis of gene-expression. *Science*, 270(5235):484-7, 1995.
9. Brenner, S., S. R. Williams, E. H. Vermaas, et al. In vitro cloning of complex mixtures of DNA on microbeads: physical separation of differentially expressed cDNAs. *Proceedings of the National Academy of Sciences USA*, 97(4):1665–70, 2000.
10. Reinartz, J., E. Bruyns, J.-Z. Lin, et al. Massively parallel signature sequencing (MPSS) as a tool for in-depth quantitative gene expression profiling in all organisms. *Briefings in Functional Genomics and Proteomics*, 1(1):95–104, 2002.
11. Novak, J. P., R. Sladek and T. J. Hudson. Characterization of variability in large-scale gene expression data: implications for study design. *Genomics*, 79(1):104–13, 2002.
12. Tu, Y., G. Stolovitzky and U. Klein. Quantitative noise analysis for gene expression microarray experiments. *Proceedings of the National Academy of Sciences USA*, 99(22):14031–6, 2002.
13. Stolovitzky, G., A. Kundage, G. A. Held, et al. Statistical analysis of MPSS measurements: application to the study of LPS activated macrophage gene expression. *Proceedings of the National Academy of Sciences USA*, 102(5): 1402–7, 2005.
14. GeneChip Expression Analysis. *Affymetrix Technical Manual*, 2001. http://affymetrix.com/support/technical/manual/expression_manual.affx
15. Lemon, W. J., J. J. T. Palatini, R. Krahe, et al. Theoretical and experimental comparisons of gene expression indexes for oligonucleotide arrays. *Bioinformatics*, 18(11):1470–6, 2002.
16. Sugarman, B. J., B. B. Aggarwal, P. E. Hass, et al. Recombinant human-tumor necrosis factor-alpha: effects on proliferation of normal and transformed-cells in vitro. *Science*, 230(4728):943–5, 1985.
17. Hoth, S., Y. Ikeda, M. Morgante, et al. Monitoring genome-wide changes in gene expression in response to endogenous cytokinin reveals targets in *Arabidopsis thaliana*. *FEBS Letters*, 554(3):373–80, 2003.

18. Hekstra, D., A. R. Taussig, M. Magnasco, et al. Absolute mRNA concentrations from sequence-specific calibration of oligonucleotide arrays. *Nucleic Acids Research*, 31(7):1962–8, 2003.
19. Held, G. A., G. Grinstein and Y. Tu. Modeling of DNA microarray data by using physical properties of hybridization. *Proceedings of the National Academy of Sciences USA*, 100(13):7575–80, 2003.
20. Statistical Algorithms Reference Guide. *Affymetrix Technical Note*, 2001. http://www.affymetrix.com/support/technical/technotes/statistical_reference_guide.pdf

5

Mapping the Genotype–Phenotype Relationship in Cellular Signaling Networks: Building Bridges Over the Unknown

Jason A. Papin, Erwin P. Gianchandani, & Shankar Subramaniam

The central challenge in postgenomic biology is the mapping of the genotype–phenotype relationship. For cellular signaling networks, the functional connections between genotype and phenotype are deciphered through an ever-growing toolbox of high-throughput experimental technologies. Each technology can be used to generate a vast amount of data particular to some aspect of a given cellular signaling network. However, despite this growing experimental toolbox and the associated repositories of data, there remains the challenge to connect these disparate data points to reconstruct the continuum from genotype to phenotype.

Cellular signaling network functions can be discretized into three tiers (see figure 5.1). The first tier involves the transcriptional activity of a genome. The proteins that result from the transcription and translation of genes orchestrate the responses to a signaling event. The second tier is composed of the chemical transformations in which these proteins are involved. This set of chemical reactions and cellular events are here called "intermediate phenotypes." The third tier is composed of the subsequent cellular behaviors that are a function of the given genotype and collection of intermediate phenotypes. These events have been called "endpoint phenotypes." This third tier of events in a cellular signaling network occurs on a much slower time scale than events that comprise "intermediate phenotypes."

Experimental technologies used to decipher cellular signaling networks can be grouped according to the data they generate within these three tiers. For example, genomic sequencing efforts decode information in the first tier, delineating the "parts list" of a given cellular signaling network. Furthermore, chromatin immunoprecipitation (ChIP) chips systematically identify which genes are regulated by a given regulatory protein under specified conditions, providing further information regarding this first tier. The intermediate phenotypes are characterized

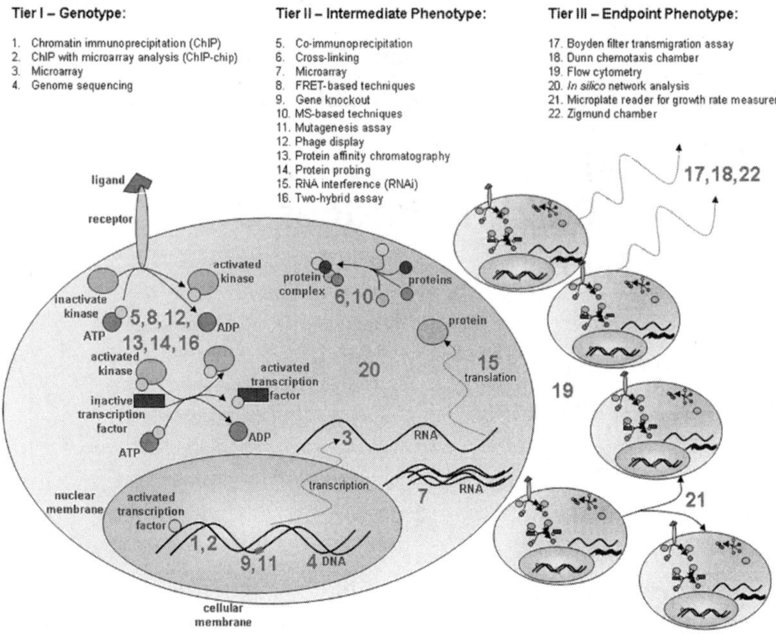

Figure 5.1 Mapping genotype–phenotype relationships in cellular signaling networks. Various experimental technologies have been developed to elucidate information about the properties of cellular signaling networks. Here a portrait is provided of the specific activities and reactions that are characterized by different techniques. Specifically, in Tier I, four examples of technologies that enable the identification of the genotype are illustrated: chromatin immunoprecipitation (ChIP) (1) and chromatin immunoprecipitation with microarrays (ChIP-chip) (2) indicate which proteins bind to DNA sequences in vivo; microarrays (3) show which genes are expressed, i.e., transcribed into RNA; and genome sequencing (4) identifies all the base pairs contained in a genome. Tier II encompasses the technologies that characterize intermediate phenotypes, or the chemical transformations that comprise a cellular signaling network: coimmunoprecipitation (5), fluorescent resonance energy transfer (FRET)-based techniques (8), and (yeast) two-hybrid assays (16) indicate which proteins interact with one another; crosslinking (6) and mass spectrometry (MS)-based techniques (10) specify the composition of protein complexes; phage display (12), protein affinity chromatography (13), and protein probing (14) indicate which proteins bind to specific targets; microarrays (7) indicate the phenotypic function of a set of genes; and gene knockouts (9), mutagenesis assays (11), and RNA interference (RNAi) (15) generate data about various genes by altering normal cellular activity through gene deletions, genetic mutations, and inhibition of RNA transcription, respectively. Finally, Tier III consists of technologies that provide data about the endpoint phenotypes, or the transcriptional and cellular modifications as a result of intermediate phenotypes: the Boyden filter transmigration assay (17), Dunn chemotaxis chamber (18), and Zigmund chamber (22) offer quantitative assessment of cellular migration, a cell-scale property; flow cytometry (19) is a technique for counting the number of cells within a given population (e.g., to assess the rate of cell division); and the microplate reader (21) provides dynamic growth-rate measurements of many different cell cultures simultaneously. Assimilating all of the data generated by these experimental technologies may provide understanding and enable modeling of cellular signaling networks (20).

with fluorescence resonance energy transfer (FRET)-based technologies, two-hybrid assays, RNA interference (RNAi)-based methodologies, and other approaches that delineate the chemical reactions that comprise signaling networks. Endpoint phenotypes are measured by migration and growth assays, flow cytometry, and other such techniques that quantify global cell behaviors.

Although each of these technologies generates specific types of data, these resulting data sets are not yet continuous. For example, there exist data on what genes are differentially regulated by interferon alpha, beta, and gamma stimulation, and on what intracellular kinases are activated by the corresponding cell surface receptors [1]. However, how combinations of these stimuli affect cellular phenotype or how different cell types respond to the same stimuli remains to be discovered. Thus, mathematical frameworks that can quickly assess a variety of conditions and account for inherent "unknowability" would be of tremendous value.

This chapter describes the promise and limitations of the experimental technologies that are used to characterize each of these levels of the genotype–phenotype relationship in cellular signaling networks. In addition, the need for mathematical techniques to both integrate these data sets and to connect the disconnected parts of a network reconstruction will be discussed.

DECIPHERING GENOTYPES

A genome annotation provides a "parts list" of all the potential protein components in a cellular system. Thus, the genotype places constraints on a signaling network by restricting what proteins are even present. The differentiated states of a cell further restrict which genes are expressed and thus which associated proteins are synthesized. To date, many cellular signaling network reconstructions are therefore "generalities" in which a given cell type may only possess a subset. Table 5.1 indicates various experimental techniques that are used to characterize the genotype, including the advantages and disadvantages of each.

Existing genome sequencing techniques have led to a tremendous advance in genomic data with the genomes of more than 250 organisms now available [17]. Further advances in sequencing techniques, called Ultra-Low Cost Sequencing (ULCS) methods, will result in an explosion of even more sequence data [18]. Such methods will lead to an increased resolution in cellular signaling network reconstructions. For example, cost-effective sequencing may allow a researcher to evaluate the genotype of a variety of cancer tissue samples leading to a catalog of all mutations that generate cancerous phenotypes. As another example particularly relevant to cellular signaling, such ULCS methods would also allow for the profiling of B-cell and T-cell receptor diversity.

Table 5.1 Genotype characterization: Experimental technologies for characterizing genotypes and the set of activated genes of a genome

Name	Description	Advantages	Disadvantages	References
Chromatin immuno-precipitation (ChIP)	Indicates whether a specific protein binds to a specific DNA sequence in vivo. Cells are grown under native conditions, and are subsequently fixed with formaldehyde so that the transcription factor proteins crosslink with the chromatin. The genomic DNA, bound to the transcription factors, is then sheared into appropriately sized fragments. These are immunoprecipitated to determine the binding domains.	• Is performed in vivo.	• Precipitation is often inefficient, i.e., insufficient protein is recovered. • Crosslinking may require purification of the DNA-protein complex, which can be tedious and expensive. • Crosslinking may fix interactions that are transient and/or of minor significance. • Suspected genetic targets must be predefined. • Significant quantities of fragmented DNA must be available.	[2,3]
Chromatin immunoprecipitation with microarray analysis (ChIP-chip)	Provides the comprehensive identification of all targets of a transcription factor within a genome. Bound DNA fragments are recovered and hybridized to a chip consisting of all the regulatory regions within the genome.	• Enables the identification, in a single experiment, of all the targets of a transcription factor. • Prior knowledge of potential binding sites is not needed, since	• Various technical aspects of the procedure remain difficult to control, including fixation, epitope accessibility, antibody specificity, and choice of microarray content.	[4–8]

		the entire cDNA genome is mapped onto the microarray.		
Microarray	A cDNA or oligonucleotide library is printed on a high-density microchip. Fluorescently tagged mRNA, previously extracted from a cell, hybridizes with the cDNA to which it is complementary. The resulting array provides an expression profile of those genes that may be regulated under the given experimental conditions.	• Provides a complete list of all genes within a cell whose expression levels are significantly altered under certain conditions.	• Does not indicate the mechanism of gene regulation, only the correlative characteristics.	[9–15]
Genome sequencing	Provides the comprehensive identification of the base pair of a genome. Several techniques have been developed and used, including whole-genome shotgun sequencing, and involve automated sequencers.	• Enables the identification of the base pair sequence, upon whichmost other experimental platforms are based.		[16]

The expression of each of these catalogued genes is regulated by a complex interplay of transcription factors and other cellular proteins. Chromatin immunoprecipitation is used to identify sites on a genome to which regulatory proteins bind. The proteins are crosslinked to DNA with formaldehyde and small fragments are precipitated, amplified, and identified. Recently, a high-throughput implementation of this technique (called genome-wide location analysis [19]) has enabled extensive characterizations of the regulatory patterns in yeast [20], and it is being developed for application to mammalian systems [21]. For example, Zeitlinger et al. found that different conditions result in the binding of yeast transcription factor Ste12 to distinct promoters in vivo, which, in turn, cause different sets of genes to be expressed, and different developmental programs to be specified [22].

In addition, cDNA microarrays can indicate which genes are expressed under specified conditions. Microarray technology has progressed to the point where community standards regarding the depositing and sharing of data have emerged [23]. Although initially there was criticism regarding the low signal to noise ratio, technical standards have significantly improved. Microarray data coupled with other data sources can generate remarkably accurate pictures of network function [24]. Microarrays complete the picture regarding the genotype of a given signaling network. Genome sequencing, coupled with expression arrays, delineates the protein participants. Genome-wide location analysis indicates which regulatory proteins control the expression of the given genes.

The first stage in mapping the genotype–phenotype relationship is a precise characterization of the genotype of interest. The methods highlighted above delineate the genotype, or "available parts list," for a given signaling network. Rapid and cost-effective sequencing efforts may lead to genotyping of individual cells in a population of highly mutable cells. This precise characterization may generate further understanding of a disease etiology as complex and diverse as cancer. Although many efforts to reconstruct signaling networks piece together global signaling reactions, only a subset of these reactions is ever expressed in any given system. Genome-wide location analysis may lead to signaling network reconstructions that correspond to the regulatory protein–gene interactions that occur in vivo in a given system. Advances in sequencing technologies and methods for identifying regulatory protein–gene interactions connect the signaling network to the regulatory program.

What questions do these technologies fail to answer? They do not indicate whether any of these proteins actually interact with each other in vivo. The approaches described above cannot generate hypotheses regarding the physical mechanism of their interaction. Furthermore, they do not account for nonprotein components of signaling networks, like lipids (e.g., PIP3) and metabolites (e.g., ATP). Rather, genomic technologies identify which proteins participate in a given cellular signaling network.

COMPONENT INTERACTION: BRIDGING GENOTYPES TO INTERMEDIATE PHENOTYPES

Relatively quick (on the order of seconds to minutes), measurable responses to a signaling event are here called "intermediate phenotypes" (see figure 5.1). For example, the rapid increase in intracellular calcium concentrations is an intermediate phenotype because it is an observable response to a triggering signaling event. These intermediate phenotypes may be transient responses or may lead to irreversible changes in gene expression (an endpoint phenotype). Most technologies developed to date to characterize cellular signaling networks, which are elaborated in table 5.2, measure these intermediate phenotypes.

Perhaps the most widely used technique to characterize signaling functions is the yeast two-hybrid approach and its associated variations [55]. Simply, the genes for target and bait proteins are fused to DNA binding and transcriptional activity protein domains so that if the target and bait proteins interact, the expression of a corresponding gene occurs. While this approach has been criticized for its lack of contextual specificity (e.g., two proteins might never be expressed at the same time, and yet they may interact in a two-hybrid screen), tremendous amounts of data have generated novel hypotheses regarding signaling network function. For example, yeast two-hybrid screens were recently used to construct a proteome-wide protein–protein interaction, or "interactome," map of *Caenorhabditis elegans* [56], including a network of 71 interactions among 59 proteins for the *C. elegans* DAF-7/TGF-β signal transduction pathway [57].

FRET-based technologies underpin another set of methods for measuring intermediate phenotypes [36]. This technology involves the fusion of two fluorophores to two proteins. Each fluorophore emits light at a distinct wavelength; however, when two proteins associate, the wavelength of the emitted light is different. Thus, associating proteins and their dynamic characteristics can be identified. For instance, a genetically encoded fluorescent indicator called a phocus, a mutant of the traditional green fluorescent protein (GFP), was recently developed specifically for the purpose of visualizing signal transduction based upon protein phosphorylation in vivo [58]. A recent study also used FRET to observe real-time changes in Ras signaling, which functions as a molecular switch in many signaling cascades [59]. Fluorescent technologies have also been used for high-throughput identification of cellular localization as well.

Table 5.2 Intermediate phenotype characterization: Experimental technologies for characterizing intermediate phenotypes, or the chemical transformations that comprise a network

Name	Description	Advantages	Disadvantages	References
Bioinformatics sequence analysis	Indicates the function of specific proteins by searching for sequence homologies. An increasing number of approaches synthesize multiple data sources to predict protein function.	• Takes advantage of the large amount of data that is already available, and the vast supply of computational resources. • Very efficient.	• Relies upon existing data, which is not necessarily always accurate, i.e., the prediction is only as good as the original assignment. • Does not provide any "new" discoveries, in the sense that proteins with no sequence homologies or structure similarities remain difficult to reconcile.	[25,26]
Coimmunoprecipitation	Indicates which proteins interact with one another. Like immunoprecipitation, this technique is based on the formation of immune complexes: a cell lysate is produced; the protein of interest is tagged by an antibody; the antigen is added and allowed to precipitate with the protein antibody (and thereby "coprecipitate" with any	• Enables the unbiased testing of multiple ligands against a specific target protein. • Protein concentrations, posttranslational modifications, and other in vivo characteristics are more likely to be maintained.	• May generate false positives for proteins in a complex that do not directly interact. • Level of sensitivity is less than that of other methods because the antigen concentration is inherently lower to avoid banding of the antibodies. There are ways to overcome this pitfall, but these usually result in perturbations of the simulated natural environment.	[27–29]

	proteins with which the target interacts); and the mixture is washed, such that the bound proteins are eluted and analyzed.			
Crosslinking	Indicates (1) the structural characteristics of proteins or complexes isolated from within a cell; and (2) which proteins interact with a specific test target. Proteins are covalently crosslinked with agents that bind to reactive groups of amino acid residues. The resulting complex is analyzed for protein interactions, interaction sites, and residue contacts.	• Capable of identifying weak interactions that are not identified with other methods. • Transient contacts between proteins at various stages in a dynamic process may be detected by crosslinking and freezing the process at different time points. • Can be performed in vivo with membrane permeable crosslinking reagents. • Is often combined with sequence analysis for improved results.	• Often detects a neighboring protein that may not be in direct contact with the target of interest.	[27–32]
Microarray	Can indicate the phenotypic function of a set of genes. A cDNA or oligonucleotide library is printed on a high-density microchip, and fluorescently tagged mRNA	• Can be used to characterize a complete cellular phenotype, i.e., a profile of all the genes that are activated and inactivated under certain cellular conditions.	• Results observed are dependent on the elements present on the chip, to the extent that extensive homology or antisense transcripts may cause cross-hybridization.	[9–12, 33–35]

Continued

Table 5.2 (*Continued*)

Name	Description	Advantages	Disadvantages	References
	extracted from the cell is hybridized on it. The resulting array identifies genes that are up- and down-regulated under the given experimental conditions.		• The probe is affected by the method of (1) RNA isolation and (2) incorporation of the available labeling agents.	
Fluorescent resonance energy transfer (FRET)	Identifies proteins that interact with each other. Proteins are tagged with fluorescent markers that generate visible fluorescent resonant energy transfer (FRET) when they come into close proximity with one another.	• Capable of yielding real-time data, including precise location and time of interaction. • Effectively detects intramolecular activity, i.e., interactions wherein the donor and acceptor fluorescent protein are fused to the same host protein.	• May not accurately identify interaction if the hybrid proteins fail to attain appropriate orientation and separation distance. • Subject to noise from unpaired proteins that are colocalized.	[36,37]
Gene knockout	Indicates the phenotypic function of a specific gene. The gene of interest is replaced with a null mutation, and the resulting cellular behavior is observed to identify functions of the deleted gene.	• Effective tool for investigating global effects of individual genes.	• Difficult to characterize functions of lethal knockouts.	[38,39]

Mass spectrometry (MS)	Identifies proteins that interact with one another. Protein complexes, which are often isolated from cell culture using peptide tags that target specific protein sequences, are purified with 2D gel electrophoresis or liquid chromatography. Individual protein constituents are then identified by molecular weight.	• Several members of a complex can be tagged, which allows for an internal consistency check. • Detects real complexes in physiological settings. • Has a low false positive rate.	• Tagging may disturb complex formation. • May miss complexes that are not present under the experimental conditions. • Loosely associated components may be lost during purification.	[29,40–43]
Mutagenesis assay	Indicates gene function through the exploration of phenotypic effects of gene mutations. Short exogenous gene sequences are randomly inserted into a genome, and can mutate a functional gene. The mutation itself serves as a molecular marker for the subsequent identification of the disrupted gene. A variant of this method involves creating a site-specific insertional mutation.	• Applicable to any organism for which transformation and chromosomal integration of DNA can occur. • Inserted DNA can be tailored to meet experimental goals, e.g., determining expression and protein localization. • Does not depend on predetermined and possibly erroneous assignments of gene. position since it can be a random approach. • Easy to determine which gene is mutated by virtue of the insertion sequence, which serves as a molecular marker in the disrupted gene.	• Is redundant, i.e., multiple hits can occur in a single gene, due to the intrinsic randomness of insertion events. • Is incapable of generating a complete collection of gene functionality, though a spectrum of alleles that can display different phenotypic and/or localization characteristics is often generated.	[25,39]

Continued

Table 5.2 (*Continued*)

Name	Description	Advantages	Disadvantages	References
Phage display	Indicates which proteins bind to specific targets. The protein of interest is introduced into a bacteriophage, which, upon infecting *E. coli*, replicates and produces phage particles that display the hybrid protein. The bacteriophage can then be used to search for binding partners. Alternatively, a library of fusion proteins is generated and screened for binding to a target protein, e.g., using an affinity column. Those proteins that bind are eluted and replicated in *E. coli*, and their DNA is ultimately recovered and sequenced to identify the binding protein.	• Enables the screening of very large numbers of proteins for binding to selected targets. • Can be very rapid.	• Requires proteins to be secreted from *E. coli*. • The use of a bacterial host may preclude certain proteins from correctly folding or being normally modified. Further, all phage-encoded proteins are fusion proteins, and so their normal activity may be hindered. • The reactions are completed in vitro, and therefore are incapable of simulating in vivo conditions. Thus, false positives or false negatives are possible.	[27,39,44]
Protein affinity chromatography	Indicates which proteins interact with a specific test protein target.	• Enables the equal testing of multiple ligands against a specific target protein,	• Requires protein of interest to be completely purified to avoid the binding of nontargets to	[27,45]

	A protein is covalently coupled to a matrix column, and only those proteins that bind to it are retained from an appropriate extract (that is ultimately eluted by, e.g., a high-salt concentration), whereas all others pass through the affinity chromatography column or are readily washed off under low-salt conditions.	enabling, e.g., less abundant proteins to be equally considered. • Highly sensitive to even the weakest of protein–protein interactions. • Enables the detection of both the domains and critical residues of a protein that are responsible for a specific interaction. • Interactions that depend upon multi-subunit tethering can be detected.	contaminants in the matrix preparation. • False negatives, because complexes may not form due to a lack of appropriate protein modifications, or due to the associated resin. • Amount of extract applied is highly critical: too little of one kind of protein may cause interactions to go unnoticed, whereas too much may cause minor ligand species to be missed.	
Protein probing	Indicates which proteins in an expression library bind to a specific probe ligand. A labeled protein is used as a probe to screen an expression library, and interactions between an immobilized (library) protein and the labeled probe protein occur on an intermediary nitrocellulose filter. The filter is subsequently washed of nonbinding compounds	• Enables the unbiased testing of multiple ligands against a specific target protein, enabling, e.g., less abundant proteins to be equally considered. • The protein probe may be manipulated in vitro to introduce a specific posttranslational modification that is essential for the ability of the probe to bind to ligand(s) in the library.	• Proteins encoded by the library may not fold or undergo appropriate posttranslational modifications in *E. coli* and/or on a nitrocellulose structure. • Binding conditions are arbitrarily imposed by the investigator, as opposed to mimicking the native environment. • All possible combinations of protein–protein interactions are assayed, including those that might never occur in vivo.	[27,46]

Continued

Table 5.2 (*Continued*)

Name	Description	Advantages	Disadvantages	References
		• Any protein or protein domain can be labeled for use as a probe, and the preparation of the probes is a relatively straightforward task.		
RNA interference (RNAi)	Relatively new technique to indicate the phenotypic function of a specific gene. A double-stranded RNA (dsRNA) sequence that corresponds to the gene of interest is synthesized. The dsRNA is introduced in vivo and disrupts the function of the gene that it encodes through normal cellular mechanisms.	• Functions like a gene knockout, but does not result in a hereditary functional gene loss. • Very specific. Sequence similarity in excess of 80% is a proven requisite for gene inactivation. • Remarkably potent, i.e., only a few dsRNA strands are required to cause interference. • Can act on cells and tissues far removed from the original introduction site. • RNAi has met with success in a wide variety of organisms.	• The precise mechanism underlying RNAi is not yet well understood. • Compensatory effects have not yet been characterized.	[29,47–51]

(Yeast) two-hybrid assay	Indicates which pairs of proteins may interact with one another. Proteins are tagged such that interactions result in the expression of a reporter gene.	• Widely used technique for determining protein–protein interactions. • Very sensitive, and therefore capable of detecting interactions that are not observed through other methods. • Highly scalable, enabling the rapid generation of large amounts of data. • Some variants of this technique can be used for membrane proteins.	• Only a small degree of congruence exists between different data sets. • Exhibits high degree of false positives, as some proteins are spatially or temporally segregated in vivo. • Limited to proteins that can be localized to the nucleus such that the reporter gene is expressed.	[27,40,52–54]

Mass spectrometry has enjoyed tremendous success in the characterization of intermediate phenotypes by delineating the composition of protein complexes and posttranslational modifications on proteins [60]. After protein purification with two-dimensional electrophoresis or liquid chromatography, mass spectrometry can be used to precisely measure the mass of peptide fragments to identify protein complex composition. In tandem, two mass spectrometers can be used to identify the amino acid composition of peptide fragments and thus characterize posttranslational modifications like phosphorylations and methylations. As an example of the utility of this technique, it was recently used to characterize the dynamic phosphorylations of proteins associated with T-cell activation and the inhibition of the overactive BCR-ABL kinase in response to the cancer therapeutic STI571 (Gleevec) [61]. This study identified 64 phosphorylation sites on 32 proteins.

Array-profiling has recently been applied to characterize intermediate phenotypes. Protein arrays are based on the ordering of protein-binding compounds (primarily antibodies) to indicate the presence and intensity of proteins in a given sample [62]. However, since protein binding typically requires the native conformation of a given protein to be effective, such array approaches have been fraught with difficulty. Nevertheless, there is emerging evidence of the utility of such approaches. For example, 96-well microtiter plates were used to characterize ErbB receptor tyrosine kinases in human tumor cells [63].

Together, the technologies that identify the presence of proteins and measure their concentrations generate so-called "proteomic" data. Additional technologies provide what has been called "metabolomic" data, or the identification of metabolites in a cell and associated concentration measurements. What questions do these technologies answer? Under the *given* experimental conditions, two proteins do or do not interact with each other. These methods provide some indication of the mechanisms of interactions. For example, mass spectrometry can identify phosphorylation sites on a given protein. The final result of all such techniques is the identification of which proteins are present, with which cellular constituents they interact, and the concentrations of some of these components. However, these data alone do not indicate how individual signaling events or intermediate phenotypes contribute to network properties.

FUNCTIONS OF INTERACTING COMPONENTS: BRIDGING INTERMEDIATE PHENOTYPES TO ENDPOINT PHENOTYPES

Methods for characterizing the function of interacting components generate data on how intermediate phenotypes contribute to global properties and endpoint phenotypes. Perhaps one of the most common techniques for measuring the global function of network components involves the use of microarrays, as mentioned above. With microarrays, genome-scale gene expression profiles are obtained and used to characterize global function. For example, genes that are expressed together are likely functionally related [33]. Simply, cellular mRNA transcripts are extracted from cells and hybridize to cDNA or oligonucleotides corresponding to the genes within the genome on extremely dense arrays. The resulting data provide details of which genes are activated or inactivated under various experimental conditions. Although this approach is affected by the method of mRNA isolation, among other factors, it has proven to be effective in constructing complete phenotypic states [34,35]. Previous studies, for instance, have been able to distinguish between different types of lymphoma via expression profiling [13].

Experimental techniques based on the principle of RNAi may be the most promising for perturbing a signaling network and measuring subsequent intermediate and endpoint phenotypes [64]. These techniques take advantage of the cell's natural response to double-stranded RNA (dsRNA) (e.g., from a virus). After introduction of dsRNA with a sequence similar to the target mRNA, an RNA-inducing silencing complex (RISC) is activated and cleaves mRNA molecules with a complementary sequence. Consequently, the function of a specific protein corresponding to the targeted mRNA can be inhibited. The endpoint phenotypes are then measured without the function of the target protein. This technique allows for tailored investigation of a signaling network. One recent application of this technique categorized the growth and viability of 91% of *Drosophila* genes [65].

Several of the technologies listed above can be integrated to evaluate intermediate and endpoint phenotypes and characterize cellular signaling networks. Two examples are presented here. First, the yeast galactose utilization system was perturbed with a series of environmental (e.g., growth with and without galactose) and genetic (e.g., gene deletion) modifications [66]. Subsequently, global measurements of network function were made (e.g., analysis of corresponding galactose utilization genes with mRNA expression arrays). This integrated approach led to several novel hypotheses and an extensive refinement of the galactose utilization pathway in yeast. The second example integrated RNA interference, mass spectrometry, and other techniques to generate a model of the human TNF-α/NF-κB signaling reactions that accounted for 221 molecular associations [29]. These studies are examples of how several techniques described above can be integrated to probe cellular signaling networks and how they are coupled with endpoint phenotypes.

What questions do these methods answer? All these techniques measure global effects of a protein's function, that is, how a signaling network responds to the absence of a given protein. Certainly, combinations of proteins may be perturbed leading to characterizations of how groups of proteins affect network function. However, the combinatorial possibilities would be tremendous.

ELUCIDATING ENDPOINT PHENOTYPES

Endpoint phenotypes are measurable, global characteristics of cellular signaling network function. For example, whether a cell undergoes apoptosis as a result of a series of signaling events is an endpoint phenotype. These global behaviors (endpoint phenotypes) are readily measured because they involve cell-scale properties (e.g., growth, apoptosis, differentiation, and migration). Furthermore, recent technologies, shown in table 5.2 and discussed above, allow for precise control of signaling network components that influence systems-level properties.

The measurement of cell migration is well developed. In table 5.3, three such techniques are listed. Each approach uses a concentration gradient to induce cell migration. After a defined time period, the numbers of cells that have migrated to a particular location are counted. These techniques measure the strength of a response to a given chemokine or the associated conditions and can be used to evaluate the effects of different genetic modifications or environmental conditions on migration processes.

The additional techniques listed in table 5.3 include flow cytometry and cell growth assays. Flow cytometry is a highly quantitative technique that counts the number of cells with a given fluorescent tag. This fluorescent marker can be associated with cell division, or a variety of intermediate phenotypes. Similarly, growth assays based on the optical density of a cell culture have been implemented in high-throughput experiments, enabling the simultaneous measurement of cell growth under a variety of conditions.

Global measurements of cellular signaling network function are also used to characterize cellular differentiation, including the identification of original and differentiated populations of stem cells. For instance, hematopoietic stem cells are known to express a unique surface marker, CD34, that distinguishes them from the remaining bone marrow cells, and from any differentiated cells [74]. Gene expression profiling or antibody screening via protein affinity chromatography for CD34 therefore enables identification of stem or differentiated cell populations. Similarly, identifying the derivatives of mesenchymal stem cells is made possible via expression profiling or screening for protein markers. Bone tissue, for example, may be screened for markers like osteopontin or osteocalcin. An investigation of rat central nervous system tissue illustrated the utility of global network function measurements in characterizing differentiated states; gene expression profiling enabled the identification of five basic "waves" of expression, each corresponding to a distinct development phase [75].

Table 5.3 Endpoint phenotype characterization: experimental technologies for characterizing endpoint phenotypes, or the transcriptional and cellular modifications as a result of intermediate phenotypes

Name	Description	Advantages	Disadvantages	References
Boyden filter transmigration assay	Provides a measure of the migration characteristics of populations of cells. A porous filter separates two compartments, with the cells of interest placed in the upper compartment, and a chemoattractant placed in the lower one. The chemoattractant forms a concentration gradient as it is allowed to diffuse up through the pores of the filter, and the cells may then respond and migrate. After a period of time, the assay is stopped, the filter membrane is fixed and stained, and migrating cells are counted.		• Original version consisted of a limited number of wells (48), though several recent attempts have expanded the capacity of the assay to 96 wells. • The assessment of the number of migrating cells is laborious and subjective. • High-throughput chemotaxis assay formats using cell lines have not yet been described in sufficient detail. • Cannot observe the cells during the assay. • The stability of the chemokine gradient is often compromised.	[67,68]

Continued

Table 5.3 (*Continued*)

Name	Description	Advantages	Disadvantages	References
Dunn chemotaxis chamber	Provides a measure of the migration characteristics of populations of cells. Like the Boyden and Zigmund microchambers, a gradient is formed by a chemoattractant diffusing between two compartments. In this design, however, the wells are arranged in a concentric layout. Additionally, the wells are precisely cut into a glass slide, which enables better transmission properties for light microscopy.	• Offers the best means to view the cells during the chemotaxis assay: a glass slide is used, enabling microscopy, and the wells are cut precisely into the glass slide, ensuring better transmission properties for many different types of light microscopy. • The use of a concentric layout maximizes the stability of the chemokine gradient; the edge effects associated with the Boyden chamber and the linear bridge in the Zigmund chamber are avoided.	• The assessment of the number of migrating cells is laborious and subjective. • High-throughput chemotaxis assays have not yet been developed.	[67,69,70]
Flow cytometry	Allows large numbers of cells to be analyzed rapidly and automatically for physiological characteristics. Cells are stained with fluorescent dyes, suspended in solution, and passed in a stream before a laser or other light source. The subsequent measurements that categorize the cells are based on how the light-sensitive dye reacts to the light source. This technique offers real-time, quantitative measurements of characteristics	• Highly sensitive and quantitative, i.e., capable of precise, discrete measurements of fluorescent signals. • Enables the simultaneous quantitative analysis of multiple optical markers.	• Severely limited in its ability to handle multiple discrete samples of cells, though recent variations of flow cytometry, including plug flow cytometry, have attempted to solve this problem. • Informatic systems, i.e., tools for accurately tracking data, have yet	[71]

	like a cell's temporal location in the cell cycle or biochemical interactions, e.g., ligand–receptor reactions.		to be fully integrated with flow cytometry data.	
In silico network analysis	Provides models of large-scale cellular behavior, as an assimilation of the functions of many individually characterized proteins. An increasing number of computational approaches are being developed to generate these kinds of models, which are based on genome data and experimental evidence.	• Takes advantage of the large amount of data that is already available. • Can predict large-scale cellular behavior. • Offers an iterative approach to discovery: global observations are matched against model predictions, leading to the formation of new models, new predictions, and new experiments to test them.	• To date, relatively few proven computational approaches exist.	[66,72,73]
Microplate reader for growth rate measurements	Provides dynamic growth rate measurements of many different cell cultures simultaneously. Several different systems have been developed recently, including BioScreen. These systems are typically comprised of microtiter plates containing the cell culture samples, as well as a microtiter plate reader and associated software for quantification of growth.	• Can track the growth of many different cell culture samples simultaneously. • Can provide dynamic, real-time data.		[67]

Continued

Table 5.3 (*Continued*)

Name	Description	Advantages	Disadvantages	References
Zigmund chamber	Provides a measure of the migration characteristics of cell populations. Like the Boyden chamber, a gradient is formed by a chemoattractant diffusing between two compartments. In this case, however, the compartments are represented by two rectangular wells that are cut into a Plexiglas slide and separated by a narrow bridge. The wells are filled with medium, one of which contains the chemoattractant, cells are added, and a gradient forms as the chemokine diffuses across the bridge.	• Because a glass slide is used, it is possible to view the cells during the assay as they migrate across the bridge in time, unlike in the Boyden microchamber. • The physical separation between the wells increases the stability of the chemokine gradient as compared to the Boyden chamber.	• The assessment of migrated cells is laborious and subjective. • High-throughput chemotaxis assays using cell lines have not yet been developed.	[67,69]

TEMPORAL REGULATION OF CELLULAR SIGNALING NETWORKS

In addition to the three tiers of information in a cellular signaling network, across all these scales there is a temporal component. For example, the binding of a ligand to its receptor may result in the activation of a transcription factor. This transcription factor may induce the expression of a gene whose corresponding protein inhibits the activity of the original ligand–receptor complex. The inhibition of the signaling function may in turn arrest some migratory activity. An understanding of the function of the ligand–receptor complex and the inhibitory protein does not reflect the temporal elements of how the components are related to each other, that is, the "sequence of events." Temporal components of cellular signaling systems have been perhaps most characterized in autocrine signaling loops for which temporal characteristics are critical elements (for an example, see [76]). While temporal regulation is critical to cellular signaling network function, its analysis at a system level is only beginning as methods for quantitative mapping of components, interactions, and resultant phenotypes are emerging.

QUANTITATIVE MAPPING OF GENOTYPE–PHENOTYPE RELATIONSHIPS

There are efforts to provide quantitative descriptions of components and their relationships in each tier of cellular signaling networks. With respect to the genotype component of cellular signaling, extensive bioinformatics analysis has identified the complete set of protein kinases in the human genome [77]. These bioinformatics analyses provide the foundation for more comprehensive quantitative analysis of relationships between components.

Three poignant examples illustrate the quantitative mapping to cell signaling phenotypes from the cataloging of genotype. First, an extensive model incorporating G-protein coupled receptors, G-proteins, and GTPase-activating proteins (GAPs) was analyzed to characterize 16 distinct "states" or intermediate phenotypes [78]. These "states" allowed for the clarification of previously paradoxical information regarding the lack of GAP regulation of current amplitude of G-protein-activated ion channels. Second, a mathematical representation of NF-κB and IκB dynamics clarified the role of various IκB isoforms [79]. These results were experimentally verified with murine cell knockouts of the associated IκB isoforms. Third, an exhaustive stoichiometric reconstruction of JAK-STAT signaling in the human B-cell resulted in a novel, quantitative analysis of crosstalk and pathway redundancy [72].

Quantitative models of endpoint phenotypes have also generated novel characterizations of genotype–phenotype relationships in cellular signaling networks. Cell migration is the endpoint phenotype that has been perhaps most thoroughly analyzed with quantitative models.

Initial work on quantifying cell motility in bacteria [80] resulted in molecular descriptions of motility mechanisms [81], in addition to sophisticated analyses of stochasticity of leukocyte cellular responses to chemokine concentration gradients [82].

WHERE ARE THE KNOWLEDGE GAPS?

While we may measure the function of an individual protein and perhaps even how that protein affects network-level behavior, a cellular signaling network is a function of the simultaneous interaction of thousands of proteins. To use the colloquialism, the whole is greater than the sum of its parts. There is emerging evidence of such integrative functions. For example, cellular responses to combinations of kinase inhibitors cannot be understood by merely the summation of individual inhibitor–kinase interactions. In fact, a recent study in *Saccharomyces cerevisiae* concluded that multiplex inhibition of cyclin-dependent kinases Cdk1 and Pho85, and *not* the inhibition of either kinase alone, controlled the expression of 76 genes involved in cell growth [83]. Thus, one knowledge gap lies in the characterization of integrated functions.

As discussed in the introduction, systems biology is not simply the analysis of a "large" number of cellular constituents. Rather, it is the study of cellular constituents at a genome scale. Systems biology is not the analysis of the *individual* function of all cellular proteins at a genome scale. Rather, it is the *integrated* analysis of all the cellular proteins at a genome scale [84]. Combinatorics of experimental conditions may reveal novel features elucidating in part the integrated properties of these cellular networks. However, it is impossible to generate all possible combinations of all possible experimental conditions. Thus, another knowledge gap lies in the inability to test all possible experimental conditions. The question remains: at what point is the repertoire of tested experimental conditions sufficient?

These principles are complicated by the fact that there is a degree of evolvability in biology. Biological systems adapt to their environment. Consequently, a signaling network may respond in a particular fashion at one point in time, but environmental pressures may select for a different signaling mechanism at another point in time. For example, while interactions between a peptide fragment of the yeast Pbs2 protein and the Sh3 domain of Sho1 protein are highly specific, this Pbs2 peptide fragment interacts with Sh3 domains from proteins of other species [85]. Thus, small sequence variations can lead to differing interaction specificities. Cells under highly selective pressures may readily mutate to initiate novel signaling pathways, as has been mechanistically demonstrated with metabolic pathways [86,87]. The evolvability of biological systems is thus another knowledge gap.

BUILDING BRIDGES OVER THE UNKNOWN

One of the simplest ways to address gaps in biological knowledge is to further develop the genotype–phenotype experimental technologies. For instance, antibody specificity in protein microarrays needs to be addressed to successfully monitor global network properties using this technique. A recent study showed that only a small number of antigen–antibody pairs exhibited the kind of specificity that is required in order to perform highly parallel quantitative analysis, that is, in order to monitor global network properties [88,89]. Similarly, protein microarrays are believed to be capable of mapping signaling pathways and their individual components, but they are hampered by the fact that they cannot yet replicate the unique conditions that are necessary to preserve proper protein folding and enzyme activity, among other in vivo properties [89].

Additionally, new tools for managing, integrating, and understanding the wealth of data that is being amassed would be very beneficial [9]. For example, a recent study analyzed the galactose metabolic pathway with microarrays coupled with mass spectrometry and identified new possibilities for the regulation of galactose and interpathway interactions [9,66]. Such novel techniques provide the means for building bridges over certain information "disconnects."

However, as was discussed above, there are types of information that may never be deciphered from experiment. To compensate for these kinds of gaps, several different research avenues are currently being pursued. Among these, a number of mathematical approaches that account for the degree of unknowability have been developed. For instance, the "constraint-based" approach is grounded in the fact that all expressed phenotypes must satisfy basic constraints that are imposed upon the molecular functions of all cells [90]. Physical laws like the conservation of mass and energy must be upheld, just as environmental factors, including nutrient availability, pH, temperature, and osmolarity, must be satisfied. These constraints have been described mathematically through the use of "balances" (constraints that are associated with conserved quantities or with phenomena like osmotic pressure) and "bounds" (constraints that limit the numerical ranges of individual variables and parameters like concentration or flux), and the solution space for these mathematical descriptions provides the range of valid states of a reconstructed network. Thus, being able to identify constraints and state them mathematically helps narrow the spectrum of possible phenotypes, and therefore provides an approach to enhancing the understanding of cellular network function that would otherwise be difficult. These approaches have had success with metabolism and regulation, and initial analysis of signaling networks shows promise [72].

Another example of a research avenue for handling the gaps in knowledge involves reverse-engineering biological networks. As discussed previously, cDNA and oligonucleotide microarrays provide gene expression profiles that help provide information about the underlying network structure. However, characterizing large-scale networks is much more challenging, simply because of the computationally intensive task that is associated with cataloging the activities of the thousands of genes within the genome. Furthermore, certain assumptions, including the hypothesis that genes that perform similar functions are expressed together, while likely to hold locally, are much more suspect when applied to global network function [91]. What if mRNA and protein expressions are not correlated? Singular value decomposition (SVD) is a technique that has been applied to biological network reconstruction with the aim of resolving these issues [91]. SVD is capable of using microarray data and generating a set of candidate biological networks. Further, based on the empirical observation that gene regulatory networks are usually large and sparse, the sparsest such candidate network is chosen as the likely biological outcome. A recent approach of reverse-engineering was successfully applied to gene regulatory networks [91].

CONCLUDING REMARKS

Cellular signaling network functions can be discretized into three distinct tiers, namely, genotypes, "intermediate phenotypes," and "endpoint phenotypes." The first tier involves the transcriptional activity of a genome, that is, all the proteins that result from the transcription and translation of genes. Indeed, the genotype effectively delineates the "parts list" of a given cellular network. The next tier, called "intermediate phenotypes," is composed of the chemical transformations in which these "parts" are involved, including, for example, the various protein–protein interactions that mediate responses to signaling events. Finally, "endpoint phenotypes" are the subsequent transcriptional and cellular events, such as apoptosis, that ultimately drive cellular phenotypes.

Numerous experimental technologies have emerged to decipher cellular signaling networks, and these are easily classified according to the three-tiered structure. For instance, ChIP-chip methodologies explain the genotype by identifying which genes are regulated by various regulatory proteins under specific conditions. Intermediate phenotypes are characterized by methods such as FRET-based technologies that delineate which proteins interact with each other. Techniques like RNAi-based approaches, which indicate the phenotypic function of a specific transcript, elucidate how particular components of a network affect global cellular responses.

Despite the enormous wealth of data that each of these technologies generates, these data are not yet continuous. For example, although it is possible to measure the function of an individual protein and perhaps even how that protein affects network-level behavior, it remains a challenge to describe how the network is a function of the simultaneous interaction of thousands of such proteins. Indeed, the understanding of integrated functions remains elusive. Furthermore, systems biology, which attempts to model the integrated properties of cellular networks, cannot adequately generate all the possible combinations of all the possible experimental conditions; realizing the point at which the repertoire of tested experimental conditions becomes sufficient is therefore an essential property that is not yet defined. The evolvability of biological systems, an important property of livelihood, is similarly not well characterized.

Nevertheless, a number of approaches are being developed in an attempt to bridge these knowledge gaps. The simplest of these have involved enhancing the experimental technologies that are used to decipher the genotype–phenotype mapping. Improving microarrays so that they map signaling networks and their individual components, for instance, is an ongoing research avenue. More significantly, to compensate for the types of information that may never be deciphered from experiment, several mathematical approaches that account for the degree of unknowability have been developed. For instance, it has been shown that the spectrum of possible cellular phenotypes can be narrowed by virtue of a "constraint-based" approach, which involves identifying constraints and stating them mathematically. In addition, reverse-engineering biological networks (e.g., generating a set of candidate networks from microarray data) has had success in describing metabolism and regulation and has shown signs of initial promise in signaling networks.

Nonetheless, direct relations between external factors, or even specific internal factors, and internal responses are rare. For instance, although specific factors are believed to uniquely specify particular responses in the cell-fate control of stem cells, few such factors have been identified [92]. Instead, stem cell-fate control relies principally on common, developmentally conserved signaling processes that are involved in multiple stem cell-fate decisions. Thus, the challenge remains to discriminate the combinations of inputs to signaling networks and the resulting phenotypes that drive higher-level behavior.

ACKNOWLEDGMENTS

We thank the National Institutes of Health for financial support.

REFERENCES

1. Der, S. D., A. Zhou, B. R. Williams, et al. Identification of genes differentially regulated by interferon alpha, beta, or gamma using oligonucleotide arrays. *Proceedings of the National Academy of Sciences USA*, 95(26):15623–8, 1998.
2. Orlando, V. Mapping chromosomal proteins in vivo by formaldehyde-crosslinked-chromatin immunoprecipitation. *Trends in Biochemical Sciences*, 25(3):99–104, 2000.
3. Orlando, V., H. Strutt and R. Paro. Analysis of chromatin structure by in vivo formaldehyde cross-linking. *Methods*, 11(2):205–14, 1997.
4. Buck, M. J. and J. D. Lieb. ChIP-chip: considerations for the design, analysis, and application of genome-wide chromatin immunoprecipitation experiments. *Genomics*, 83(3):349–60, 2004.
5. Dunn, K. L., P. S. Espino, B. Drobic, et al. The Ras-MAPK signal transduction pathway, cancer and chromatin remodeling. *International Journal of Biochemistry and Cell Biology*, 83(1):1–14, 2005.
6. Kirmizis, A. and P. J. Farnham. Genomic approaches that aid in the identification of transcription factor target genes. *Experimental Biology and Medicine*, 229(8):705–21, 2004.
7. Hanlon, S. E. and J. D. Lieb. Progress and challenges in profiling the dynamics of chromatin and transcription factor binding with DNA microarrays. *Current Opinion in Genetic Development*, 14(6):697–705, 2004.
8. Banerjee, N. and M. Q. Zhang. Identifying cooperativity among transcription factors controlling the cell cycle in yeast. *Nucleic Acids Research*, 31(23):7024–31, 2003.
9. Patterson, S. D. and R. H. Aebersold. Proteomics: the first decade and beyond. *Nature Genetics*, 33(Suppl):311–23, 2003.
10. Chittur, S. V. DNA microarrays: tools for the 21st century. *Combinatorial Chemistry and High Throughput Screening*, 7(6):531–7, 2004.
11. DeRisi, J. L., V. R. Iyer and P. O. Brown. Exploring the metabolic and genetic control of gene expression on a genomic scale. *Science*, 278(5338):680–6, 1997.
12. Nagaraj, V. H., R. A. O'Flanagan, A. R. Bruning, et al. Combined analysis of expression data and transcription factor binding sites in the yeast genome. *BMC Genomics*, 5(1):59, 2004.
13. Alizadeh, A. A., M. B. Eisen, R. E. Davis, et al. Distinct types of diffuse large B-cell lymphoma identified by gene expression profiling. *Nature*, 403(6769):503–11, 2000.
14. Schena, M., D. Shalon, R. W. Davis, et al. Quantitative monitoring of gene expression patterns with a complementary DNA microarray. *Science*, 270(5235):467–70, 1995.
15. Pease, A. C., D. Solas, E. J. Sullivan, et al. Light-generated oligonucleotide arrays for rapid DNA sequence analysis. *Proceedings of the National Academy of Sciences USA*, 91(11):5022–6, 1994.
16. Venter, J. C., M. D. Adams, G. G. Sutton, et al. Shotgun sequencing of the human genome. *Science*, 280(5369):1540–2, 1998.
17. Bernal, A., U. Ear and N. Kyrpides. Genomes OnLine Database (GOLD): a monitor of genome projects world-wide. *Nucleic Acids Research*, 29(1):126–7, 2001.

18. Shendure, J., R. D. Mitra, C. Varma, et al. Advanced sequencing technologies: methods and goals. *Nature Reviews Genetics*, 5(5):335–44, 2004.
19. Ren, B., F. Robert, J. J. Wyrick, et al. Genome-wide location and function of DNA binding proteins. *Science*, 290(5500):2306–9, 2000.
20. Lee, T. I., N. J. Rinaldi, F. Robert, et al. Transcriptional regulatory networks in *Saccharomyces cerevisiae*. *Science*, 298(5594):799–804, 2002.
21. Ren, B. and B. D. Dynlacht. Use of chromatin immunoprecipitation assays in genome-wide location analysis of mammalian transcription factors. *Methods in Enzymology*, 376:304–15, 2004.
22. Zeitlinger, J., I. Simon, C. T. Harbison, et al. Program-specific distribution of a transcription factor dependent on partner transcription factor and MAPK signaling. *Cell*, 113(3):395–404, 2003.
23. Brazma, A., P. Hingamp, J. Quackenbush, et al. Minimum information about a microarray experiment (MIAME)—toward standards for microarray data. *Nature Genetics*, 29(4):365–71, 2001.
24. Herrgard, M. J., M. W. Covert and B. O. Palsson. Reconciling gene expression data with known genome-scale regulatory network structures. *Genome Research*, 13(11):2423–34, 2003.
25. Ross-Macdonald, P. Functional analysis of the yeast genome. *Functional and Integrative Genomics*, 1(2):99–113, 2000.
26. Marcotte, E. M., M. Pellegrini, H. L. Ng, et al. Detecting protein function and protein-protein interactions from genome sequences. *Science*, 285(5428):751–3, 1999.
27. Phizicky, E. M. and S. Fields. Protein-protein interactions: methods for detection and analysis. *Microbiology Review*, 59(1):94–123, 1995.
28. Fremont, J. J., R. W. Wang and C. D. King. Coimmunoprecipitation of UDP-glucuronosyltransferase isoforms and cytochrome P450 3A4. *Molecular Pharmacology*, 67(1):260–2, 2005.
29. Bouwmeester, T., A. Bauch, H. Ruffner, et al. A physical and functional map of the human TNF-α/NF-κ B signal transduction pathway. *Nature Cell Biology*, 6(2):97–105, 2004.
30. de Gunzburg, J., R. Riehl and R. A. Weinberg. Identification of a protein associated with p21ras by chemical crosslinking. *Proceedings of the National Academy of Sciences USA*, 86(11):4007–11, 1989.
31. Friedhoff, P. Mapping protein-protein interactions by bioinformatics and cross-linking. *Analytical and Bioanalytical Chemistry*, 381(1):78–80, 2005.
32. Back, J. W., L. de Jong, A. O. Muijsers, et al. Chemical cross-linking and mass spectrometry for protein structural modeling. *Journal of Molecular Biology*, 331(2):303–13, 2003.
33. Fraser, A. G. and E. M. Marcotte. Development through the eyes of functional genomics. *Current Opinion in Genetic Development*, 14(4):336–42, 2004.
34. Clarke, P. A., R. te Poele and P. Workman. Gene expression microarray technologies in the development of new therapeutic agents. *European Journal of Cancer*, 40(17):2560–91, 2004.
35. Walker, M. G., W. Volkmuth, E. Sprinzak, et al. Prediction of gene function by genome-scale expression analysis: prostate cancer-associated genes. *Genome Research*, 9(12):1198–203, 1999.
36. Truong, K. and M. Ikura. The use of FRET imaging microscopy to detect protein-protein interactions and protein conformational changes in vivo. *Current Opinion in Structural Biology*, 11(5):573–8, 2001.

37. Kenworthy, A. K. Imaging protein-protein interactions using fluorescence resonance energy transfer microscopy. *Methods*, 24(3):289–96, 2001.
38. Holmes, A., J. E. Lachowicz and D. R. Sibley. Phenotypic analysis of dopamine receptor knockout mice; recent insights into the functional specificity of dopamine receptor subtypes. *Neuropharmacology*, 47(8):1117–34, 2004.
39. Alberts, B. *Molecular Biology of the Cell*, 4th ed. Garland Science, New York, 2002.
40. von Mering, C., R. Krause, B. Snel, et al. Comparative assessment of large-scale data sets of protein-protein interactions. *Nature*, 417(6887):399–403, 2002.
41. Rigaut, G., A. Shevchenko, B. Rutz, et al. A generic protein purification method for protein complex characterization and proteome exploration. *Nature Biotechnology*, 17(10):1030–2, 1999.
42. Mann, M., R. C. Hendrickson and A. Pandey. Analysis of proteins and proteomes by mass spectrometry. *Annual Review of Biochemistry*, 70:437–73, 2001.
43. Puig, O., F. Caspary, G. Rigaut, et al. The tandem affinity purification (TAP) method: a general procedure of protein complex purification. *Methods*, 24(3):218–29, 2001.
44. Koolpe, M., R. Burgess, M. Dail, et al. EphB receptor-binding peptides identified by phage display enable design of an antagonist with ephrin-like affinity. *Journal of Biological Chemistry*, 280(17):17301–11, 2005.
45. Takeuchi, R., M. Oshige, M. Uchida, et al. Purification of *Drosophila* DNA polymerase zeta by REV1 protein-affinity chromatography. *Biochemical Journal*, 382(Pt 2):535–43, 2004.
46. Young, R. A. and R. W. Davis. Yeast RNA polymerase II genes: isolation with antibody probes. *Science*, 222(4625):778–82, 1983.
47. Hannon, G. J. RNA interference. *Nature*, 418(6894):244–51, 2002.
48. Friedman, A. and N. Perrimon. Genome-wide high-throughput screens in functional genomics. *Current Opinion in Genetic Development*, 14(5):470–6, 2004.
49. Maeda, I., Y. Kohara, M. Yamamoto, et al. Large-scale analysis of gene function in *Caenorhabditis elegans* by high-throughput RNAi. *Current Biology*, 11(3):171–6, 2001.
50. Fire, A., S. Xu, M. K. Montgomery, et al. Potent and specific genetic interference by double-stranded RNA in *Caenorhabditis elegans*. *Nature*, 391(6669):806–11, 1998.
51. Hannon, G. J. *RNAi: A Guide to Gene Silencing*. Cold Spring Harbor Laboratory Press, Cold Spring Harbor, N.Y., 2003.
52. Walhout, A. J., S. J. Boulton and M. Vidal. Yeast two-hybrid systems and protein interaction mapping projects for yeast and worm. *Yeast*, 17(2):88–94, 2000.
53. Fields, S. and O. Song. A novel genetic system to detect protein-protein interactions. *Nature*, 340(6230):245–6, 1989.
54. Ito, T., T. Chiba, R. Ozawa, et al. A comprehensive two-hybrid analysis to explore the yeast protein interactome. *Proceedings of the National Academy of Sciences USA*, 98(8):4569–74, 2001.
55. Phizicky, E., P. I. Bastiaens, H. Zhu, et al. Protein analysis on a proteomic scale. *Nature*, 422(6928):208–15, 2003.
56. Li, S., C. M. Armstrong, N. Bertin, et al. A map of the interactome network of the metazoan *C. elegans*. *Science*, 303(5657):540–3, 2004.
57. Tewari, M., P. J. Hu, J. S. Ahn, et al. Systematic interactome mapping and genetic perturbation analysis of a *C. elegans* TGF-β signaling network. *Molecular Cell*, 13(4):469–82, 2004.

58. Sato, M., T. Ozawa, K. Inukai, et al. Fluorescent indicators for imaging protein phosphorylation in single living cells. *Nature Biotechnology*, 20(3):287–94, 2002.
59. Mochizuki, N., S. Yamashita, K. Kurokawa, et al. Spatio-temporal images of growth-factor-induced activation of Ras and Rap1. *Nature*, 411(6841):1065–8, 2001.
60. Aebersold, R. and M. Mann. Mass spectrometry-based proteomics. *Nature*, 422(6928):198–207, 2003.
61. Salomon, A. R., S. B. Ficarro, L. M. Brill, et al. Profiling of tyrosine phosphorylation pathways in human cells using mass spectrometry. *Proceedings of the National Academy of Sciences USA*, 100(2):443–8, 2003.
62. Zhu, H. and M. Snyder. "Omic" approaches for unraveling signaling networks. *Current Opinion in Cell Biology*, 14(2):173–9, 2002.
63. Nielsen, U. B., M. H. Cardone, A. J. Sinskey, et al. Profiling receptor tyrosine kinase activation by using Ab microarrays. *Proceedings of the National Academy of Sciences USA*, 100(16):9330–5, 2003.
64. Dykxhoorn, D. M., C. D. Novina and P. A. Sharp. Killing the messenger: short RNAs that silence gene expression. *Nature Reviews Molecular Cell Biology*, 4(6):457–67, 2003.
65. Boutros, M., A. A. Kiger, S. Armknecht, et al. Genome-wide RNAi analysis of growth and viability in *Drosophila* cells. *Science*, 303(5659):832–5, 2004.
66. Ideker, T., V. Thorsson, J.A. Ranish, et al. Integrated genomic and proteomic analyses of a systematically perturbed metabolic network. *Science*, 292(5518):929–34, 2001.
67. Palsson, B. and S. Bhatia. *Tissue Engineering*. Pearson Prentice Hall, Upper Saddle River, N.J., 2004.
68. Ott, T. R., A. Pahuja, F. M. Lio, et al. A high-throughput chemotaxis assay for pharmacological characterization of chemokine receptors: utilization of U937 monocytic cells. *Journal of Pharmacological and Toxicological Methods*, 51(2):105–14, 2005.
69. Zigmond, S. H. Ability of polymorphonuclear leukocytes to orient in gradients of chemotactic factors. *Journal of Cell Biology*, 75(2 Pt 1):606–16, 1977.
70. Zicha, D., G. A. Dunn and A. F. Brown. A new direct-viewing chemotaxis chamber. *Journal of Cell Science*, 99(Pt 4):769–75, 1991.
71. Edwards, B. S., T. Oprea, E. R. Prossnitz, et al. Flow cytometry for high-throughput, high-content screening. *Current Opinion in Chemical Biology*, 8(4):392–8, 2004.
72. Papin, J. A. and B. O. Palsson. The JAK-STAT signaling network in the human B-cell: an extreme signaling pathway analysis. *Biophysical Journal*, 87(1):37–46, 2004.
73. Ideker, T., T. Galitski and L. Hood. A new approach to decoding life: systems biology. *Annual Review of Genomics and Human Genetics*, 2:343–72, 2001.
74. Epstein, R. J. *Human Molecular Biology: An Introduction to the Molecular Basis of Health and Disease*. Cambridge University Press, New York, 2003.
75. Wen, X., S. Fuhrman, G. S. Michaels, et al. Large-scale temporal gene expression mapping of central nervous system development. *Proceedings of the National Academy of Sciences USA*, 95(1):334–9, 1998.
76. Shvartsman, S. Y., H. S. Wiley, W. M. Deen, et al. Spatial range of autocrine signaling: modeling and computational analysis. *Biophysical Journal*, 81(4):1854–67, 2001.

77. Manning, G., D. B. Whyte, G. Martinez, et al. The protein kinase complement of the human genome. *Science*, 298(5600):1912–34, 2002.
78. Bornheimer, S. J., M. R. Maurya, M. G. Farquhar, et al. Computational modeling reveals how interplay between components of the GTPase-cycle module regulates signal transduction. *Proceedings of the National Academy of Sciences USA*, 101(45):15899–904, 2004.
79. Hoffmann, A., A. Levchenko, M. L. Scott, et al. The IκB-NF-κB signaling module: temporal control and selective gene activation. *Science*, 298(5596):1241–5, 2002.
80. Berg, H. C. and D. A. Brown. Chemotaxis in *Escherichia coli* analysed by three-dimensional tracking. *Nature*, 239(5374):500–4, 1972.
81. Shimizu, T. S., N. Le Novere, M. D. Levin, et al. Molecular model of a lattice of signalling proteins involved in bacterial chemotaxis. *Nature Cell Biology*, 2(11):792–6, 2000.
82. Tranquillo, R. T. and D. A. Lauffenburger. Stochastic model of leukocyte chemosensory movement. *Journal of Mathematical Biology*, 25(3):229–62, 1987.
83. Kung, C., D. M. Kenski, S. H. Dickerson, et al. Chemical genomic profiling to identify intracellular targets of a multiplex kinase inhibitor. *Proceedings of the National Academy of Sciences USA*, 102(10):3587–92, 2005.
84. Palsson, B. O. Two-dimensional annotation of genomes. *Nature Biotechnology*, 22(10):1218–19, 2004.
85. Zarrinpar, A., S. H. Park and W. A. Lim. Optimization of specificity in a cellular protein interaction network by negative selection. *Nature*, 426(6967):676–80, 2003.
86. Fong, S. L. and B. O. Palsson. Metabolic gene-deletion strains of *Escherichia coli* evolve to computationally predicted growth phenotypes. *Nature Genetics* 36(10):1056–8, 2004.
87. Raghunathan, A. and B. O. Palsson. Scalable method to determine mutations that occur during adaptive evolution of *Escherichia coli*. *Biotechnology Letters*, 25(5):435–441, 2003.
88. Haab, B. B., M. J. Dunham, and P. O. Brown. Protein microarrays for highly parallel detection and quantitation of specific proteins and antibodies in complex solutions. *Genome Biology*, 2(2):RESEARCH0004, 2001.
89. Kabuyama, Y., K. A. Resing and N. G. Ahn. Applying proteomics to signaling networks. *Current Opinion in Genetic Development*, 14(5):492–8, 2004.
90. Price, N. D., J. L. Reed and B. O. Palsson. Genome-scale models of microbial cells: evaluating the consequences of constraints. *Nature Reviews Microbiology*, 2(11):886–97, 2004.
91. Yeung, M. K., J. Tegner and J. J. Collins. Reverse engineering gene networks using singular value decomposition and robust regression. *Proceedings of the National Academy of Sciences USA*, 99(9):6163–8, 2002.
92. Davey, R. E. and P. W. Zandstra. Signal processing underlying extrinsic control of stem cell fate. *Current Opinion in Hematology*, 11(2):95–101, 2004.

6

Integrating Innate Immunity into a Global "Systems" Context: The Complement Paradigm Shift

Dimitrios Mastellos & John D. Lambris

The Human Genome Project has pioneered scientific discovery by eloquently demonstrating that the basic constituents of a biological system (namely, the genome, transcriptome, and proteome) can be precisely defined, measured, and quantified at predetermined stages or time intervals [1]. The genome carries the entire digital (DNA) and inheritable code of life which remains largely unaffected by external stimuli, while the transcriptome and proteome reflect the dynamic and complex nature of biological information, as they are subject to fine regulatory control (e.g., transcriptome) and chemical modifications (e.g., proteome) that together dictate the actual behavior of cells and organisms. The deciphering of the entire human genomic sequence has provided the essential driver for developing high-throughput screening strategies and methodologies for the measurement of the entire RNA and protein output of a cell or organism [2]. This breakthrough has led to the development of *proteomics* and the unraveling of the dynamic protein profile of different tissues and organisms as it constantly adapts to discreet environmental perturbations.

It should be noted that the genome analysis of various organisms (e.g., yeast, insects, mammals) has revealed that what actually defines the differences between species, and favors natural selection across the evolutionary ladder, is not the gene content per se, but the fine interplay of signaling pathways and inducible gene regulatory networks that are activated in response to various molecular signals and environmental stimuli [1]. In this respect, *systems biology* has emerged as the field that studies these regulatory networks in an integrated and comprehensive manner. The "systems" approach attempts to "reconstruct" biological networks by integrating data produced on a multidisciplinary platform and aims at designing "model biological systems" with potentially new properties by exploiting the vast computing capability of modern bioinformatics [1].

Immunological processes epitomize in many ways the inherent complexity of biological systems. Random or stochastic signals are often responsible for triggering diversity in biological responses and can also serve as a factor that increases the adaptability and complexity of an immune response [1]. Indeed, it is often a case of minimizing a noise/signal ratio, and being able to convert "noise" into meaningful signals. For example, stochastic processes govern many of the genetic mechanisms that underlie the generation of antibody diversity. B-cells producing high-affinity antibodies undergo expansion, while low-affinity antibody-producing cells are subject to negative selection and deletion. In such a way, the signal (high affinity) is distinguished from the noise (low affinity). Similar stochastic events dictate various immunological processes such as antibody gene recombination and hypermutation, T-cell selection in the thymus, B-cell maturation in germinal centers, and immunological tolerance to self-antigens.

Even though the immune response has been extensively studied over the years, very little is known about the way in which different components of this response interact in a "systems" perspective. Our knowledge of the system's properties, and the pathway associations that are formed within the immune response, as well as the study of its dynamics, are still in an early stage. This can be partly attributed to the fact that researchers have been investigating this complex system only one gene, or one protein, at a time. It is now becoming evident that the enormous growth in structural and biomolecular data, generated by high-throughput methodologies, will force scientists to address immunology in an integrated "systems" context.

Complement, a pivotal system of the innate immune response, serves as an ideal paradigm of how traditional immunology is revisiting basic immune processes in a "systems" approach. In the postgenomic era, our knowledge of this innate immune system is enriched by findings that point to novel functions that do not strictly correlate with immunological defense and surveillance, immune modulation, or inflammation [3]. Several studies indicate that complement proteins exert functions that are either more complex than previously thought, or go well beyond the innate immune character of the system.

As we depart from the traditional hallmarks of molecular biology, such as the genome and the transcriptome, and begin to appreciate more the "proteome" as the dynamic expression profile of all organisms, novel associations between complement-modulated pathways and apparently unrelated biological processes are constantly being revealed. In this respect, recent evidence produced by our laboratory (and others) suggests that complement components can modulate diverse biological processes by closely interacting with other intra- and intercellular networks [3]. Furthermore, the structure and functions of several complement proteins, as well as the protein–protein interactions that underlie these functions, are now being investigated with the aid of cross-disciplinary approaches ranging from mathematics and biophysics to comparative phylogenesis, molecular modeling, and proteomics.

The complement system has long been appreciated as a major effector arm of the innate immune response. It consists of a complex group of serum proteins and glycoproteins and soluble or membrane-bound receptors, which play an important role in host defense against infection [4] (figure 6.1). Complement, a phylogenetically conserved arm of innate immunity, functions together with the adaptive immune response by serving as an important inflammatory mediator of antigen–antibody interactions. It also provides an interface between the innate and adaptive immune response by contributing to the enhancement of the humoral response mounted against specific antigens [5].

Complement can be activated through the classical, alternative, or lectin pathways (figure 6.1). Antigen–antibody complexes initiate the activation of the classical pathway, whereas the alternative and lectin pathways are activated in an antibody-independent fashion through interaction of complement components with specific carbohydrate groups and lipopolysaccharides present on the surface of foreign pathogens (e.g., bacteria) [6]. Complement activation proceeds in a sequential fashion, through the proteolytic cleavage of a series of proteins, and leads to the generation of active products that mediate various biological activities through their interaction with specific cellular receptors and other serum proteins. During the course of this cascade, a number of biological processes are initiated by the various complement components, including inflammation, leukocyte migration, and phagocytosis of complement-opsonized particles and cells. The end result of these complement-mediated events is a direct lysis of target cells and microorganisms as a consequence of membrane-penetrating lesions (pores).

Currently over 35 complement proteins have been identified, and deficiencies in any particular components have been frequently associated with a diminished ability to clear circulating immune complexes or fight bacterial and viral infection. Here we present a comprehensive account of how an integrated "systems" approach has contributed to the elucidation of the structural-functional aspects of C3–ligand interactions and the rational design of small-size complement inhibitors. We also present critical aspects of our studies on viral molecular mimicry and immune evasion as well as evolutionary aspects of complement biology and diversity that are integral to a "systems" overview of the complement system. Furthermore, we discuss novel associations of various complement components with developmental pathways and present our research on the role of complement in tissue regeneration and early hematopoietic development using an array of genomic, biochemical, and proteomic approaches.

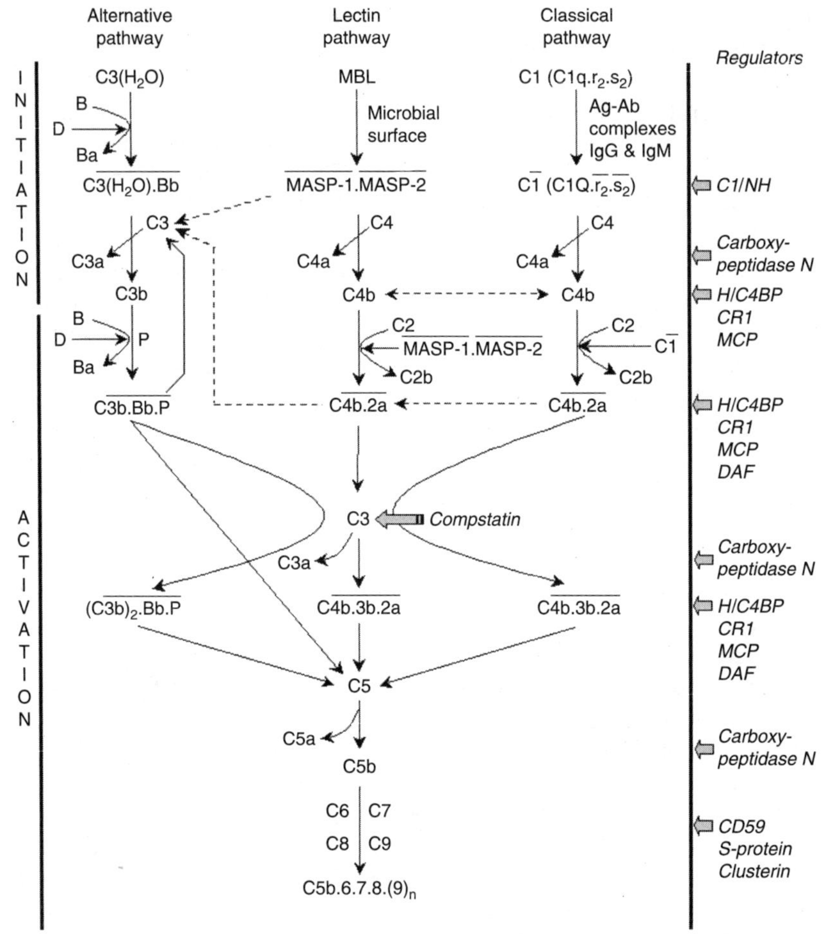

Figure 6.1 Schematic illustration of the complex network of protein–protein interactions that regulate the activation state of the complement system. This innate immune system comprises more than 30 soluble proteins and receptors that play important roles in innate and adaptive immune pathways and also participate in various immunoregulatory processes. Complement activation initiates a cascade of proteolytic cleavages that lead to the release of several bioactive fragments that mediate diverse functions in normal physiology and disease. All activation pathways converge at the level of C3, a multifunctional protein that constitutes the central component of the system. Recent evidence suggests that several key components of the system interact with cellular networks and signaling pathways that modulate cell differentiation, cell survival, hematopoietic development, and organ/tissue regeneration. The involvement of specific complement proteins in such noninflammatory pathways is discussed in various sections of this chapter.

It is our conviction that biomedical discovery will be spearheaded in the next decade by combinatorial and cross-disciplinary approaches (such as those outlined in the present chapter) that will address basic biological networks in a global and integrated manner. Furthermore, the mining of biomolecular and textual databases will essentially complement these experimental strategies and enable scientists to form the integrative context for hypothesis-driven scientific discovery.

STUDIES ON PROTEIN–PROTEIN INTERACTIONS OF COMPLEMENT COMPONENTS

Biophysical and Computational Studies of Protein–Protein Interactions in the Complement System

Biomolecular interactions (DNA–protein or protein–protein) are the core component of complex biological networks and define at a molecular level the nature of diverse cellular regulatory networks. Computer-assisted modeling is a means by which such complex interactions can be simulated, optimized, or even predicted in a quantitative and dynamic manner. The use of mathematical algorithms coupled to the power of supercomputing is integral to elucidating structure/function relations within these complex interactions. Computational studies greatly rely on the fine resolution of three-dimensional structures at an atomic level, which is achieved through the use of x-ray crystallographic or NMR approaches. To gain insight into the transitional structural changes that occur during association, recognition, or ligand binding, it is imperative to elucidate first the structure of the free components and compare this to the structure of the complex. The formation of interfaces in protein complexes is achieved by hydrophobic interactions among nonpolar side chains, hydrogen bonding interactions, electrostatic interactions (e.g., salt bridges), and steric (van der Waals) interactions. Furthermore, the electrostatic nature and shape constraints of the interacting partners within a complex dictate to a great extent the mechanism by which the optimum and more stable configuration is selected for recognition and binding. In addition, the stability of the biomolecular association is dependent on the inclusion or exclusion of solvent molecules within the interacting interface.

In this respect, studies have been designed to elucidate the dynamics that govern complex interactions between various complement components, using a cross-disciplinary platform that integrates biochemical, physicochemical, and computational methods. The interaction between complement components iC3b/C3d and complement receptor type 2 (CR2) was used as a research prototype. Integrating available crystallographic data [7–9] with experimental findings that indicate a strong dependence of this interaction on the solution pH and ionic strength [10–14], we went on to perform electrostatic calculations in an effort to understand the nature of the C3d/CR2 association at a theoretical level [15].

This was achieved by analyzing the overall electrostatic field of each protein both in free form and in complex, and also by measuring the contribution of charged residues within each protein to the binding reaction [15]. To gain further insight into the involvement of individual C3d residues to the C3d/CR2 association, theoretical site-directed mutations were designed [15] on the basis of available crystallographic structures, and their electrostatic potentials were subsequently calculated. It should be noted that these theoretical calculations showed great correlation to the experimental data produced by in vitro mutagenesis studies [12], and supported a two-step association model comprising recognition and binding [15]. An integrative survey addressing the complex dynamics and biophysical nature of the C3d/CR2 interaction should provide a comprehensive platform for the design of more potent complement therapeutics.

Interestingly, a similar biophysical approach was recently employed to determine the role of electrostatic potential in predicting the functional activity of two viral complement inhibitory proteins, VCP and SPICE [16]. These two orthopoxvirus complement regulators differ only in 11 amino acids. Yet, SPICE exhibits a 100-fold more potent inhibitory activity on complement [17]. Electrostatic modeling clearly predicted that switching critical residues between these two viral proteins would result in a VCP protein with enhanced complement regulatory activity. Furthermore, this study concluded that electrostatic forces are a major determinant of VCP binding to C3b, suggesting a two-step association model that involves electrostatically driven recognition and enhanced binding, as in the case of the C3d/CR2 association.

Mass Spectrometry and Hydrogen/Deuterium Exchange Studies

Hydrogen/deuterium exchange has traditionally been used to understand the formation of protein core or stable intermediate or transient states in pathways of protein folding, because it provides a noninvasive method for identifying protected (or deprotected) exchanging amides. The same principles can be applied to studies of protein–protein association, where the loss in solvent-accessible surface area upon association can be correlated with amide protection from exchange for the amides that lose their contact with solvent. Recent advances in the use of mass spectrometry allow for rapid collection of data of free and complexed proteins [18–20]. Comparison of mass spectra of free and complexed proteins provides the sites of interaction without the need for previously available structural data. The efficiency and rapidity of the technique makes it a suitable method for the study of the multiplicity of interactions of complement components within the complement system and with other serum or membrane-bound proteins.

Hydrogen/deuterium exchange coupled to mass spectrometry has recently been used to probe the conformational changes of the C3 molecule in its transition from native to hydrolyzed state [21], and it is becoming clear that such a methodology could provide valuable insight into the structural determinants that govern the interaction of C3 with various ligands and receptors (e.g., C3d/CR2).

Rational and Combinatorial Design of Complement Inhibitors: The Case of Compstatin

Complement has been primarily associated with the propagation of proinflammatory responses in the context of human disease. Complement activation has been implicated in the induction of acute inflammatory reactions leading to complications such as acute graft rejection, local tissue injury, and multiple organ failure [22]. Bedside therapies that target these harmful proinflammatory properties of complement activation are not yet available and considerable effort has been devoted in recent years to the development of effective drugs that inhibit complement activation. Complement inhibitors that are currently under development include small-size organic compounds, synthetic peptides, and also large monoclonal antibodies [23,24]. Our laboratory has focused its research efforts on the discovery of complement inhibitors that bind to complement component C3. Potential C3-binding inhibitors are considered a more effective means of blocking complement activation since all complement pathways converge to the cleavage and activation of C3.

High-throughput combinatorial screening strategies have been employed in conjunction with fine structural and computational approaches for the identification and subsequent optimization of potential C3-binding inhibitors. The screening of a phage-displayed random peptide library led to the identification of a 27-residue peptide that binds to C3 and inhibits complement activation [25]. This peptide was truncated to a 13-residue cyclic species that maintained complete activity, and was hereafter named *compstatin*. Compstatin blocks complement activation by preventing the C3 convertase-dependent proteolytic cleavage of C3 and the release of its bioactive fragments, C3b and the anaphylatoxin C3a. The activity of compstatin has been successfully tested in a series of in vitro, in vivo, ex vivo, and in vivo/ex vivo interface studies [25–34]. Compstatin constitutes an ideal lead compound for drug development because of its low toxicity in vivo and its small size, which allows for rapid, cost-effective, and large-scale synthesis.

To gain insight into the structural basis of the inhibitory activity of compstatin, we have employed a combination of NMR-based strategies, positional scanning, site-directed mutagenesis, and computational modeling approaches [35,36]. This multidisciplinary approach resulted in the resolution of the structure of compstatin and helped us elucidate several structure/function aspects of its activity. NMR studies of compstatin's structure revealed a molecular surface that comprises a polar patch and a nonpolar region. The polar part (residues 5–11) includes a type I β-turn and the nonpolar part (residues 1–4 and 12–13) includes the disulfide bridge that cyclizes the peptide.

Compstatin was subsequently subjected to several rounds of sequence and structure optimization using a comprehensive and integrative strategy comprising biophysical, molecular, and computational methods [25, 37–42]. The structure/activity relations determined in this structure-based rational design opened the way for a second round of phage-displayed random peptide library design, which produced a peptide that was 4-fold more active than the parent peptide [40].

Extending on our experimental high-throughput screening strategy, we decided to use compstatin as a model peptide for in silico combinatorial design, using a novel two-step computational optimization methodology. This round of design yielded a 16-fold more active analog than the parent peptide with sequence Ac-I[CVYQDW-GAHRC]T-NH_2 [43–46].

In an effort to integrate the data derived from our structure/function studies of compstatin into a new generation of C3-inhibitory peptides, we have recently produced new compstatin analogs, some of which were also expressed *in E. coli* using an intein-mediated expression system [47,48]. This approach has yielded an analog that is about 220 times more active than the original compstatin [48]. Finally, a different round of in silico studies was performed using molecular dynamics simulations, in our effort to understand the dynamic character of compstatin [49]. Collectively, these studies underscore the concept that bioactive peptides should be regarded as dynamic and flexible molecules rather than rigid structures when studying their binding properties. This concept is integral to our research for more effective complement inhibitors and should be taken into consideration in any drug design effort that involves peptide screening, synthesis, and structure manipulation.

Studies on the C5a/C5aR Interaction Using Synthetic C5aR Antagonists

Considerable effort has been devoted to characterizing the interaction of the complement anaphylatoxin C5a with its receptor C5aR, in several biological contexts. In this direction, an array of synthetic and modified peptides corresponding to the C-terminal sequence of the C5a anaphylatoxin have been developed and tested as potent antagonists of binding to the C5a receptor. Great promise has been shown by a potent, chemically synthesized C5a antagonist developed by Dr. S. M. Taylor's group (AcF-[OPdChaWR]) [50,51]. This cyclic hexapeptide has been successfully used as a selective inhibitor of C5a-mediated functions in animal models of disease and as a means to discern the involvement of C5a in several pathophysiological processes. In C5aR blockade studies using the acetylated species of the antagonist, it was demonstrated that C5aR is critically involved in the development of sepsis [52], in liver regeneration [53], and in fetal loss associated with the antiphospholipid antibody syndrome [54]. Recently, this antagonist was also used in blockade studies that demonstrated an essential role of C5a in the modulation of $CD8^+$ T-cell responses during acute influenza infection in mice [55].

Proteomic Approaches for Studying Protein–Protein Interactions and Profiling Global Protein Expression

Proteomics is the study of the protein content ("proteome") of an organism or a given tissue sample [56]. The basic principle relies on separating a complex protein/peptide mixture into its components and then analyzing quantitative changes or identifying certain proteins of interest. The information about a cell's proteome at a specific time point can then be correlated with the genomic approach that has been successfully applied to the study of cell physiology in recent years. Furthermore, the proteome contains additional functional information about the investigated tissue or cells that cannot be retrieved from the corresponding genome or transcriptome (e.g., posttranslational modifications of proteins) [57].

Appreciating the vast capabilities offered by this novel, high-throughput technology in global protein expression profiling, our laboratory has pursued a proteomic analysis of liver regeneration, in an effort to identify crucial signaling pathways, and changes in acute-phase response and lipid metabolism after partial hepatectomy in mice.

Liver regeneration is a complex physiological process that recruits multiple and redundant molecular pathways in order to ensure effective restoration of the hepatic architecture and function [58]. Identifying hepatocellular targets that are subject to posttranslational modifications and could thereby modulate or enhance the regenerative potential of the liver is of great therapeutic benefit for liver transplant recipients and living donors. In this respect, a recent study using a broad-range proteomic approach, coupled to mass spectrometry, has identified several proteins, including acute-phase and metabolic gene products, that are noticeably affected during liver regeneration [59]. Future investigations will address the functional relevance of the identified proteins for the integrity of the regenerative process and their potential use as therapeutic targets.

In the postgenomic era, it is anticipated that the integrated use of proteomic technologies will serve as a powerful analytical tool in the study of C3–ligand interactions by helping researchers to better define and monitor the dynamic protein changes that underlie various complement-dependent biological responses.

Thermodynamic Studies of Complement Protein–Protein Interactions

Isothermal titration calorimetry (ITC) is a method used to study energetics of the formation of macromolecular complexes in solution that have association constants in the range $\sim 10^3$–10^9 M^{-1}. It measures the electric power that is required to reequilibrate the temperature of a binding reaction cell with respect to a reference cell as a function of time, upon binding [60], and yields information on the stoichiometry, enthalpy, association constant, and free energy of binding. ITC has recently been applied to the study of energetics of the interaction of C3 with its inhibitor, compstatin. Thermodynamic measurements have indicated that the binding of compstatin to C3 is 1:1 and occurs through hydrophobic interactions with possible conformational changes in C3 or compstatin. Some protonation changes, occurring at the binding interface, have also been observed by ITC analysis [61]. This analysis will be extended to the energetics of various protein–protein interactions, with the goal of obtaining the energetic parameters of complement activation and regulation pathways.

The Complement System and Viral Molecular Mimicry

The complement system serves as both an innate and an acquired defense against viral infection. Activation of the complement (C′) system in the presence or absence of antibodies can lead to virus neutralization, phagocytosis of C3b-coated viral particles, lysis of infected cells, and generation of inflammatory and specific immune responses. To circumvent these defenses, viruses have not only developed mechanisms to control C′ but have also turned these interactions to their own advantage. Given the clinical importance of devising effective antiviral therapeutics, considerable attention has been devoted in recent years to the structural and functional analysis of several viral proteins (such as, VCP/SPICE, gC, and kaposica) that are involved in C′ evasion.

VACCINIA VIRUS IMMUNE EVASION STRATEGIES: STUDIES ON THE C3b/VCP INTERACTION

Vaccinia virus complement control protein (VCP) is a 35 kDa secretory protein of vaccinia virus that contains four short consensus repeats (SCRs) [62]. The protein shows homology to C4-binding protein (C4Bp) and many other members of the CCP group. Culture supernatant containing VCP and partially purified VCP has previously been shown to inhibit C′-mediated lysis of sheep erythrocytes, bind to C4b and C3b, and decay the classical and alternative pathway C3 convertases [63,64]. Further studies have indicated that VCP acts as a cofactor in the proteolytic inactivation of C4b and C3b by factor I [64].

It has been suggested that all four SCRs are required for VCP's complement-neutralizing activity. To elucidate which SCR domains are involved in abolishing complement-enhanced neutralization of vaccinia virus virions and further define the mechanisms of complement inactivation by VCP, we have developed monoclonal antibodies that react with VCP [65]. We used the recombinant VCP expressed in the yeast *Pichia pastoris* to vaccinate mice for the development of VCP-specific hybridomas. Ten Mabs were isolated and all recognized VCP on Western blots under reducing conditions as well as native-bound VCP in a sandwich enzyme-linked immunosorbent assay. Three of the 10 MAbs (2E5, 3D1, and 3F11) inhibited VCP's abolition of complement-enhanced neutralization of vaccinia virus virions. These MAbs blocked the interaction of VCP with C3b/C4b. The seven remaining MAbs did not alter VCP function in the complement neutralization assay and recognized VCP bound to C3b/C4b.

To understand MAb specificity and mode of interaction with VCP, we mapped the MAb binding regions on VCP. The seven nonblocking MAbs all bound to the first SCR of VCP. One of the blocking MAbs recognized SCR 2 while the other two recognized either SCR 4 or the junction between SCRs 3 and 4, indicating that structural elements involved in the interaction of VCP with C3b/C4b are located within SCR domains 2 and 3 and 4. These anti-VCP MAbs may have important clinical applications serving as potential therapeutic inhibitors of VCP's complement control activity, and may also offer a novel therapeutic platform for managing vaccinia virus vaccine complications that occur from smallpox vaccination. We also hope to use these monoclonals to further characterize the contribution of VCP to viral pathogenesis.

STUDIES ON THE VCP/SPICE INTERACTION WITH COMPLEMENT COMPONENT C3b

The smallpox inhibitor of complement enzymes (SPICE) is encoded by variola virus, the causative agent of smallpox, and was found to have complement-modulating activity that is similar to that of VCP [17]. Strikingly, despite the fact that it is 100-fold more potent than VCP in inactivating human C3b, SPICE differs from VCP in only 11 amino acid residues. In order to identify the amino acid residues that account for the significant difference in function between the two molecules, chimeric proteins consisting of VCP and SPICEs CCP modules were expressed in an *E. coli* expression system. All proteins retained functional activity and were able to inhibit classical pathway-mediated complement activation on an immunocomplex ELISA set.

The amino acids that are associated with the enhanced activity of SPICE were identified by testing the ability of appropriate VCP/SPICE chimeras to inhibit alternative pathway-mediated complement activation [16]. Furthermore, it was demonstrated that despite this enhanced activity, SPICE cofactor activity in the factor I-mediated degradation of C3b does not result in the cleavage of the α'-chain of C3b between residues 956 and 957 to generate C3c and C3dg. The multi-$\log_{10}$ increase in the complement regulatory activity of SPICE when compared to VCP permitted the systematic study of the domains and amino acid changes that contributed to this enhanced activity and demonstrated the role of electrostatics in the interaction of C3b with VCP and SPICE. That is, a predominantly negative C3b and a predominantly positive VCP variant favor their electrostatically driven recognition and enhance their association. Increase in the positive charge of VCP variants occurs by mutations of acidic amino acids, which reduce the negative character of the electrostatic potential at the vicinity of SCR-2 and SCR-3 and enhance the positive character of the electrostatic potential at SCR-1. This model has important implications in both our understanding of how RCA proteins interact with C3b and in the rational design of potent complement inhibitors that might be used as therapeutic agents [66].

To define the mechanism by which other viral homologs of complement regulatory proteins interact with C3b and contribute to immune evasion and pathogenesis, research efforts have focused on defining the complement-neutralizing function of two viral proteins of the herpesvirus family: gC, a glycoprotein encoded by herpes simplex virus type-1 (HSV-1) [67], and kaposica, a complement inhibitor encoded by Kaposi's sarcoma-associated herpesvirus (KSHV) [68]. Studies using animal gene knockout models and expressed recombinant proteins have revealed several aspects of the molecular interaction between C3b and gC/kaposica and will hopefully contribute to developing effective therapeutics for the treatment of these viral diseases.

COMPLEMENT EVOLUTION AND DIVERSITY

Studies of evolutionary processes that give rise to living organisms as they are known today constitute an integral part of modern biology. A comprehensive understanding of any complex biological system such as the complement system requires a thorough knowledge of its origin and evolution, development, and diversity.

In this respect, it has recently been shown that teleost fish possess multiple isoforms of various complement components, as a means of expanding their innate immune capabilities and also compensating for their weak adaptive immune response and poor and undiversified Ig repertoire. In an effort to characterize the phylogeny of C3 in invertebrates and lower vertebrates, recent studies have documented the existence of multiple C3 isoforms in two teleost fish species: the rainbow trout, *Oncorhynchus mykiss*, and the gilthead seabream, *Sparus aurata* [69,70]. In all other animals studied so far, C3 was shown to exist as a single gene product. However, the three trout C3 isoforms (C3-1, C3-3, C3-4) are products of different genes and significantly differ in their binding efficiencies to various C′-activating surfaces [69]. Such findings have favored the hypothesis that teleost fish have developed a unique approach to expand their innate immunity by duplicating their C3 genes. Thus, the structural and functional C3 diversity observed in fish may have important consequences for our understanding of the evolution of the C3 molecule, and other innate immune processes.

The Search for Complement Receptors in Lower Vertebrates

Complement anaphylatoxins mediate their proinflammatory functions by binding to G-protein-coupled transmembrane receptors that are expressed mainly on blood-derived leukocytes. Their function has been—thus far—characterized only in mammalian systems, since no such receptors have been cloned or isolated in lower vertebrates. Recently, evidence was produced suggesting that inflammatory pathways mediated by anaphylatoxic peptides are present both in teleost fish and urochordates, an invertebrate species.

The presence of two C3-like genes in the invertebrate urochordate *Ciona intestinalis* [71] was recently demonstrated. To define the functions mediated by these innate immune molecules in urochordates, the C3a moiety of *Ciona* C3-1 (CiC3-1a) was expressed in *E. coli* and shown to stimulate the dose-dependent directional migration of granular amebocytes (hemocytes), in a manner similar to mammalian C3a. The chemotactic activity of this peptide was localized to the C-terminus, and pretreatment of *Ciona* hemocytes with pertussis toxin abolished the chemotactic response to C3a, suggesting that this effect is mediated by a Gi-protein-coupled receptor. This is the first report documenting the presence of a C3aR-like receptor in an invertebrate species [72].

Furthermore, recent studies have demonstrated the presence of a functional C5aR receptor in a teleost species, the rainbow trout. Recombinant trout C5a was able to induce the chemotactic response of trout granulocytes isolated from a head kidney cell population in a dose-dependent fashion, and C5aR receptor expression was localized in granulocyte/macrophage cell suspensions of the head kidney as well as in trout peripheral blood leukocytes [73].

Taken together, these results strongly indicate that the inflammatory pathways mediated by C3a and C5a are also present in lower vertebrates, further underlying the fact that throughout evolution complement anaphylatoxins and their receptor-mediated pathways have retained a high degree of structural conservation and functional homology.

COMPLEMENT MEDIATES NOVEL FUNCTIONS IN DEVELOPMENTAL PROCESSES

Complement has long been recognized as an arm of innate immunity that mediates strictly immunological functions, by maintaining host defense against invading pathogens, and by mediating local and systemic inflammatory responses under various pathophysiological settings. Recently, however, it has become evident that several complement components exert novel functions that are associated with normal biological and developmental processes in various tissues and cannot be clearly placed in an "inflammatory" context [3,74].

The role of various complement components has been rigorously investigated in three distinct developmental processes. These include limb and lens regeneration in urodeles, liver regeneration in mammals, and stem cell differentiation during hematopoietic development.

Complement Components Promote Hepatocyte Regeneration

The liver is one of the few organs in mammals that have retained the ability to regenerate, restoring its functional integrity after partial hepatectomy, or viral or acute toxic injury [75]. The hepatic parenchyma reacts to these insults by eliciting a robust proliferative response that causes previously quiescent hepatocytes to reenter the cell cycle and divide. Several cytokine, hormonal, and growth factor-dependent pathways have been implicated in triggering liver regeneration [76–78].

However, to this date, the potential interaction of such hepatocellular regenerative pathways with components of the innate immune response has not been addressed. It has been previously shown that complement is critical for the normal recovery of the liver after acute toxic challenge [79]. Recently, it was established that complement components are essential priming partners in the early growth response of the liver by showing that complement-mediated pathways are coupled to the cytokine network that drives the regenerative response of hepatocytes [53]. This study revealed a pivotal role for both anaphylatoxins C3a and C5a in promoting the growth response of regenerating hepatocytes. Furthermore, it was shown that blockade of C5aR leads to abrogation of liver regeneration by affecting the normal induction of cytokines and the early activation of hepatic transcription factors that are essential for the "priming" of quiescent hepatocytes [53].

Further delineating the mechanisms by which complement proteins and receptors interact with other signaling networks in the regenerating liver will provide further insight into the molecular pathways that drive the early growth response of the liver and "prime" quiescent hepatocytes to reenter the cell cycle and proliferate.

Complement and Regenerative Pathways in Lower Vertebrates

The ability to regenerate complex structures and reconstruct entire body parts from damaged tissues is a trait widely encountered among invertebrates (e.g., annelids, hydroids, etc.) and in lower vertebrates such as amphibians [80]. In urodele amphibians (axolotls) the process of regeneration is quite prominent in the limb, tail, and in structures of the eye (retinal epithelium and lens) [81,82]. Limb regeneration in urodeles entails the activation of complex developmental pathways that act in concert to promote dedifferentiation, proliferation, and redifferentiation of mesenchymal cells into the specific cell types that comprise the various tissues of the regenerating limb.

The molecular pathway(s) that underlie these developmental stages are largely unknown. Surprisingly, a recent study implicated complement component C3 in urodele regeneration by showing specific expression of C3 in the blastema cells located in the regenerative zone of the amputated limb [83].

To determine the impact of complement in urodele regeneration and to dissect the specific involvement of the critical components C3 and C5 in limb and lens regeneration, studies were performed using the newt (*Notophthalmus viridescens*) as a model organism that possesses extensive regerative capacity in both these tissues. To monitor the expression of complement proteins during newt lens and limb regeneration, newt cDNAs for complement C3 and C5 were isolated and used for the generation of antibodies against C3a and C5a; these antibodies were shown to be specific for C3 and C5, respectively, and found to inhibit their activation [84]. Expression of both proteins was demonstrated in limb and lens structures during regeneration by immunostaining using the respective polyclonal antibodies. The expression of C3 and C5 was also confirmed by in situ hybridization.

To assess the in vivo role of complement in regeneration, cobra venom factor was injected into newts before amputation and found to cause a significant delay in limb regeneration. In contrast, similar treatment before lentectomy resulted in bigger fiber formation in the lens. To dissect the role of C3 and C5 in regeneration, further studies are in progress to analyze the effect of anti-C3a and C5a antibodies in both limb and lens regeneration [85].

A Role for Complement in Hematopoietic Development and Stem Cell Engraftment

The role of various complement regulatory molecules and receptors in protecting blood cells from complement-mediated lysis and promoting their inflammatory recruitment and activation during the course of infection [86] is well appreciated. Very little is known, however, about the distribution of complement components in early hematopoietic progenitor cells, their potential role in hematopoietic development, and the complement-mediated interactions that influence the homing of lymphoid progenitors to various tissues.

In this respect, a recent study has demonstrated that the G-protein-coupled receptors for both C3a and C5a anaphylatoxins are expressed by human clonogeneic $CD34^+$ cells, and that both complement components C3 and C5 are locally synthesized by the bone marrow stroma [87]. In addition, stimulation of the C3a receptor (C3aR) appears to regulate the chemotaxis of human $CD34^+$ cells by synergistically increasing the migration of these cells in the presence of *a*-chemokine stromal-derived factor-1 (SDF-1). Moreover, C3a has been shown to modulate various homing activities of stem cells by increasing their sensitivity to low doses of SDF-1 [87,88]. These striking observations have laid the groundwork for further investigation of the hypothesis that a functional cross-talk between the C3aR and CXCR4 signaling pathways may play an important role in the homing of human stem/progenitor cells to the bone marrow hematopoietic niches.

Corroborating this hypothesis, recent studies have shown that C3a and its receptor C3aR promote the retention of hematopoietic progenitor/stem cells in the bone marrow during stem cell mobilization in mice [89]. C3–/– or C3aR–/– mice exhibit increased release of bone marrow progenitors into circulation following GCSF-induced mobilization. Furthermore, mice treated with a specific C3aR antagonist show accelerated mobilization of hematopoitic stem cells (HSC) to the periphery, suggesting that antagonists of the C3aR can potentially serve as mobilizing agents for HSC transplantation (http://www.lambris.com/papers/C3a-patent-application.pdf).

A Novel Bioinformatics Platform for Text-Based Knowledge Discovery and "Mining" of Complement-Related "Systems" Associations

Biomolecular (structural and sequence) data generated by means of high-throughput screening techniques play an integral role in the generation of new biomedical knowledge and in the interpretation of complex biological systems. However, this enormous growth of experimental data has led to a reciprocal increase of the volume of scientific literature available in the databases. In this respect, researchers are now faced with the ultimate challenge of integrating both experimental and textual information in order to create a comprehensive context for interpreting and predicting gene and pathway associations and also generating new knowledge in a systematic, hypothesis-driven way. New text-mining algorithms are being developed in an effort to enable scientists to efficiently extract biological information from text databases, refine their search queries, manage complex ontogenies, and cluster biological entities in a meaningful manner that can shed light onto novel systems associations [90].

Recently, a novel text-mining approach, called Systems Literature Analysis [91], was adopted in an effort to retrieve from bibliographical databases global pathway associations of complement proteins. [92]. Biovista's BioLab Experiment Assistant (BEA^{TM}) tool, which supports literature analysis and hypothesis generation, was used in this study as a literature mining platform. This survey documented a broad range of biological processes that are modulated by complement proteins and presents for the first time an integrated map of complement-mediated networks that incorporates well over 85 diverse biological pathways.

The use of this novel mining platform has expanded the complement ontology beyond its ~35 designated components by identifying putative protein–protein interactions involving novel, noncomplement ligands. Furthermore, unusual associations of complement with signaling cascades and cellular networks that affect both inflammatory and noninflammatory processes were revealed through text-based correlations of complement genes and multiple pathways [92]. Integrating complement within a unified "systems" framework underscores the concept that innate immunity goes well beyond the maintenance of host defense, extending previously underappreciated links to critical developmental, homeostatic, and metabolic processes.

Complement essentially serves as a paradigm of how a defined knowledge space can be manipulated in a cross-disciplinary fashion, and thus be expanded by integrating both experimental and textual information dispersed across the databases.

PERSPECTIVES

In recent years, it has become evident that "cutting edge" biomedical research cannot be conducted with the exclusive use of traditional experimental approaches. The enormous amount of raw data accumulating in nucleotide and protein databases has urged the contemporary scientist to adopt a more global and cross-disciplinary approach to "old" scientific questions. Resolving the fine structure and biochemical properties of proteins may still contribute to addressing functions that underlie complex processes, provided that these research components are placed into a wider context of interacting systems and pathways.

In this respect, the investigation of complement structure/function relationships and protein–protein interactions that underlie critical biological processes is now being integrated in a systems-wide context that requires the cooperative application of different disciplines, resources, and experimental platforms. Using a combination of biophysical and computational approaches, together with high-throughput screening technologies and in vivo animal models, we have been able to verify and also predict the activity of a new generation of complement inhibitors and provide a versatile platform for developing effective complement therapeutics. Furthermore, unique, species-specific complement reagents have been developed that will help us study C3-mediated effects in mouse models of disease and determine those structural modules that are responsible for these functions [93]. The use of proteomic analysis, as a novel tool for probing protein–protein interactions, has produced evidence that complement can interact with networks that affect complex developmental processes, such as liver regeneration. In conclusion, all these diverse studies are integrated under the unifying theme of evolution that provides further insight into the molecular aspects of complement function and underscores the fact that complement—despite its ancient origin—has evolved into a versatile and yet "unpredictable" innate immune system.

ACKNOWLEDGMENTS

This research was supported by National Institutes of Health Grants AI 30040, GM-56698, GM-62134, and DK-059422.

Limited portions of this material appeared in *Molecular Immunology*, 41 (2004) 153–64 (reproduced by permission of Elsevier), and in a special issue of *Immunological Research*: Immunology at Penn, 7 (2003) 367–85.

REFERENCES

1. Hood, L. and D. Galas. The digital code of DNA. *Nature*, 421(6921):444–8, 2003.
2. Hood, L. Systems biology: integrating technology, biology, and computation. *Mechanisms of Ageing and Development*, 124(1):9–16, 2003.
3. Mastellos, D. and J. D. Lambris. Complement: more than a "guard" against invading pathogens? *Trends in Immunology*, 23(10):485–91, 2002.
4. Lambris, J. D. The multifunctional role of C3, the third component of complement. *Immunology Today*, 9(12):387–93, 1988.
5. Sahu, A. and J. D. Lambris. Structure and biology of complement protein C3, a connecting link between innate and acquired immunity. *Immunological Reviews*, 180:35–48, 2001.
6. Muller-Eberhard, H. J. Molecular organization and function of the complement system. *Annual Reviews in Biochemistry*, 57:321–47, 1988.
7. Nagar, B., R. G. Jones, R. J. Diefenbach, et al. X-ray crystal structure of C3d: a C3 fragment and ligand for complement receptor 2. *Science*, 280(5367): 1277–81, 1998.
8. Prota, A. E., D. R. Sage, T. Stehle, et al. The crystal structure of human CD21: implications for Epstein-Barr virus and C3d binding. *Proceedings of the National Academy of Sciences USA*, 99(16):10641–6, 2002.
9. Szakonyi, G., J. M. Guthridge, D. Li, et al. Structure of complement receptor 2 in complex with its C3d ligand. *Science*, 292(5522):1725–8, 2001.

10. Moore, M. D., R. G. DiScipio, N. R. Cooper, et al. Hydrodynamic, electron microscopic, and ligand-binding analysis of the Epstein-Barr virus/C3dg receptor (CR2). *Journal of Biological Chemistry*, 264(34):20576–82, 1989.
11. Diefenbach, R. J. and D. E. Isenman. Mutation of residues in the C3dg region of human complement component C3 corresponding to a proposed binding site for complement receptor type 2 (CR2, CD21) does not abolish binding of iC3b or C3dg to CR2. *Journal of Immunology*, 154(5):2303–20, 1995.
12. Clemenza, L. and D. E. Isenman. Structure-guided identification of C3d residues essential for its binding to complement receptor 2 (CD21). *Journal of Immunology*, 165(7):3839–48, 2000.
13. Guthridge, J. M., J. K. Rakstang, K. A. Young, et al. Structural studies in solution of the recombinant N-terminal pair of short consensus/complement repeat domains of complement receptor type 2 (CR2/CD21) and interactions with its ligand C3dg. *Biochemistry*, 40(20):5931–41, 2001.
14. Sarrias, M. R., S. Franchini, G. Canziani, et al. Kinetic analysis of the interactions of complement receptor 2 (CR2, CD21) with its ligands C3d, iC3b, and the EBV glycoprotein gp350/220. *Journal of Immunology*, 167(3):1490–9, 2001.
15. Morikis, D. and J. D. Lambris. The electrostatic nature of C3d-complement receptor 2 association. *Journal of Immunology*, 172(12):7537–47, 2004.
16. Sfyroera, G., M. Katragadda, D. Morikis, et al. Electrostatic modeling predicts the activities of orthopoxvirus complement control proteins. *Journal of Immunology*, 174(4):2143–51, 2005.
17. Rosengard, A. M., Y. Liu, Z. Nie, et al. Variola virus immune evasion design: expression of a highly efficient inhibitor of human complement. *Proceedings of the National Academy of Sciences USA*, 99(13):8808–13, 2002.
18. Smith, D. L., Y. Z. Deng and Z. Q. Zhang. Probing the non-covalent structure of proteins by amide hydrogen exchange and mass spectrometry. *Journal of Mass Spectrometry*, 32(2):135–46, 1997.
19. Mandell, J. G., A. M. Falick and E. A. Komives. Identification of protein–protein interfaces by decreased amide proton solvent accessibility. *Proceedings of the National Academy of Sciences USA*, 95(25):14705–10, 1998.
20. Mandell, J. G., A. M. Falick and E. A. Komives. Measurement of amide hydrogen exchange by MALDI-TOF mass spectrometry. *Analytical Chemistry*, 70(19):3987–95, 1998.
21. Winters, M. S., D. S. Spellman and J. D. Lambris. Solvent accessibility of native and hydrolyzed human complement protein 3 analyzed by hydrogen/deuterium exchange and mass spectrometry. *Journal of Immunology*, 174(6):3469–74, 2005.
22. Sahu, A. and J. D. Lambris. Complement inhibitors: a resurgent concept in anti-inflammatory therapeutics. *Immunopharmacology*, 49:133–48, 2000.
23. Sahu, A. and J. D. Lambris. Structure and biology of complement protein C3, a connecting link between innate and acquired immunity. *Immunological Reviews*, 180:35–48, 2001.
24. Sahu, A. and J. D. Lambris. Complement inhibitors: a resurgent concept in anti-inflammatory therapeutics. *Immunopharmacology*, 49:133–48, 2000.
25. Sahu, A., B. K. Kay and J. D. Lambris. Inhibition of human complement by a C3-binding peptide isolated from a phage displayed random peptide library. *Journal of Immunology*, 157(2):884–91, 1996.

26. Soulika, A. M., M. M. Khan, T. Hattori, et al. Inhibition of heparin/protamine complex-induced complement activation by compstatin in baboons. *Clinical Immunology*, 96(3):212–21, 2000.
27. Nilsson, B., R. Larsson, J. Hong, et al. Compstatin inhibits complement and cellular activation in whole blood in two models of extracorporeal circulation. *Blood*, 92(5):1661–7, 1998.
28. Fiane, A. E., T. E. Mollnes, V. Videm, et al. Prolongation of ex vivo-perfused pig xenograft survival by the complement inhibitor compstatin. *Transplantation Proceedings*, 31(1-2):934–5, 1999.
29. Fiane, A. E., T. E. Mollnes, V. Videm, et al. Compstatin, a peptide inhibitor of C3, prolongs survival of ex vivo perfused pig xenografts. *Xenotransplantation*, 6:52–65, 1999.
30. Fiane, A. E., V. Videm, J. D. Lambris, et al. Modulation of fluid-phase complement activation inhibits hyperacute rejection in a porcine-to-human xenograft model. *Transplantation Proceedings*, 32(5):899–900, 2000.
31. Mollnes, T. E., O. L. Brekke, M. Fung, et al. Essential role of the C5a receptor in *E. coli*-induced oxidative burst and phagocytosis revealed by a novel lepirudin-based human whole blood model of inflammation. *Blood*, 100(5):1869–77, 2002.
32. Klegeris, A., E. A. Singh and P. L. McGeer. Effects of C-reactive protein and pentosan polysulphate on human complement activation. *Immunology*, 106(3):381–8, 2002.
33. Furlong, S. T., A. S. Dutta, M. M. Coath, et al. C3 activation is inhibited by analogs of compstatin but not by serine protease inhibitors or peptidyl alpha-ketoheterocycles. *Immunopharmacology*, 48(2):199–212, 2000.
34. Sahu, A., D. Morikis and J. D. Lambris. Compstatin, a peptide inhibitor of complement, exhibits species-specific binding to complement component C3. *Molecular Immunology*, 39(10):557–66, 2003.
35. Morikis, D., N. Assa-Munt, A. Sahu, et al. Solution structure of compstatin, a potent complement inhibitor. *Protein Science*, 7(3):619–27, 1998.
36. Morikis, D., A. Sahu, W. T. Moore and J. D. Lambris. Design, structure, function and application of compstatin. In J. Matsukas and T. Mavromoustakos (Eds.), *Bioactive Peptides in Drug Discovery and Design: Medical Aspects* (pp. 235–46). Ios Press, Amsterdam, 1999.
37. Sahu, A., A. M. Soulika, D. Morikis, et al. Binding kinetics, structure-activity relationship, and biotransformation of the complement inhibitor compstatin. *Journal of Immunology*, 165(5):2491–9, 2000.
38. Furlong, S. T., A. S. Dutta, M. M. Coath, et al. C3 activation is inhibited by analogs of compstatin but not by serine protease inhibitors or peptidyl alpha-ketoheterocycles. *Immunopharmacology* 48(2):199–212, 2000.
39. Morikis, D., M. Roy, A. Sahu, et al. The structural basis of compstatin activity examined by structure-function-based design of peptide analogs and NMR. *Journal of Biological Chemistry*, 277(17):14942–53, 2002.
40. Soulika, A. M., D. Morikis, M. R. Sarrias, et al. Studies of structure-activity relations of complement inhibitor compstatin. *Journal of Immunology*, 171(4):1881–90, 2003.
41. Sahu, A., A. M. Soulika, D. Morikis, et al. Binding kinetics, structure-activity relationship, and biotransformation of the complement inhibitor compstatin. *Journal of Immunology*, 165(5):2491–9, 2000.

42. Morikis, D., M. Roy, A. Sahu, et al. The structural basis of compstatin activity examined by structure-function-based design of peptide analogs and NMR. *Journal of Biological Chemistry*, 277(17):14942–53, 2002.
43. Klepeis, J. L., C. A. Floudas, D. Morikis, et al. Integrated computational and experimental approach for lead optimization and design of compstatin variants with improved activity. *Journal of the American Chemical Society*, 125(28):8422–3, 2003.
44. Klepeis, J. L., C. A. Floudas, D. Morikis and J. D. Lambris. Predicting peptide structures using NMR data and deterministic global optimization. *Journal of Computational Chemistry*, 20:1354–70, 1999.
45. Klepeis, J. L. and C. A. Floudas. Ab initio tertiary structure prediction of proteins. *Journal of Global Optimization*, 25:113–40, 2003.
46. Klepeis, J. L., H. D. Schafroth, K. M. Westerberg, et al. Deterministic global optimization and ab initio approaches for the structure prediction of polypeptides, dynamics of protein folding, and protein-protein interactions. *Computational Methods for Protein Folding–Advances in Chemical Physics*, 120:265–457, 2002.
47. Chong, S., G. E. Montello, A. Zhang, et al. Utilizing the C-terminal cleavage activity of a protein splicing element to purify recombinant proteins in a single chromatographic step. *Nucleic Acids Research*, 26(22):5109–15, 1998.
48. Mallik, B., M. Katragadda, L. A. Spruce, et al. Design and NMR characterization of active analogues of compstatin containing non-natural amino acids. *Journal of Medicinal Chemistry*, 48(1):274–86, 2005.
49. Mallik, B., J. D. Lambris and D. Morikis. Conformational interconversion in compstatin probed with molecular dynamics simulations. *Proteins*, 53(1):130–41, 2003.
50. Paczkowski, N. J., A. M. Finch, J. B. Whitmore, et al. Pharmacological characterization of antagonists of the C5a receptor. *British Journal of Pharmacology*, 128(7):1461–6, 1999.
51. Morikis, D. and J. D. Lambris. Structural aspects and design of low-molecular-mass complement inhibitors. *Biochemical Society Transactions*, 30(6):1026–36, 2002.
52. Huber-Lang, M. S., N. C. Riedeman, J. V. Sarma, et al. Protection of innate immunity by C5aR antagonist in septic mice. *FASEB Journal*, 16(12):1567–74, 2002.
53. Strey, C. W., M. Markiewski, D. Mastellos, et al. The proinflammatory mediators C3a and C5a are essential for liver regeneration. *Journal of Experimental Medicine*, 198(6):913–23, 2003.
54. Girardi, G., J. Berman, P. Redecha, et al. Complement C5a receptors and neutrophils mediate fetal injury in the antiphospholipid syndrome. *Journal of Clinical Investigation*, 112(11):1644–54, 2003.
55. Kim, A. H., I. D. Dimitriou, M. C. Holland, et al. Complement C5a receptor is essential for the optimal generation of antiviral $CD8^+$ T cell responses. *Journal of Immunology*, 173(4):2524–9, 2004.
56. Pennington, S. R., M. R. Wilkins, D. F. Hochstrasser, et al. Proteome analysis: from protein characterization to biological function. *Trends in Cell Biology*, 7(1):168–73, 1997.
57. Anderson, L. and J. Seilhamer. A comparison of selected mRNA and protein abundances in human liver. *Electrophoresis*, 18(3-4):533–7, 1997.

58. Michalopoulos, G. K. and M. C. DeFrances. Liver regeneration. *Science*, 276(5309):60–6, 1997.
59. Strey, C. W., M. S. Winters, M. M. Markiewski and J. D. Lambris. Partial hepatectomy induced liver proteome changes in mice. *Proteomics*, 5(1):318–25, 2005.
60. Cooper, A. and C. M. Johnson. Isothermal titration microcalorimetry. *Methods in Molecular Biology*, 22:137–50, 1994.
61. Katragadda, M., D. Morikis and J. D. Lambris. Thermodynamic studies on the interaction of the third complement component and its inhibitor, compstatin. *Journal of Biological Chemistry*, 279(53):54987–95, 2004.
62. Kotwal, G. J. and B. Moss. Vaccinia virus encodes a secretory polypeptide structurally related to complement control proteins. *Nature*, 335(6186): 176–8, 1988.
63. Kotwal, G. J., S. N. Isaacs, R. Mckenzie, et al. Inhibition of the complement cascade by the major secretory protein of vaccinia virus. *Science*, 250(4982):827–30, 1990.
64. Mckenzie, R., G. J. Kotwal, B. Moss, et al. Regulation of complement activity by vaccinia virus complement-control protein. *Journal of Infectious Diseases*, 166(6):1245–50, 1992.
65. Isaacs, S. N., E. Argyropoulos, G. Sfyroera, et al. Restoration of complement-enhanced neutralization of vaccinia virus virions by novel monoclonal antibodies raised against the vaccinia virus complement control protein. *Journal of Virology*, 77(15):8256–62, 2003.
66. Mizuno, M. and B. P. Morgan. The possibilities and pitfalls for anti-complement therapies in inflammatory diseases. *Current Drug Targets–Inflammation and Allergy*, 3(1):87–96, 2004.
67. Lubinski, J., L. Wang, D. Mastellos, et al. In vivo role of complement-interacting domains of herpes simplex virus type 1 glycoprotein gC. *Journal of Experimental Medicine*, 190(11):1637–46, 1999.
68. Mullick, J., J. Bernet, A. K. Singh, et al. Kaposi's sarcoma-associated herpesvirus (human herpesvirus 8) open reading frame 4 protein (kaposica) is a functional homolog of complement control proteins. *Journal of Virology*, 77(6):3878–81, 2003.
69. Sunyer, J. O., I. K. Zarkadis, A. Sahu and J. D. Lambris. Multiple forms of complement C3 in trout, that differ in binding to complement activators. *Proceedings of the National Academy of Sciences USA*, 93(16): 8546–51, 1996.
70. Sunyer, J. O., L. Tort and J. D. Lambris. Structural C3 diversity in fish: characterization of five forms of C3 in the diploid fish *Sparus aurata*. *Journal of Immunology*, 158(6):2813–21, 1997.
71. Marino, R., Y. Kimura, R. De Santis, et al. Complement in urochordates: cloning and characterization of two C3-like genes in the ascidian *Ciona intestinalis*. *Immunogenetics*, 53(12):1055–64, 2002.
72. Pinto, M. R., C. M. Chinnici, Y. Kimura, et al. CiC3-1a-mediated chemotaxis in the deuterostome invertebrate *Ciona intestinalis* (Urochordata). *Journal of Immunology*, 171(10):5521–8, 2003.
73. Holland, M. C. and J. D. Lambris. A functional C5a anaphylatoxin receptor in a teleost species. *Journal of Immunology*, 172(1):349–55, 2004.

74. Lee Y. L., K. F. Lee, J. S. Xu, et al. The embryotrophic activity of oviductal cell derived complement C3b and iC3b—a novel function of complement protein in reproduction. *Journal of Biological Chemistry*, 279(13):12763–8, 2004.
75. Fausto, N. and E. M. Webber. Liver regeneration. In I. M. Arias, J. L. Boyer, N. Fausto, W. B. Jacoby, D.A. Schachter and D. A. Shafritz (Eds.), *The Liver: Biology and Pathobiology* (pp. 1059–84). Raven Press, New York, 1994.
76. Cressman, D. E., L. E. Greenbaum, R. A. DeAngelis, et al. Liver failure and defective hepatocyte regeneration in interleukin-6-deficient mice. *Science*, 274:1379–83, 1996.
77. Yamada, Y., I. Kirillova, J. J. Peschon and N. Fausto. Initiation of liver growth by tumor necrosis factor: deficient liver regeneration in mice lacking type I tumor necrosis factor receptor. *Proceedings of the National Academy of Sciences USA*, 94(4):1441–6, 1997.
78. Taub, R., L. E. Greenbaum and Y. Peng. Transcriptional regulatory signals define cytokine-dependent and -independent pathways in liver regeneration. *Seminars in Liver Disease*, 19(2):117–27, 1999.
79. Mastellos, D., J. C. Papadimitriou, S. Franchini, et al. A novel role of complement: mice deficient in the fifth component of complement (C5) exhibit impaired liver regeneration. *Journal of Immunology*, 166(4):2479–86, 2001.
80. Brockes, J. P. Amphibian limb regeneration: rebuilding a complex structure. *Science*, 276(5309):81–7, 1997.
81. Call, M. K. and P. A. Tsonis. Vertebrate limb regeneration. *Advances in Biochemical Engineering/Biotechnology*, 93:67–81, 2005.
82. Tsonis, P. A. Regeneration in vertebrates. *Developmental Biology*, 221(2):273–84, 2000.
83. Del Rio-Tsonis, K., P. A. Tsonis, I. K. Zarkadis, et al. Expression of the third component of complement, C3, in regenerating limb blastema cells of urodeles. *Journal of Immunology*, 161(12):6819–24, 1998.
84. Kimura, Y., M. Madhavan, M. K. Call, et al. Expression of complement 3 and complement 5 in newt limb and lens regeneration. *Journal of Immunology*, 170(5):2331–9, 2003.
85. Madhavan, M. C., Y. Kimura, M. K. Call, et al. Do complement molecules play a role in lens regeneration? *Investigative Ophthalmology and Visual Science*, 44(Suppl 1):1241, 2003.
86. Sun, X., C. D. Funk, C. Deng, et al. Role of decay-accelerating factor in regulating complement activation on the erythrocyte surface as revealed by gene targeting. *Proceedings of the National Academy of Sciences USA*, 96(2):628–33, 1999.
87. Reca, R., D. Mastellos, M. Majka, et al. Functional receptor for C3a anaphylatoxin is expressed by normal hematopoietic stem/progenitor cells, and C3a enhances their homing-related responses to SDF-1. *Blood*, 101(10):3784–93, 2003.
88. Zou, Y. R., A. H. Kottmann, M. Kuroda, et al. Function of the chemokine receptor CXCR4 in haematopoiesis and in cerebellar development. *Nature*, 393(6685):595–9, 1998.
89. Ratajczak, J., R. Reca, M. Kucia, et al. Mobilization studies in mice deficient in either C3 or C3a-receptor (C3aR) reveal a novel role for complement in

retention of hematopoietic stem/progenitor cells in bone marrow. *Blood*, 103(6):2071–8, 2004.

90. Mack, R. and M. Hehenberger. Text-based knowledge discovery: search and mining of life-sciences documents. *Drug Discovery Today*, 7:S89–98, 2002.
91. Persidis, A., S. Deftereos and A. Persidis. Systems literature analysis. *Pharmacogenomics*, 5(7):943–7, 2004.
92. Mastellos, D., C. Andronis, A. Persidis and J. D. Lambris. Novel biological networks modulated by complement. *Clinical Immunology*, 115(3):225–35, 2005.
93. Mastellos, D., J. Prechl, G. László, et al. Novel monoclonal antibodies against mouse C3 interfering with complement activation: description of fine specificity and applications to various immunoassays. *Molecular Immunology*, 40(16):1213–21, 2004.

7

Systems Biotechnology: Combined in Silico and Omics Analyses for the Improvement of Microorganisms for Industrial Applications

Sang Yup Lee, Dong-Yup Lee, Tae Yong Kim, Byung Hun Kim, & Sang Jun Lee

Biotechnology plays an increasingly important role in the healthcare, pharmaceutical, chemical, food, and agricultural industries. Microorganisms have been successfully employed for the production of recombinant proteins [1–4] and various primary and secondary metabolites [5–8]. As in other engineering disciplines, one of the ultimate goals of industrial biotechnology is to develop lower-cost and higher-yield processes. Toward this goal, fermentation and downstream processes have been significantly improved thanks to the effort of biochemical engineers [9]. In addition to the effort of making these mid- to downstream processes more efficient, microbial strains have been improved by recombinant and other molecular biological methods, leading to the increase in microbial metabolic activities toward desired goals [10]. However, these conventional attempts have not always been successful owing to unexpected changes in the physiology and metabolism of host cells. Alternatively, rational metabolic and cellular engineering approaches have been tried to solve such problems in a number of cases, but they were also still limited to the manipulation of only a handful of genes encoding enzymes and regulatory proteins. In this regard, systematic approaches are indeed required not only for understanding the global context of the metabolic system but also for designing better metabolic engineering strategies.

Recent advances in high-throughput experimental techniques have resulted in rapid accumulation of a wide range of biological data and information at different levels: genome, transcriptome, proteome, metabolome, and fluxome [11–17]. This technology-driven biology is now allowing us to study large-scale sets and networks of components and molecules of a biological system, thus building a foundation for the global understanding of biological processes.

Concurrently with high-throughput experiments, it is increasingly accepted that in silico or dry experiments of modeling and simulation improve our ability to analyze and predict the cellular behavior of microorganisms under any perturbations, for example, the genetic modifications and/or environmental changes [18–20]. Thus, metabolic engineering of microorganisms can be best performed by integrating wet experiments (high-throughput experiments) and dry experiments (in silico modeling and simulation) at the systems level (figure 7.1); the results of genomic, transcriptomic, proteomic, metabolomic, and fluxomic studies, information and data available in public and in-house databases, and those generated by in silico analysis can be integrated within the global context of the metabolic system. This leads to the generation of new knowledge that can facilitate strain development suitable for biotechnological processes with the highest efficiencies and productivities. Hereafter, such a systems approach is referred to

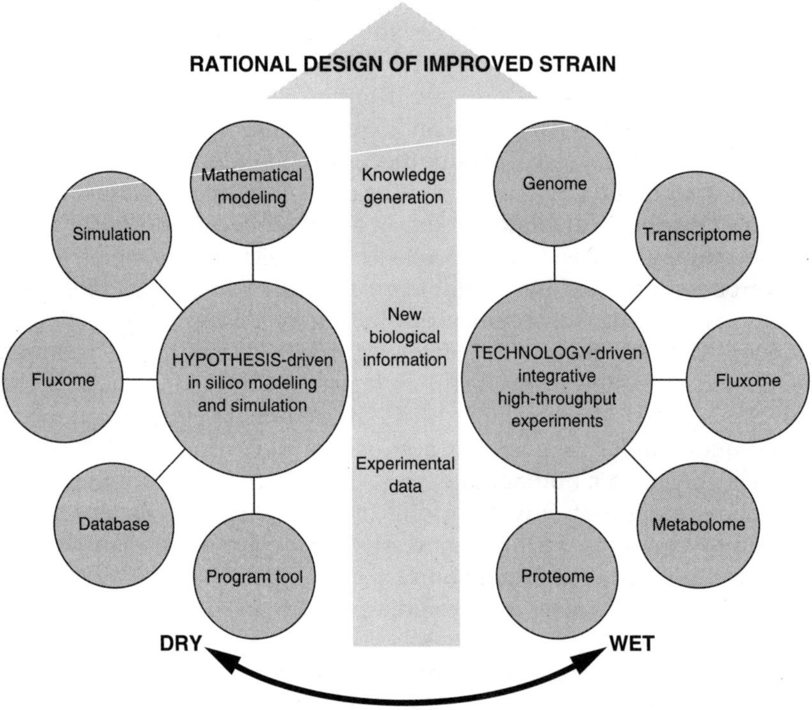

Figure 7.1 Overview of systems biotechnology. High-throughput experiments generating large amounts of genome, transcriptome, proteome, metabolome, and fluxome data can be analyzed in combination with in silico modeling and simulation results in order to develop improved organisms for applications in biotechnology industries.

as "systems biotechnology." Thus, biotechnology processes can be developed in a much more rational and systematic way by following systems biotechnological strategies rather than traditionally used "trial and error" type approaches.

In this chapter, high-throughput experimental techniques and in silico modeling and simulation methods for strain development are reviewed in the context of systems biotechnology. This is followed by strategic procedures and relevant case studies addressing how wet and dry experiments can be interactively combined to achieve the improvement of microbial strains. Such an integrated strategy based on systems-level quantitative analysis of cellular phenotypes in conjunction with real experiments is the main focus of this chapter. Future perspectives on systems biotechnology are also discussed to suggest new opportunities and challenges in this field.

HIGH-THROUGHPUT X-OMIC EXPERIMENTS

Biology has evolved rapidly during the last decade, and so has biotechnology. This rapid development is being even more accelerated by the advent of high-throughput experimental techniques, allowing the generation of various biological data at unprecedentedly high rates. These revolutionary techniques for the rapid acquisition of genome, transcriptome, proteome, metabolome, and fluxome data are changing the traditional paradigm of biological and biotechnological research [21–23]. These technological advances naturally led to the birth of "systems biology," which aims to elucidate biological mechanisms and phenomena in a quantitative manner at a holistic and genome-scale level (figure 7.2). Applying this approach to biotechnological development leads to "systems biotechnology," which enables improvement of organisms and bioprocesses employing them at systems level.

Before describing some successful stories on the use of high-throughput x-omic experiments to design improved organisms, we wish to insert a few words of caution. Transcriptome profiling, which allows measurement of the levels of all transcripts in a cell, provides a global view on gene expression under particular circumstances. However, it should be noted that gene expression is a necessary but not a sufficient condition for the corresponding protein production, since there exist translational regulation, uncharacterized RNA and protein stability, and sometimes posttranslational modification. This is further complicated by the fact that protein abundance does not necessarily correlate with the high activity of that protein due to several factors such as varying substrate concentration, cofactor abundance, and feedback inhibition. Therefore, even though proteome profiling by two-dimensional gel electrophoresis (2DE) reveals a relative abundance of proteins, we still have limited information on the

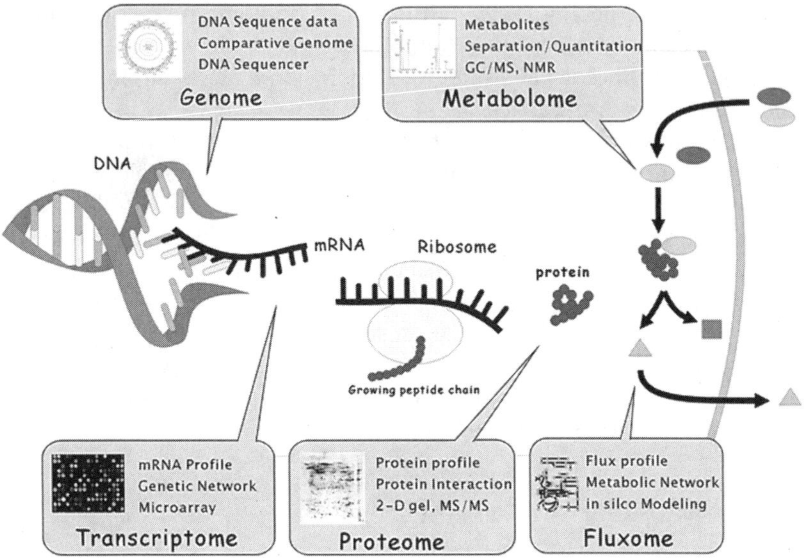

Figure 7.2 Postgenomic omics research. High-speed DNA sequencing has allowed completion of a number of genome sequencing projects. Having complete genome sequences for many microorganisms in our hands, comparative genomic studies can be performed. Development of high-density DNA microarray allowed profiling of the transcriptome of cells under investigation, thus allowing us to examine the expression levels of multiple genes simultaneously. Two-dimensional gel electrophoresis (2DE) combined with mass spectrometry made it possible to obtain complete profiling of cellular or subcellular proteins, which complements the results of transcriptome profiling. Metabolome profiling is also becoming increasingly popular as better GC/MS, LC/MS, and NMR techniques are developed. Fluxome profiling is still mostly based on computation, but is supplemented with the results of isotopomer profiling for better estimation. Much effort is being devoted to integrate all these omics data toward better understanding of global cellular physiology, metabolism, and regulation, and consequently to use the results for strain improvement.

activities of these enzymes. Metabolome profiling can complement some of these limitations, but is currently hampered by the absence of robust methods for sample preparation and metabolites analysis. Complementary to this, fluxome profiling allows determination of metabolic fluxes by metabolic reaction network simulation constrained by the measurement of net excretion rates of extracellular metabolites, uptake rates of metabolites, and/or isotopomer distributions following the use of isotopically labeled substrates.

It is clear that we are still far from understanding cellular behavior at the systems level. Nevertheless, the combination of these x-omes profiling will suggest holistic insight into the whole-cell metabolism and regulation, and the interaction of cells with the environment. The benefits of having global x-omic data will be truly realized when we can correctly integrate all these data using well-validated in silico modeling and simulation tools. Again, it will take a while to see this achieved. In the meantime, we can still enjoy developing improved organisms based on these large-scale data. One can extract interesting knowledge on local pathways from global-scale data sets, and use it for altering metabolism toward the desired goals. It should be emphasized that the generation of new knowledge on the local (rather than global) reactions and pathways to be manipulated was possible because large-scale data on the genome, transcriptome, proteome, metabolome, and fluxome are available. In what follows, we describe how each x-ome's data were generated and their combinations were used for strain improvement.

Genome Analysis

Comparative analysis of genomes is a relatively simple yet powerful way of extracting information necessary for understanding differences in metabolism and identifying targets for strain improvement. One can identify unnecessary genes or operons that are not beneficial or even harmful for producing the desired biochemical products, and those genes that need to be newly introduced, amplified, or modified to establish new pathways or to enhance the pathway fluxes [24,25]. The genomes can be compared among different organisms as well as between the wild type and its mutant strains. This can obviously be extended to comparing the control strain with the engineered strains. Engineering of microorganisms based on comparative genomics has recently been successfully demonstrated by Ohnishi et al. [25]. The genome sequence of the lysine-overproducing *Corynebacterium* strain was compared with that of the wild-type strain to identify genes with point mutations that might be beneficial for the overproduction of L-lysine. Consequently, the point mutations in five genes responsible for improved lysine production could be identified. Introduction of new genes and/or knockout of undesirable genes have been frequently practiced in metabolic engineering. Having the complete genome sequences for a number of microorganisms, we can identify many more candidate genes to be manipulated based on genome-scale comparison data.

Transcriptome Analysis

Development of high-density DNA microarrays is allowing us to simultaneously measure relative mRNA abundance in multiple samples.

By comparing transcriptome profiles between different strains of interest or between cells cultured under different conditions, one can elucidate possible regulatory circuits and identify potential target genes to be manipulated for strain improvement. There have been numerous papers on the use of DNA microarrays to obtain global transcriptome profiles and consequently to understand cellular gene expression regulations and physiological changes in microbial strains of industrial relevance. Among them, Gonzalez et al. [26] carried out transcriptome profiling of two ethanologenic *Escherichia coli* strains that show different levels of ethanol tolerance. From the comparison of transcriptome profiles, several genes and potential mechanisms responsible for ethanol tolerance, such as loss of FNR function and enhanced metabolism of glycine, serine, and pyruvate, were identified. This example demonstrates the usefulness of transcriptome profiling in understanding the mechanisms behind the observed expression and physiological changes. Several excellent review papers have been published [27,28].

However, there are not yet many examples of strain improvement based on transcriptome profiling. Nevertheless, several examples have recently appeared, and more examples will likely follow. In one such example, transcriptome profiles of recombinant *E. coli* producing human insulin-like growth factor I fusion protein (IGF-I_f) were analyzed. Recombinant *E. coli* was cultured in an industrially relevant fed-batch mode during which samples were taken before and after induction for transcriptome profiling. The results of comparative transcriptome analyses were used to identify target genes that may improve IGH-I_f production [29]. Among the ~200 genes that were downregulated after induction, those involved in amino acid/nucleotide biosynthetic pathways were selected as the first targets to be manipulated. As will be descried later, the expression of these genes was found to be downregulated during the high cell density culture (HCDC) of *E. coli* [30]. Amplification of one of these genes, the *prsA* gene encoding the phosphoribosyl pyrophosphate synthetase, allowed the enhanced production of IGH-I_f. However, it was found that cell growth was negatively affected by the *prsA* overexpression. The *glpF* gene encoding glycerol transporter was then selected and amplified together with the *prsA* gene, resulting in an increase of IFG-I_f production from 1.8 to 4.3 g/l [29].

As shown in this example, the target genes to be manipulated to improve a strain can be selected from the genes prescreened by the analysis of transcriptome profiles. However, one should be aware that this approach of selecting genes from transcriptome profiles is not an easy and straightforward task. In the worst case of the above example, one should try amplification of many more genes out of 200 downregulated genes in different combinations to achieve a desired goal. Often, however, this is not the case. One can reduce the number of potential target genes to 10–20 based on previous knowledge and experience (i.e., in the above example, the results from the transcriptome profiling of *E. coli* during the HCDC, as will be described later). Are 10–20 genes still too many to be manipulated? Considering the effort and money invested for rather expensive microarray experiments, this is not the case. Basically, we have to do more homework, but homework that gives higher value back to us. Furthermore, the good news is that we have other tools such as genome-scale metabolic flux analysis that can narrow down the number of candidate genes, as will be described later.

Proteome Analysis

Metabolic reactions are catalyzed by enzymes, and therefore proteome profiling takes us one step closer toward understanding cellular metabolic status. Proteome analysis typically begins with separating proteins by 2DE followed by identification of protein spots by mass spectrometry [31]. The 2DE is a rather laborious process, and its quality and reproducibility can be affected by the researchers. Furthermore, the number of protein spots that have been identified is still considerably lower than that supposedly encoded in the genome. For example, fewer than 1000 protein spots are detected in 2DE gel for *E. coli*, which has more than 4000 genes in its genome. Among those protein spots, about 200 proteins have been functionally identified. Other problems include failure to detect low-abundance proteins, difficulties in separating certain proteins, and the appearance of multiple spots for one protein. Therefore, proteome profiling has not yet gone truly global. Nonetheless, it can suggest interesting candidate proteins to be examined through the comparative analysis of protein spots showing altered intensities under two or more different conditions or genotypic backgrounds. There have been a couple of examples successfully demonstrating strain improvement by this approach.

In another example, the proteome analysis of recombinant *E. coli* overproducing human leptin, a pharmaceutical protein for treating obesity, was conducted to identify targets for the possible improvement of human leptin production [32]. During human leptin production, the levels of heat shock proteins increased while those of protein elongation factors, 30S ribosomal protein, and some enzymes in amino acid biosynthetic pathways decreased. Interestingly, the significantly decreased expression levels of some enzymes in the serine family of amino acids biosynthetic pathway indicate that leptin production can possibly be limited by serine family amino acids. This is because the serine content of leptin is 11.6%, which is much higher than the average serine content of *E. coli* proteins (5.6%). Thus, one of these downregulated genes, the *cysK* gene encoding cysteine synthase A, was amplified, which resulted in 2- and 4-fold increases in cell growth and leptin productivity, respectively. It was also found that *cysK* coexpression could improve production of another serine-rich protein, interleukin-12β chain (serine content of 11.1%), as well. These examples demonstrate that the strategy of local pathway engineering based on global information (proteome profiling in the above examples) is highly efficient for strain improvement.

Metabolome Analysis

The general aims of metabolomics are to identify metabolites and quantitatively determine their concentrations. Various quantitative tools for identifying and analyzing cellular metabolites have been developed to investigate the detailed metabolic status of cells [33]. High-throughput quantitative analysis of metabolites has become possible with the development of increasingly sophisticated gas chromatography–mass spectrometry (GC-MS), gas chromatography time-of-flight mass spectrometry (GC-TOF), liquid chromatography–mass spectrometry (LC-MS), and nuclear magnetic resonance (NMR) equipment. No single technique is suitable for the analysis of all different types of molecule, so a mixture of techniques is used. The diversity of methods ensures broad coverage of a wide range of different classes of organic compounds.

Even though much effort is needed to solve existing problems such as the limited number of metabolites detectable and the accuracy of metabolite concentrations measured, there have been some successful uses of metabolome profiling in analyzing cellular metabolism. Villas-Boas et al. [34] reported the application of a novel derivatization method for metabolome analysis in yeasts. A method was developed to simultaneously measure the metabolites throughout the central carbon metabolism and amino acid biosynthetic pathways using GC-MS. As model systems, they assayed the metabolite levels for two yeast strains and two different culture conditions; the changing levels of many metabolites caused by the genetic (*gdh1* knockout) and environmental (aerobic and anaerobic) perturbations were examined. By comparing the metabolite levels throughout the samples, it was possible to identify the activated genes or metabolic pathways. This study demonstrates the true power of metabolome analysis in understanding specific metabolic pathways and provides new insight into the integrated transcription-metabolism studies. There has been another interesting report on the combined analysis of the transcriptome and metabolome for understanding the metabolic characteristics of *Aspergillus* producing lovastatin and consequently for the development of improved strains, as will be seen below. As our ability to quantitatively analyze the metabolites increases, metabolome profiling will become an indispensable tool for the analysis and engineering of metabolic pathways.

Combined X-omes Analysis for Strain Improvements

Even though the true integration of all x-omic data is not yet possible at this time, local information extracted from these global-scale data can be exploited in understanding cellular physiology and metabolism, and consequently developing improved organisms [35]. There have been several successful examples of combining two or more high-throughput x-omic studies for analyzing and engineering cellular metabolism, and these are described below as case studies [16,30,36–38].

CASE STUDY 1: COMBINED ANALYSIS OF TRANSCRIPTOME AND PROTEOME OF *E. coli* CELLS DURING HIGH CELL DENSITY CULTURE

E. coli has been the workhorse for the production of various bio-products including recombinant proteins. In order to increase the concentration and volumetric productivity of a desired product such as recombinant protein, high cell density culture (HCDC) is often carried out. However, it is often observed that the specific productivity, defined as gram product per gram dry cell weight (DCW) per hour, decreases as cell density increases. In order to understand the physiological changes occurring as cell density increases, combined transcriptome and proteome analyses were carried out during the HCDC of *E. coli* W3110 [30]. As shown in figure 7.3, cells were grown exponentially at a specific growth rate of $0.14 h^{-1}$ to 74 g DCW/l using a chemically defined medium. Sample were taken at the cell concentrations of 0.13 (S1), 0.88 (S2), 3.5 (S3), 12.7 (S4), 24 (S5), 40 (S6), 52 (S7), 62 (S8), 72 (S9), and 74 (S10) g DCW/l. For transcriptome profiling, the cDNA made from S1 was labeled with Cy3 while all the others were labeled with Cy5. Also, proteome analyses were carried

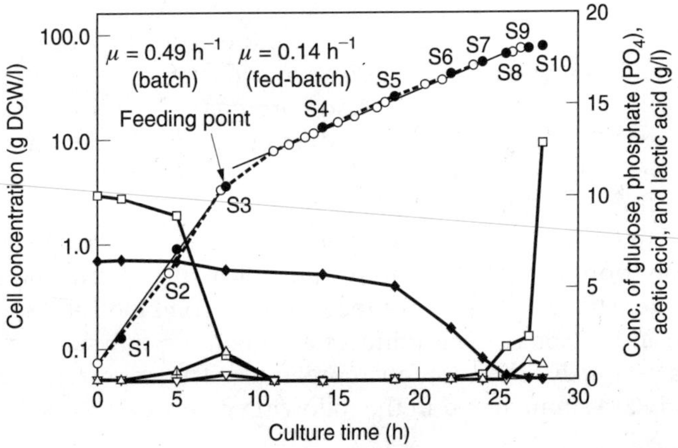

Figure 7.3 Time profiles of the concentrations of cell (○ and •; the filled circles are sampling points), glucose (□), acetic acid (△), lactic acid(▽), and phosphate (◆) during the high cell density fed-batch fermentation of *E. coli* W3110. In the batch phase, the specific growth rate was $0.49 h^{-1}$. During the fed-batch period, the specific growth rate was controlled at $0.14 h^{-1}$ by exponential feeding. Reproduced with permission from Yoon et al. [30].

out for the same samples taken during the cultivation (S1–S10). It was found that the expression of genes of TCA cycle enzymes, NADH dehydrogenase, and ATPase was upregulated during the exponential fed-batch period and downregulated afterwards. The expression of chaperone genes was upregulated as cell density increased, suggesting that cells experience stress during the HCDC. Expression of phosphate starvation genes was most strongly upregulated toward the end of cultivation, suggesting possible phosphate limitation at this point.

The most interesting finding was that the expression of most amino acid biosynthesis genes was downregulated as cell density increased. This finding immediately answers why the specific productivity of recombinant protein is reduced during the HCDC. Even though the results of transcriptome and proteome profiling were similar, there were still a number of genes showing different profiles. At this time, we do not know how to extract more information from these discrepancies, and therefore it will be one of the important tasks in the future. Instead, the "expression map," which are transcriptome and proteome profiles mapped onto the metabolic pathway network, was proposed (figure 7.4). This expression map can provide a global snapshot of metabolic and physiological changes under evaluation. In the earlier example of using transcriptome profiling results for increasing IGF-I_f production by recombinant *E. coli* [29], the target gene *prsA* could be successfully selected from many candidates and amplified based on the finding of broad downregulation of amino acid biosynthetic enzymes observed from the expression map of *E. coli* during the HCDC. These results prove the usefulness of having such an expression map.

There have recently been several more examples of combined x-omic analysis for strain characterization and engineering. Lee et al. [37] compared the transcriptome and proteome profiles of *E. coli* W3110 and its L-threonine-overproducing mutant strain. Among the 4290 genes, 54 genes exhibited meaningfully differential gene expression. Further analysis of nucleotide sequences of some of these genes revealed that mutations in the *thrA* and *ilvA* genes were responsible for L-threonine production and the physiological changes in the mutant strain. There has recently been a report on the combined analysis of the genome and fluxome toward understanding the metabolic characteristics of a relatively unknown microorganism. The approaches taken in this study are summarized in the following case study.

CASE STUDY 2: COMBINED ANALYSIS OF GENOME AND FLUXOME FOR THE ELUCIDATION OF UNKNOWN METABOLIC CHARACTERISTICS

Mannheimia succiniciproducens, which is able to produce large amounts of succinic acid along with other acids, was isolated from the rumen of Korean cows [39]. In order to engineer the metabolic pathways for enhanced succinic acid production, it is essential to understand the metabolic characteristics of this bacterium under various conditions.

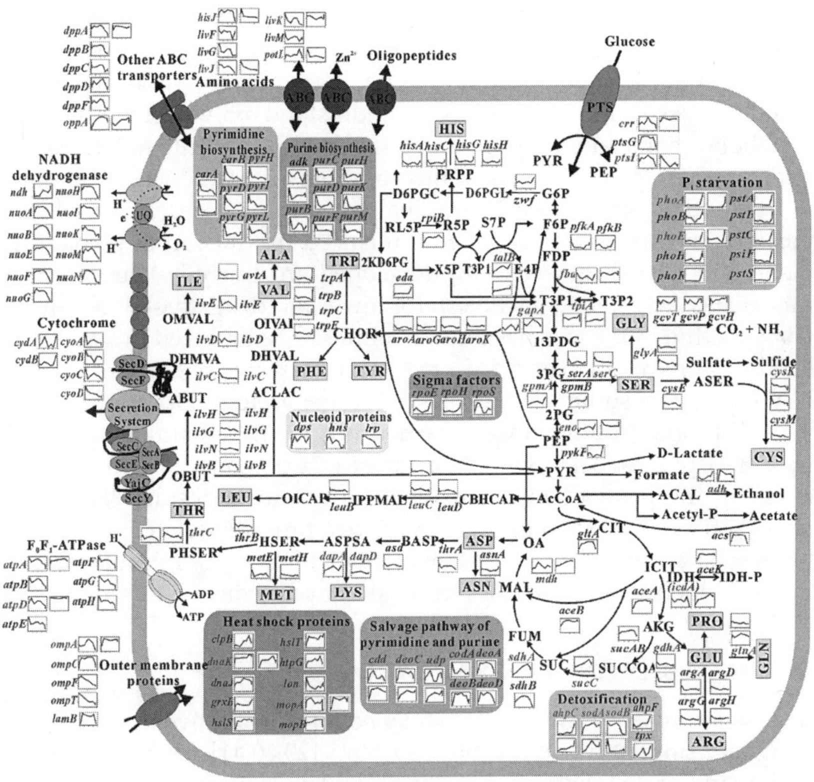

Figure 7.4 An expression map showing an integrated view of transcriptome and proteome profiles of *E. coli* during HCDC. The *x*-axis denotes cell concentration (g DCW/l), and the *y*-axis denotes expression level in $\log_2$ scale for the transcriptome and in absolute value of volume % for the proteome. The one located on either the left or the upper side of the two maps indicates the transcriptome profile while the other one is the proteome profile. Only a single map represents the transcriptome profile. A larger image is available from http://mbel.kaist.ac.kr. For more details, see Yoon et al. [30]. Reproduced with permission from Yoon et al. [30]. To view this figure in color, see the companion web site for *Systems Biology*, http://www.oup.com/us/sysbio.

Therefore, the 2.31 Mbp genome sequence of this bacterium was completely determined [38]. Based on the complete genome sequence, a genome-scale in silico metabolic model composed of 373 reactions and 352 metabolites was constructed. Metabolic flux analyses were carried out under varying pH and gas conditions. CO_2 was found to be important for cell growth as well as the carboxylation of phosphoenolpyruvate to oxaloacetate, which is then converted to succinic acid by the reductive tricarboxylic acid cycle using fumarate as a major electron acceptor. It was interesting to note that the presence of CO_2 increased the glucose uptake rate and succinic acid formation rate by 4.6- and 37-fold, respectively. Based on these findings, strategies for genome-scale metabolic engineering of *M. succiniciproducens* could be suggested. It should be emphasized that genome-scale metabolic flux analyses made it possible to understand previously unknown metabolic characteristics from a minimal number of cultivation experiments, which provided constraints for more accurate flux calculation.

In another study, combined analyses of the transcriptome and metabolome were carried out for enhanced production of secondary metabolite. Askenazi et al. [16] developed an *Aspergillus* strain overproducing lovastatin, a cholesterol-lowering drug, based on these analyses. Initially, the genes thought either to be involved in lovastatin synthesis or known to broadly affect secondary metabolites production in the parental strain were expressed to generate a library of strains. Transcriptome profilings were carried out for these strains, followed by a statistical association analysis to extract potential key parameters affecting the production of lovastatin. From these results, the target genes were identified and manipulated to improve lovastatin production by over 50%. There is no doubt that more successful examples of employing combined x-omic analyses for strain improvement will appear.

IN SILICO MODELING AND SIMULATION

Several approaches have been developed for quantitative in silico modeling and simulation of metabolic systems [20,40,41]. They can mainly be classified into two types: kinetic model-based dynamic analysis and constraints-based stationary analysis.

Kinetic Model-Based Analysis

In the dynamic approaches, kinetic and regulatory information is incorporated into a kinetic model set of continuous differential equations. The mass conservation around metabolites is derived from the detailed mechanisms and characteristics of the system; the resultant rate equations can be formulated as:

$$\frac{dc}{dt} = Sv - b \tag{1}$$

where $c \in \Re^{m\times 1}$ is the vector of the concentrations of m metabolites; $v \in \Re^{n\times 1}$, the n-dimensional rate vector, whose jth component represents the rate of reaction j as a nonlinear function of the concentrations of the metabolites involved in the reaction, $v_j = f(c)$; $S \in \Re^{m\times n}$, the stoichiometric matrix of the network; and b is the vector of the net transport of metabolites into and out of the cell and dilution due to cell growth. Thus, this kinetic model can completely describe the dynamic behavior as well as the structure of the metabolic network.

Once the kinetic model is developed, model parameters p in reaction j, $v_j = f(c, p)$, are adjusted by best fitting to in vitro kinetic experiments. Obtaining such experimental data is, however, still a laborious task. In many cases, the kinetic information on the reactions of interest can be collected from the published works and/or the available public databases [42,43]. However, one may fail to obtain reliable solutions because the parameters collected from different sources can cause system inconsistency. Thus, it is highly desirable to establish an effective method in the form of an easily usable software package for facilitating the kinetic modeling and parameter estimation. Currently, a variety of software tools and computational environments for such dynamic simulation and/or parameter estimation are available; they include GEPASI [44], DBsolve [45], E-Cell [46], SCAMP [47], Virtual Cell [48], StochSim [49], STOCKS [50], Dynetica [51], GENESIS [43], and many others (http://sbml.org/).

Constraints-Based Flux Analysis

Although the kinetic model-based dynamic modeling and simulations are most appropriate for fully characterizing the metabolic reaction systems, determining a large number of kinetic parameters is not an easy task. Moreover, the values of many of these parameters cannot be trusted since the reaction mechanisms and parameters in these models are derived from in vitro rather than in vivo measurements [52]. The stationary modeling approach is therefore a good alternative to the kinetic model for the simulation of a large metabolic reaction network. Assuming the pseudo-steady state, one can simplify the kinetic model into static representation. Unlike the dynamic approaches, the stationary model only considers the network's connectivity and capacity as time-invariant properties of the metabolic system.

The stationary approaches include stoichiometric analysis, structural or topological pathway analysis, and constraints-based flux analysis [53–58]. Of these, constraints-based flux analysis, also known as "metabolic flux analysis" (MFA), and more specifically flux balance analysis (FBA), is the most widely adopted method since it requires the least amount of information to quantify the fluxes and analyze the metabolic system [59]. Basically, under the pseudo-steady-state assumption, eq. (1) can be simplified as follows:

$$Sv = b \tag{2}$$

This matrix form can be rewritten in balance equations for I metabolites given a series of J candidate metabolic reactions:

$$\sum_{j \in J} S_{ij} v_j = b_i, \quad \forall i \in I \tag{3}$$

where S_{ij} is the stoichiometric coefficient of metabolite i in reaction j; v_j, the flux of reaction j; and b_i, the net transport flux of metabolite i. The fluxes are often calculated as mmol/g DCW·h. If this metabolite is an intermediate, b_i is zero. Extracellular metabolites serve as inputs to or outputs from the system; the former includes carbon substrates, such as glucose and fructose, while the latter includes excretory metabolic products. When metabolite i is transported into the system as a substrate uptake (input), b_i represents its consumption rate, thereby rendering its value negative. If metabolite i is secreted to the culture medium (output), b_i represents its secretion rate, thus rendering its value positive. Note that these constraints for the extracellular metabolites are imposed only when the consumption or secretion rates are experimentally observable. Otherwise, the net transport flux for each extracellular metabolite cannot be given a priori, and must be calculated just like intracellular fluxes.

In addition to the mass balance constraints of eq. (3), the reversibility of metabolic reactions and the capacity limitations for all the fluxes of these reactions can be imposed as:

$$\alpha_j \leq v_j \leq \beta_j, \quad \forall j \in J \tag{4}$$

where α_j and β_j are the lower and upper bounds, respectively, of the flux of reaction j. Here, the flux of any irreversible reaction is considered to be positive; the negative flux signifies the reverse direction of the reaction. Hence, the lower bound for the flux of irreversible reaction j, α_j, should be set to zero or a positive value whereas the flux of reversible reaction j is not constrained ($\alpha_j = -\infty$ and $\beta_j = \infty$) or can be constrained within any range according to the capacity limitation. It should be noticed that in silico fluxome analysis based on isotopomer labeling techniques can provide useful information on such capacity constraints, as will be discussed later.

The flux balance analysis by linear programming (LP) which maximizes or minimizes an objective function can be formulated as:

$$\text{Maximize/minimize } Z = \sum_{j \in J} c_j v_j \tag{5}$$

where c_j is the weight of reaction j. The objective function, Z, can vary depending on the system under analysis. It can be the growth rate, ATP usage, substrate uptake, or product formation [57]. Most frequently, the growth rate or the product formation rate is selected as an objective function. When the growth rate (g biomass/g DCW·h) is used as an objective function, the conversion of all the biosynthetic precursors (e.g., proteins, DNA, and RNA) into biomass is required.

Such biomass composition of metabolites (mmol metabolite/g biomass) can be experimentally evaluated at a given specific growth rate, or can be found from the literature [60]. For the maximization/minimization of the rate of product or byproduct formation, the objective function corresponds to its net transport flux in eq. (3), which is formulated by summing all fluxes linked to the product. For example, the maximization of the rate of succinic acid production can be formulated as follows:

$$\text{Maximize } \sum_{j \in J} S_{\text{succ},j} v_j \tag{6}$$

where $S_{\text{succ},j}$ designates the stoichiometric coefficient of reaction j involving succinic acid.

Flux Analysis Based on Labeled Substrates

The constraints-based flux analysis can be upgraded with respect to its accuracy by providing additional material balance equations through isotope balancing. Thus, more realistic internal fluxes may be predicted [61]. Using a labeled carbon source such as ^{13}C-labeled glucose as a substrate, the labeling states of the metabolites can be traced. Generally, two approaches are used for the flux interpretation of ^{13}C labeling patterns (figure 7.5). One is the identification of the flux ratio in converging reaction at the branch point using partial isotopomer data which are generated from the ^{13}C pattern of proteinogenic amino acids [62]. This approach is taken by solving the equations of metabolic reactions and optimizing the solution by error minimization. The flux ratios can be readily determined through GC/MS data without significant computational burden, but absolute flux values cannot be obtained [33]. The other iterative approach uses all available ^{13}C labeling data, extracellular material fluxes, and biomass composition for simultaneous interpretation of metabolic models of varying complexities [63]. The labeling state of metabolic intermediates is balanced within a model through an iterative fitting procedure on the isotopomer patterns of network metabolites. These approaches of employing ^{13}C data from NMR [64–66] allowed successful determination of the in vivo internal flux distributions in different microorganisms. It should be mentioned that isotope balances are bilinear with respect to its labeling and reaction rate. This is why an iterative approach of error minimization is used for flux determination. The mathematical complexity of this approach has been addressed by several researchers, who introduced concepts like exchange fluxes [67,68], isotopomer mapping matrices [64] resembling atom mapping matrices [69], cumomer balances [68], and summed fractional labeling [70].

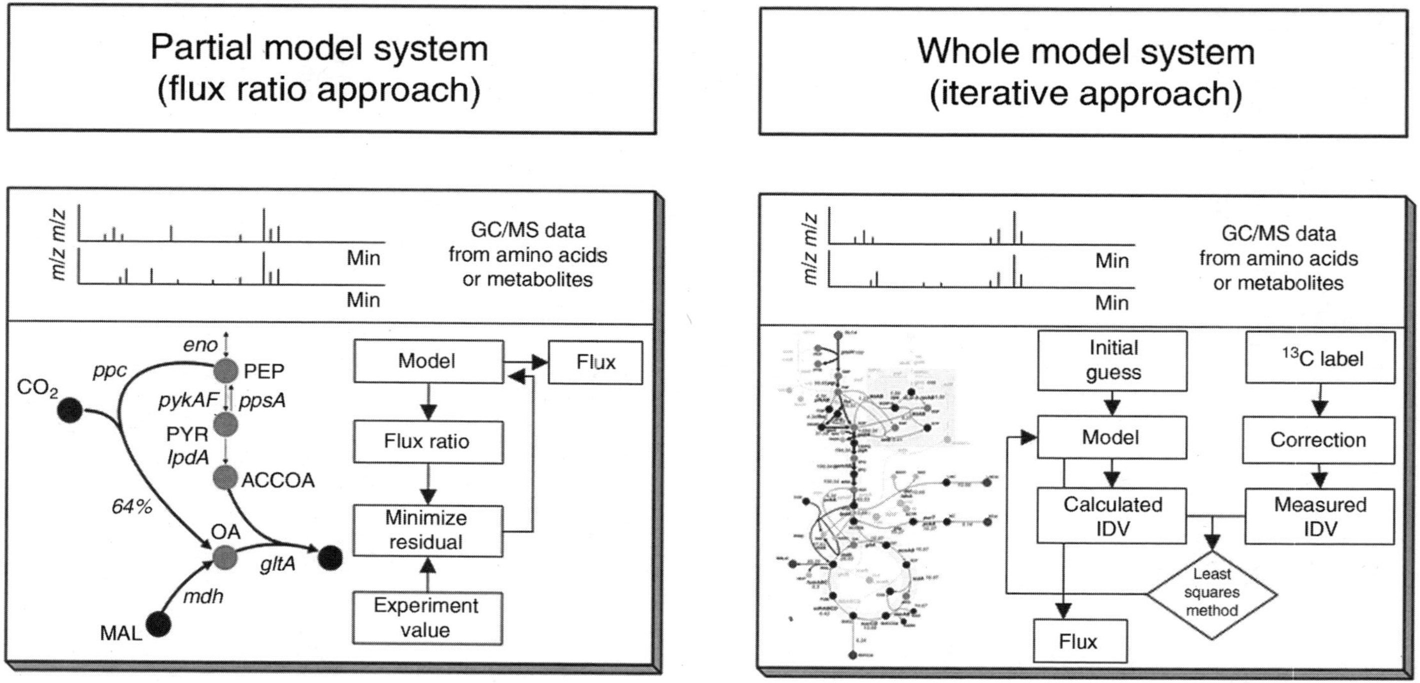

Figure 7.5 Two approaches for flux interpretation using the ^{13}C labeling patterns of metabolites [33]. The ^{13}C-labeled carbon substrate is used for the quantification of the relative amounts of the differently labeled metabolites. GC/MS and NMR can be used for determining isotopomer distribution.

DATABASES AND SUPPORTING TOOLS FOR IN SILICO MODELING AND SIMULATION

Integrated Metabolic Database System

Recent years have seen an explosion in the amount of biological data and information. This has resulted in the development of various databases containing biological information at different levels ranging from DNA sequences to metabolic pathways [71]. Among them, metabolic databases provide all available information on metabolic pathways, reactions, and enzymes that are essential to the elucidation of the metabolic and physiological characteristics of organisms. Currently available public metabolic databases are given in table 7.1. These databases, however, possess different data collections, which will become much more usable if they are unified into composite resources with common interfaces [72–74]. To properly handle such redundant and heterogeneous data sources scattered over different sites, an integrated metabolic database system, BioSilico, was developed [75]. BioSilico integrates publicly available metabolic databases for efficient navigation and analysis of metabolic pathway information.

BioSilico

BioSilico is a comprehensive, web-based database system that facilitates the search and analysis of metabolic pathways. Inhomogeneous metabolic databases including LIGAND, ENZYME, EcoCyc, and MetaCyc are compiled and integrated in a systematic way, thereby allowing users to efficiently retrieve and explore the relevant information on

Table 7.1 Representative metabolic databases useful for studying enzymes, metabolites, reactions, and pathways

Database	Corresponding URL
Biocarta	http://www.biocarta.com/genes/allPathways.asp
BioCyc	http://biocyc.org
BioSilico	http://biosilico.kaist.ac.kr
BRENDA	http://www.brenda.uni-koeln.de
EcoCyc	http://ecocyc.org
ENZYME	http://www.expasy.org/enzyme/
ERGO-LIGHT	http://www.ergo-light.com/
KEGG	http://www.genome.ad.jp/kegg
Klotho	http://www.biocheminfo.org/klotho/
LIGAND	http://www.genome.ad.jp/ligand/
MetaCyc	http://metacyc.org/
PathDB	http://www.ncgr.org/pathdb
UMBBD	http://umbbd.ahc.umn.edu/

enzymes, biochemical compounds, reactions, and pathways. In addition, a robust systematic architecture of BioSilico is extended to a customized web-start interface, the BioSilico Modeler, which provides the integrated environment to enhance the construction of in silico metabolic network models of interest.

In BioSilico, various querying logics and user-friendly web-accessible interfaces are designed to efficiently retrieve the relevant information on enzymes, compounds, reactions, and pathways from the integrated database. Queries using Enzyme Commission (EC) number, name, formula, CAS number, pathway, and organism simultaneously examine all the classes of the database; the matched entries are collected and a view page of query results is dynamically generated, allowing access to more detailed information using the hyperlinks provided. Moreover, a Java Applet for chemical structure search via the web interface allows users to either draw the compound to be searched or enter its SMILES string, giving rise to the list of matching compounds satisfying the selected search type (similarity, substructure, or exact match). Clicking on a returned structure of the compound from Chem DB, which was developed using two Marvin Java applets (MarvinSketch and MarvinView) and JChem libraries from ChemAxon (http://chemaxon.com), leads to the corresponding BioSilico compound page (figure 7.6). Consequently, one query renders it possible to efficiently search for the entire classes of the integrated database, retrieve the relevant information, and finally to display well-designed view pages interactively for more detailed information.

In addition to the web-based interface, BioSilico provides a web-client application, the BioSilico Modeler, which is launched with Java Web Start independently of the web browser. Thus, users can comprehensively build reaction network models of their own. The model composer supported in this application allows users to define components (e.g., compounds or metabolites) and their interactions (e.g., reactions) after creating a model project. Furthermore, through the customized client–server interface, a wide variety of queries retrieve the list of reactions that can be imported from the BioSilico database system on a remote server. Imported reactions are then mapped into the current model project, thus leading to the construction of the comprehensive model of a metabolic reaction network (figure 7.7). The resultant model network can be automatically visualized by several layout algorithms.

Integrated Environment for In Silico Modeling and Simulation of Metabolic Networks

Considering the importance of metabolic flux analysis in systems biotechnology, it will be extremely useful to have a user-friendly computer program for quantitatively analyzing metabolic fluxes.

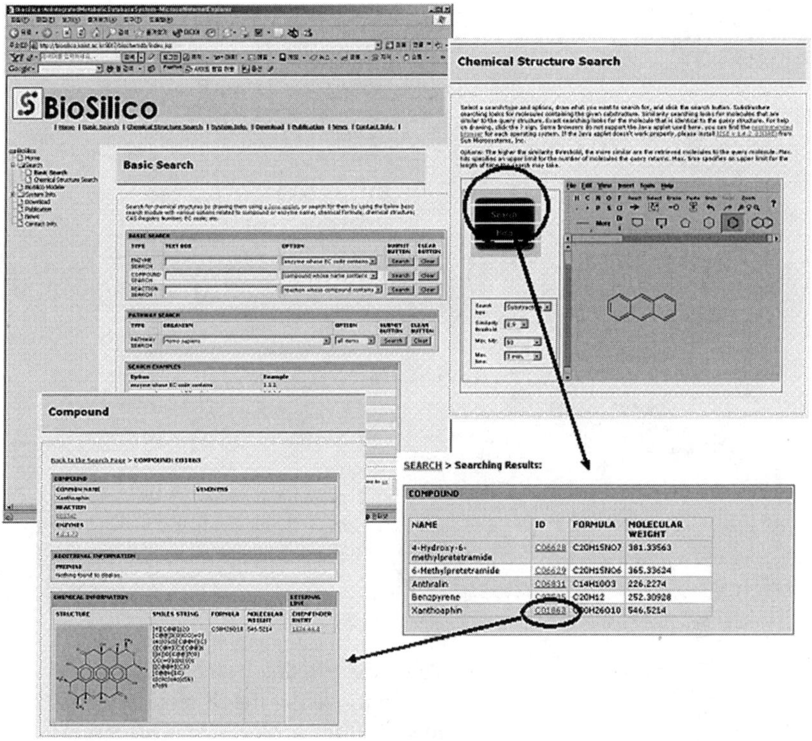

Figure 7.6 Database query interface of BioSilico.

This is indeed needed for researchers who are less familiar with the computational methods in detail. For this purpose, a program package, MetaFluxNet, was developed, which provides one of the systems biology platforms for metabolic characterization and engineering [76].

MetaFluxNet

MetaFluxNet is a program package for managing information on the metabolic reaction network and for quantitatively analyzing metabolic fluxes in an interactive and customized way. Users can interpret and examine metabolic behaviors and changes in response to genetic and/or environmental modifications. Consequently, quantitative in silico simulations of metabolic pathways can be carried out to understand the metabolic status and to design the metabolic engineering strategies. The main features of MetaFluxNet include a well-designed model construction environment, a user-friendly interface for constraints-based flux analysis, comparative flux analysis of different strains under varying environmental conditions, several options for choosing numerical solvers, and automated pathway layout creation [76].

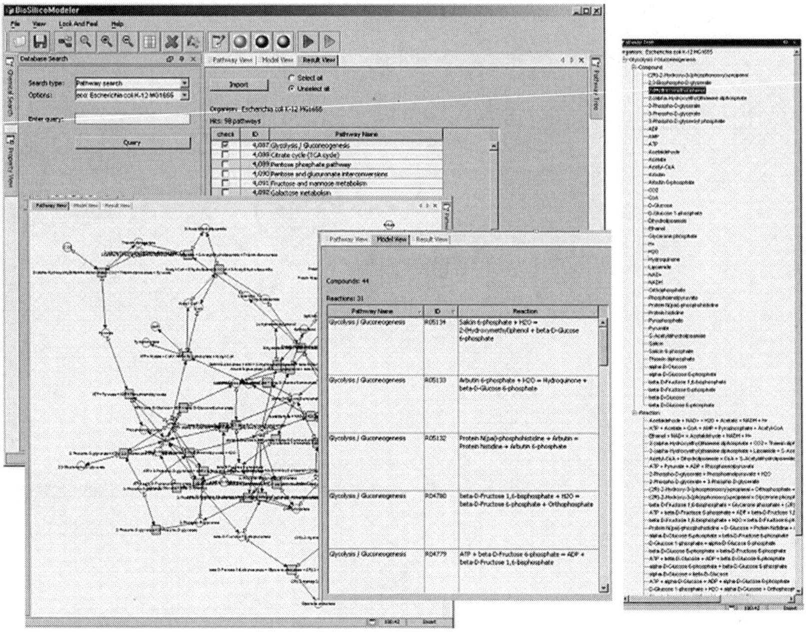

Figure 7.7 Database query interface of the BioSilico Modeler. In this screen capture, the list of metabolic reactions is retrieved as the result of a query, "Find all reactions involved in the glycolysis pathway of *E. coli* K-12 strain." The resultant network is visualized and summarized in a tabular format.

MetaFluxNet allows users to set up their own metabolic network models by registering information on two object classes, *metabolites* and *reactions*. Each class consists of several fields describing biological information. Users can freely edit and store the data contents in the fields. A stoichiometric model can be built from the metabolic reaction model under the steady-state assumption. The model system can be classified by one of four possible cases according to *determinacy* and *redundancy* [77]. In the case of the determined system, a unique solution or a least-squares solution can be obtained if the system is observable. In the case of the redundant system, the measured fluxes are reconciled to remove the inconsistency, followed by inspecting calculable fluxes which can be uniquely determined by the least-squares solution using the pseudo-inverse. If the system is underdetermined, the constraints-based flux analysis can be exploited to quantify the flux distribution by maximizing or minimizing the objective function [59].

MetaFluxNet also allows users to compare the changes in flux distribution under genetically or environmentally perturbed conditions,

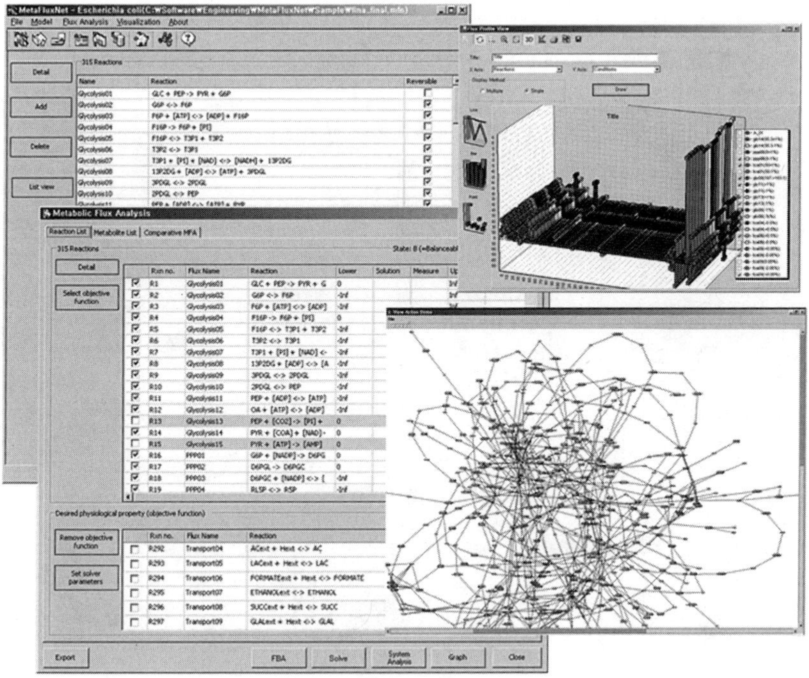

Figure 7.8 Screen shot of the flux analysis part of MetaFluxNet. Flux distributions can be interactively determined and dynamically visualized via a user-friendly interface.

which is invaluable not only for understanding the metabolic and physiological changes in cells under different conditions, but also for developing new metabolic engineering strategies to achieve desired goals. An interactive and dynamic graphical user interface is also provided to display metabolic reaction pathways with flux distribution results as depicted in figure 7.8. The pathways are automatically and dynamically visualized by the spring embedder layout algorithm supported in the program. In the most recent version released (version 1.6.9.9), systems biology markup language (SBML) [78] is supported to communicate with other systems biology platforms, thus expanding its usability.

SYSTEMS BIOTECHNOLOGICAL STRATEGY

Figure 7.9 outlines the conceptual procedure for the systems biotechnological strategy toward the development of improved strains. In silico and wet experiments are carried out on the basis of computational

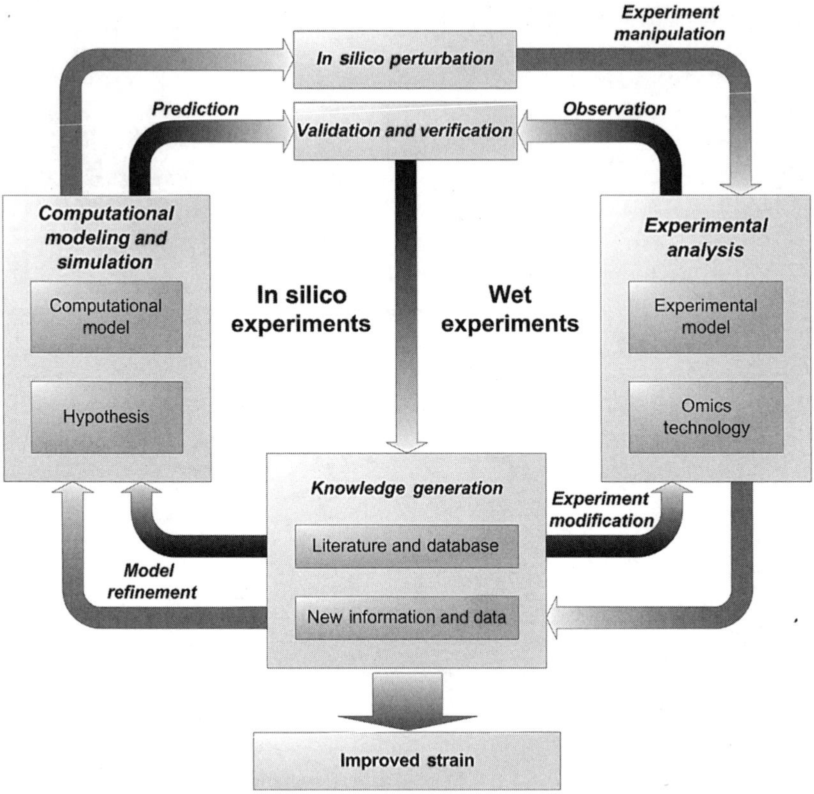

Figure 7.9 Overview of systems biotechnological research for the development of improved organisms. Hypothesis-driven computational modeling and simulation, and technology-driven high-throughput x-omic experiments, are combined to generate new knowledge, which is used to fine-tune the model and design new experiments. This process is iterated until one achieves the goal of developing improved strains having desired traits.

models of a biological system of interest and rational experimental procedures, respectively, which complementarily interact with each other for the generation of new knowledge [14,79,80]. Accumulation of ever-increasing amounts of biological data ranging from DNA and protein sequences to metabolic pathways results in the development of various databases. These databases are used not only for extracting local biological information (e.g., identifying a gene or genes of interest based on homology search) but also for larger-scale studies such as metabolic network construction and analysis in model-driven research. Even though our understanding of the genome is still far from complete, we can reconstruct genome-scale metabolic models using as much available information as possible. A good example of this approach has been reported by Segre et al. [81], who suggested an automated process to construct stoichiometric models from annotated genomes. Obviously, data contents and their quality in the metabolic databases are important in constructing reliable metabolic models. Several excellent metabolic databases are available (table 7.1) and are expected to evolve as an integrated one, as demonstrated for the BioSilico database.

Once the valid model is constructed, the system's behavior under a particular experimental situation can be better predicted by systematic perturbations encompassing genetic to environmental alterations. Accordingly, in silico experiments of the system through metabolic modeling can provide crucial information on cellular behavior under various genetic and environmental conditions, thereby giving rise to a multitude of strategies for strain improvement. For example, using the genome-scale in silico model, constraints-based flux analysis can be carried out to identify the gene knockout targets of recombinant *E. coli* by resorting to various optimization techniques, for example, iterative linear programming, mixed-integer optimization [82], quadratic programming [83], and bilevel optimization [84]. As a whole, in silico and wet experiments are interactively conducted under the perturbing conditions of biotechnological relevance. The results are compared for the generation of new knowledge, which is used to refine the model and to design new experiments. This process is iterated along systems biotechnological research cycles until the development of an improved microorganism having desired traits is accomplished.

As shown in figure 7.9, there exist two adaptive feedback loops and one interactive cycle. The in silico model and experimental design are iteratively modified through respective feedback loops [14,18,85]. This conventional iterative model-building process, however, may be insufficient to validate the hypothetical model and experimental results. Therefore, another interactive communication cycle between in silico and wet experiments makes such a process more efficient [17,79,86]. Computational models can be directly refined from new experimental data, and subsequently the resultant in silico perturbations of the refined model can suggest new experimental design. Thus, new knowledge can be effectively generated after several iterations and interactions, thereby facilitating the improvement of strains suitable for successful industrial applications. Consequently, the essence of systems biotechnology resides in the integration of wet and dry experiments to achieve a goal of rational metabolic design. In the following, two examples are presented to illustrate the truly enormous range of possibilities for the systems biotechnological strategy afforded by high-throughput experiments and in silico modeling and simulation (case studies 3 and 4).

CASE STUDY 3: PREDICTION AND VALIDATION OF THE IMPORTANCE OF THE ENTNER–DOUDOROFF (ED) PATHWAY IN PHB BIOSYNTHESIS BY METABOLICALLY ENGINEERED *E. coli*

Poly(3-hydroxybutyrate), PHB, is one of the most common members of the polyhydroxyalkanoates (PHAs) and is a biodegradable polymer that can be used as an excellent alternative to conventional petrochemical-based polymers [87,88]. However, the production cost of PHB should be considerably reduced to make it competitive with petroleum-derived plastics. To achieve this goal, much effort has been exerted to develop metabolically engineered strains for the efficient production and recovery of PHB [88–90]. In this case study, we show the systems biotechnological procedures taken to understand the metabolic characteristics of an engineered *E. coli* strain producing large amounts of PHB.

Step 1: Development of metabolically engineered E. coli strain. The PHB biosynthesis operon of *Wautersia eutropha* and *Alcaligenes latus* encodes three enzymes: β-ketothiolase, reductase, and PHB synthase [89,91,92]. β-Ketothiolase condenses two acetyl-CoA moieties to form acetoacetyl-CoA, which is then reduced to (*R*)-3-hydroxybutyryl-CoA by an NADPH-dependent acetoacetyl-CoA reductase. PHB synthase finally links (*R*)-3-hydroxybutyryl-CoA to the growing chain of PHB (figure 7.10). Plasmid-based expression of the *W. eutropha* or *A. latus* PHB biosynthesis genes in *E. coli* led to the accumulation of a large amount of PHB from glucose [92,93]. In order to understand the metabolic characteristics of the engineered *E. coli* strain, which accumulates PHB to a surprisingly high level (up to 90% of dry cell weight) inside the cell, systems biotechnological research was carried out as described below.

Step 2: Construction of in silico model. The in silico metabolic model describing recombinant *E. coli* metabolism is derived from the publicly available information and database [75,94]. The three-step PHB biosynthetic pathway, which is heterologous to *E. coli*, is also introduced in the in silico metabolic model. For simplicity, a small metabolic model consisting of 310 reactions and 295 metabolites was constructed and employed [95].

Step 3: In silico experiment. The constraints-based flux analysis was carried out to quantify flux distribution under various conditions. During the simulation, the previously reported fermentation data during PHB production were imposed as additional constraints [96,97]. Metabolic fluxes under PHB-producing conditions were determined for two different PHB contents of 49% (figure 7.11a) and 78% (figure 7.11b) of dry cell weight, and were compared with those under non-PHB-producing condition. Bold arrows indicate the increased fluxes by more than 2-fold while dotted arrows indicate the decreased fluxes to less than half under PHB-accumulating conditions.

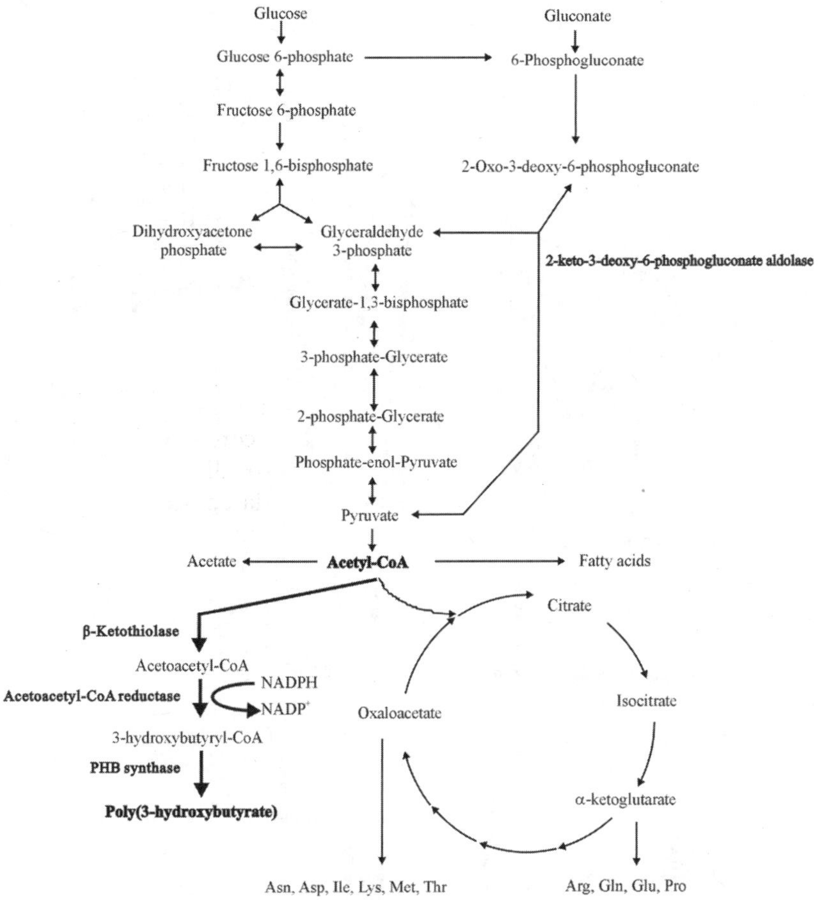

Figure 7.10 Simplified metabolic reaction network of engineered *E. coli* for the production of PHB. Bold arrows indicate three heterologous reactions that are newly introduced into *E. coli*.

First, in the case of PHB accumulation up to 49% of dry cell weight [96], it was predicted that about 24% of total carbon flux was directed to the PHB pathway, and most of it was supplied through the Entner–Doudoroff (ED) pathway (figure 7.11a). The flux of acetyl-CoA into the TCA cycle, which competes with the PHB biosynthesis pathway, decreased significantly. Next, the intracellular flux distribution in recombinant *E. coli* that accumulated PHB up to 78% of dry cell weight was simulated [97] and compared with that predicted in a non-PHB-producing strain (figure 7.11b). A significant amount of carbon flux (74%) was directed into the PHB biosynthetic pathway, and most of the other intracellular fluxes were severely reduced. Again, it was

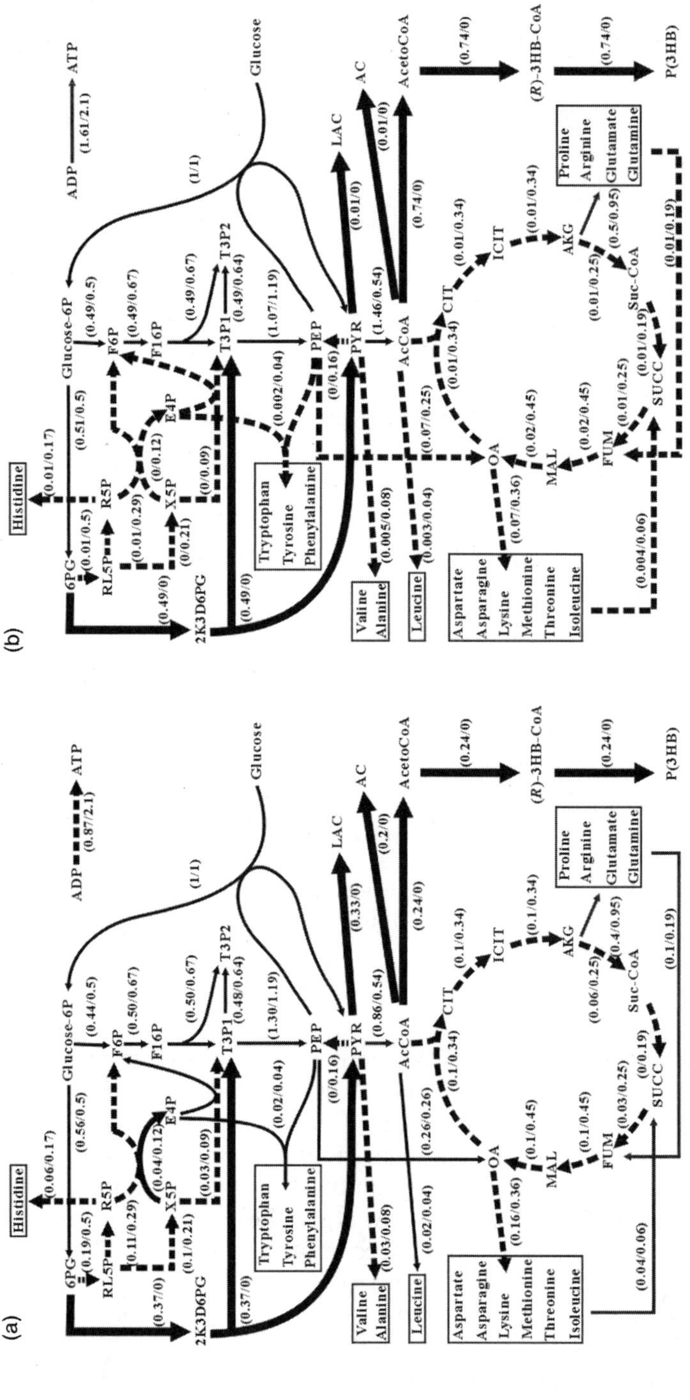

(a)
Histidine
6PG
RL5P
R5P
X5P
E4P
2K3D6PG
Glucose-6P
F6P
F16P
T3P1
T3P2
PEP
PYR
AcCoA
AcetoCoA
(R)-3HB-CoA
P(3HB)
Glucose
LAC
AC
ADP
ATP
(0.87/2.1)
(1/1)
(0.06/0.17)
(0.56/0.5)
(0.44/0.5)
(0.50/0.67)
(0.50/0.67)
(0.48/0.64)
(1.30/1.19)
(0.19/0.5)
(0.11/0.29)
(0.04/0.12)
(0.1/0.21)
(0.03/0.09)
(0.37/0)
(0.37/0)
(0.02/0.04)
(0/0.16)
(0.33/0)
(0.2/0)
(0.86/0.54)
(0.24/0)
(0.24/0)
(0.24/0)
(0.03/0.08)
(0.02/0.04)
(0.26/0.26)
(0.16/0.36)
(0.1/0.34)
(0.1/0.34)
(0.1/0.34)
(0.4/0.95)
(0.06/0.25)
(0/0.19)
(0.03/0.25)
(0.1/0.45)
(0.1/0.45)
(0.1/0.19)
(0.04/0.06)
Tryptophan
Tyrosine
Phenylalanine
Valine
Alanine
Leucine
Aspartate
Asparagine
Lysine
Methionine
Threonine
Isoleucine
Proline
Arginine
Glutamate
Glutamine
CIT
ICIT
AKG
Suc-CoA
SUCC
FUM
MAL
OA
(b)
Histidine
6PG
RL5P
R5P
X5P
E4P
2K3D6PG
Glucose-6P
F6P
F16P
T3P1
T3P2
PEP
PYR
AcCoA
AcetoCoA
(R)-3HB-CoA
P(3HB)
Glucose
LAC
AC
ADP
ATP
(1.61/2.1)
(1/1)
(0.01/0.17)
(0.51/0.5)
(0.49/0.5)
(0.49/0.67)
(0.49/0.67)
(0.49/0.64)
(1.07/1.19)
(0.01/0.5)
(0.01/0.29)
(0/0.12)
(0/0.21)
(0/0.09)
(0.49/0)
(0.49/0)
(0.002/0.04)
(0/0.16)
(0.01/0)
(0.01/0)
(1.46/0.54)
(0.74/0)
(0.74/0)
(0.74/0)
(0.005/0.08)
(0.003/0.04)
(0.07/0.25)
(0.07/0.36)
(0.01/0.34)
(0.01/0.34)
(0.01/0.34)
(0.5/0.95)
(0.01/0.25)
(0.01/0.19)
(0.01/0.25)
(0.02/0.45)
(0.02/0.45)
(0.01/0.19)
(0.004/0.06)
Tryptophan
Tyrosine
Phenylalanine
Valine
Alanine
Leucine
Aspartate
Asparagine
Lysine
Methionine
Threonine
Isoleucine
Proline
Arginine
Glutamate
Glutamine
CIT
ICIT
AKG
Suc-CoA
SUCC
FUM
MAL
OA

most notable that the ED pathway flux was highly activated under the PHB-producing condition. The fluxes of the pentose phosphate (PP) pathway and cellular material synthesis were not greatly affected when the PHB content was 49% but were significantly decreased by more than 90% when the PHB content was 78%. It has been generally accepted that the ED pathway is not functional in *E. coli* when glucose is used as a carbon source. However, the in silico simulation studies suggested that the ED pathway is important for PHB production from glucose in recombinant *E. coli*. Since the in silico prediction cannot be trusted 100%, the roles of the ED pathway on PHB biosynthesis in recombinant *E. coli* were examined experimentally.

Step 4: Experimental validation. To verify the importance of the ED pathway on PHB production in *E. coli* as suggested by in silico simulation, actual experiments including gene manipulation, cultivation, and proteome analysis were performed. The *eda* mutant strain KEDA and its parent strain KS272 were transformed with pJC4, which harbors the *A. latus* PHB biosynthesis operon, and were compared for their ability to produce PHB in LB and MR media containing 20 g/l of glucose at 30°C (table 7.2). The final PHB contents obtained with the *eda* mutant strain KEDA (pJC4 + pACYC184) were 61.9 wt% of dry cell weight in LB medium and 44.0 wt% in MR medium, which are lower than those (75.9% and 60%, respectively) obtained with KS272 (pJC4 + pACYC184). When the activity of Eda was restored by the coexpression of the *eda* gene in KEDA (pJC4 + pAC104Eda), the PHB contents increased back to 72.5 wt% of DCW in LB medium and 64.5 wt% in MR medium.

Figure 7.11 Pictorial representation of the intracellular fluxes (mM/g dry cell weight/h) in recombinant *E. coli* producing PHB versus wild-type *E. coli* not producing PHB; the flux values are given in parentheses for (a) PHB content of 49%/non-PHB-producing condition, (b) PHB content of 78%/non-PHB-producing condition. Bold arrows indicate the fluxes that are increased by more than 2-fold, while dotted arrows indicate the fluxes that are decreased to less than a half. Abbreviations: 2K3D6PG, 2-keto-3-deoxy-6-phosphogluconate; (R)-3HBCoA, (R)-3-hydroxybutyryl-CoA; 6PG, 6-phosphogluconate; AcetoCoA, acetoacetyl-CoA; AcCoA, acetyl-CoA; CIT, citrate; T3P2, dihydroxyacetone-phosphate; E4P, erythrose-4-phosphate; F16P, fructose-1,6-diphosphate; F6P, fructose-6-phosphate; FUM, fumarate; T3P1, glyceraldehyde-3-phosphate; PHB, poly(3-hydroxybutyrate); ICT, isocitrate; AKG, α-ketoglutarate; LAC, lactate; MAL, malate; OA, oxaloacetate; PEP, phosphoenolpyruvate; PYR, pyruvate; R5P, ribose-5-phosphate; RL5P, ribulose-5-phosphate; SUCC, succinic acid; Suc-CoA, succinyl-CoA; X5P, xylulose-5-phosphate. Reproduced with permission from Hong et al. [95].

Table 7.2 Results of flask cultures of recombinant *E. coli* strains producing PHB[a]

Strain (Plasmid[b])	Medium[c]	Dry cell weight (g/l)	PHB conc. (g/l)	PHB content (wt%)
KS272 (pJC4 + pACYC184)	LB	4.13	3.12	75.9
KEDA (pJC4 + pACYC184)	LB	4.40	2.70	61.9
KEDA (pJC4 + pAC104Eda)	LB	6.10	4.40	72.5
KS272 (pJC4 + pACYC184)	MR	4.30	2.50	60.0
KEDA (pJC4 + pACYC184)	MR	4.50	2.00	44.0
KEDA (pJC4 + pAC104Eda)	MR	6.20	4.00	64.5

[a]These data were taken from [95].
[b]pJC4 is a high copy number plasmid harboring the *A. latus* PHB biosynthesis operon. pAC104Eda is a pACYC184 derivative harboring the *eda* gene. Plasmids pJC4 and pACYC184 are compatible in *E. coli*.
[c]Glucose was added to 20 g/l in both media. Abbreviations are: LB medium, Luria-Bertani medium [101]; MR medium, a chemically defined medium [102].

These results verify the essential role of the ED pathway during PHB biosynthesis in recombinant *E. coli* as predicted by in silico experiment. The increase in the ED pathway flux during PHB production is also experimentally supported by the proteome analysis of a different *E. coli* strain, XL1-Blue, which showed an increase in Eda expression level under PHB-producing condition [31]. It should be noticed that there exists the possibility that metabolic pathways might be differently controlled in different *E. coli* strains, which can also be studied in a similar way. This case study demonstrates the effectiveness of the systems biotechnological approach to elucidate previously unknown metabolic characteristics.

CASE STUDY 4: INCREASING THE METABOLIC FLUX TO SUCCINIC ACID IN *E. coli* BY SYSTEMS BIOTECHNOLOGICAL RESEARCH CYCLE

There have been many successful cases of the development of metabolically engineered *E. coli* strains for the production of succinic acid (C4), which is used as a food additive, an ion chelator, and a precursor of polymers, and has been chemically produced from maleic anhydride [98]. Recently, metabolic engineering of *E. coli* by heterologous gene expression (the *pyc* gene encoding pyruvate carboxylase), amplification of an inherent gene (the *ppc* gene encoding phosphoenolpyruvate carboxylase), and deletion of several genes (*ptsG, ldhA, pfl*) has shown the possibility for the biotechnological production of succinic acid. In this case study, the aim is to develop an *E. coli* strain that is able to overproduce succinic acid by means of comparative genomics and metabolic flux prediction, followed by subsequent gene knockout and cultivation experiments. This approach follows the systems biotechnological research cycle toward stain improvement as shown in figure 7.12.

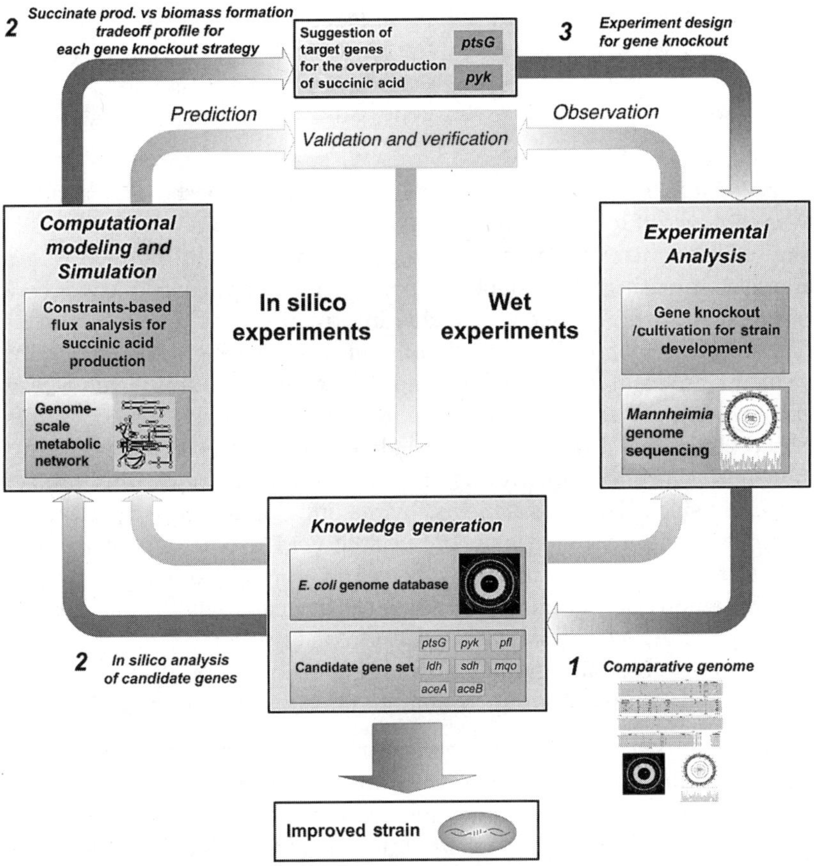

Figure 7.12 Procedures for strain improvement combined with in silico prediction. Step 1: Comparative analysis of two bacterial genomes, *E. coli* and *M. succiniciproducens*, was conducted to select knockout gene candidates for the subsequent in silico perturbation. Step 2: In silico experiments were then carried out by means of constraints-based flux analysis, suggesting target genes for the overproduction of succinic acids. Step 3: Combinatorial gene knockout experiments were carried out within the suggested candidate genes, and experimental data were obtained by actual cultivation. Finally, an engineered strain for the enhanced succinic acid production could be successfully designed.

Step 1: Comparative genomics for specifying candidate genes. Recently, the full genome sequence of *M. succiniciproducens* MBEL55E, and the genome-scale metabolic characteristics leading to high-level succinic acid productions, were published [38]. In this step, comparative genome analysis of this succinic acid-producing strain, *M. succiniciproducens*, and *E. coli* was conducted to select gene candidates for strain improvement.

The resultant genes include the *ptsG, pykF, mqo, sdhABCD,* and *aceBA* genes that were not found in the genome of *M. succiniciproducens*, and only in the central pathway of *E. coli*. In addition, the *pflB* and *ldhA* genes were selected for gene candidates from the literature information [99].

Step 2: Modeling and simulation. Next, constraints-based flux analysis was carried out to rationally select the target genes to be manipulated in the subsequent experiments. Initially, a genome-scale in silico *E. coli* model comprising 814 metabolites and 979 metabolic reactions was constructed from the publicly available information and database [75,94]. Then, a variety of in silico knockout strains were designed by fixing the relevant fluxes at zeros among candidate genes (i.e., *aceBA, ldhA, mqo, ptsG, pykFA, pfl,* and *sdh*) identified in step 1. For each knockout strain based on the in silico model, the correlation between the biomass formation and succinic acid production was examined in various mutant strains in silico. This was achieved by maximizing the growth rate while recursively limiting the level of succinic acid production in the flux model, thereby identifying gene targets for succinic acid overproduction. Finally, among all mutant strains, only the five cases (*ΔpflB, ΔptsG pykFA, ΔptsG pfl ldhA, ΔptsG pykF pykA pfl,* and *ΔptsG pykF pykA pfl ldhA*) were predicted as possible candidates capable of both overproducing succinic acid and sustaining biomass formation (table 7.3 and figure 7.13).

Step 3: Gene knockout and cultivation for strain engineering. According to the predictions in step 2, several mutant strains were constructed by gene knockout experiments and cultured to examine succinic acid production (tables 7.4 and 7.5). Interestingly, anaerobic cultivation profile of *E. coli* W3110GFA (*ptsG pykF* and *pykA* triple knockout mutant) showed about a 3.4-fold increase in succinic acid formation with significant reduction of other fermentation products (8.29-fold increase in succinic acid ratio) in 24 h of anaerobic fermentation (table 7.5). In addition, *E. coli* W3110GFA could convert 50 mmoles of glucose to 17.4 mmoles of succinic acid, while other mutants did not exceed 3 mmoles of succinic acid (data not shown). These results are consistent with in silico prediction (table 7.3), and thus prove the effectiveness of the systems biotechnological approach.

FUTURE PERSPECTIVES ON SYSTEMS BIOTECHNOLOGY

The cell is a complex system. Various types of biochemical processes are seamlessly integrated for generating mass and energy (metabolic), transmitting information (signaling), and regulating cellular and metabolic behaviors (gene regulatory) through complex networks and pathways of the molecular interactions and reactions [100]. Thus, systemic and integrative approaches to systems biotechnology allow us to understand its organization within a more global context of the

Table 7.3 In silico prediction of succinic acid fluxes in *E. coli* W3110 mutants

Disrupted genes	Maximum biomass formation rate (h^{-1})	Succinic acid production rate (mmol/g DCW/h)	Succinic acid flux ratio[a]
Wild-type	0.2156	0.1714	0.002
PykFA	0.2156	0.1714	0.002
PtsG	0.1884	0.1714	0.002
PflB	0.1882	0.857	—
ldhA	0.2156	0.1714	—
ptsG pykFA	0.1366	6.834	0.231
ptsG mqo	0.1884	0.1714	—
ptsG sdhA	0.1884	0.1714	—
ptsG aceBA	0.1884	0.1714	—
pykFA sdhA	0.2156	0.1714	—
pykFA aceBA	0.2156	0.1714	—
mqo sdhA	0.2156	0.1714	—
mqo aceBA	0.2156	0.1714	—
sdhA aceBA	0.2156	0.1714	—
ptsG pfl ldhA	0.1611	0.6856	—
ptsG pykF pykA pfl	0.1219	9.1236	0.503
ptsG pykF pykA pfl ldhA	0.1219	9.1236	0.503

[a]Calculated as succinic acid/(succinic acid + lactate + formate + acetate + ethanol).

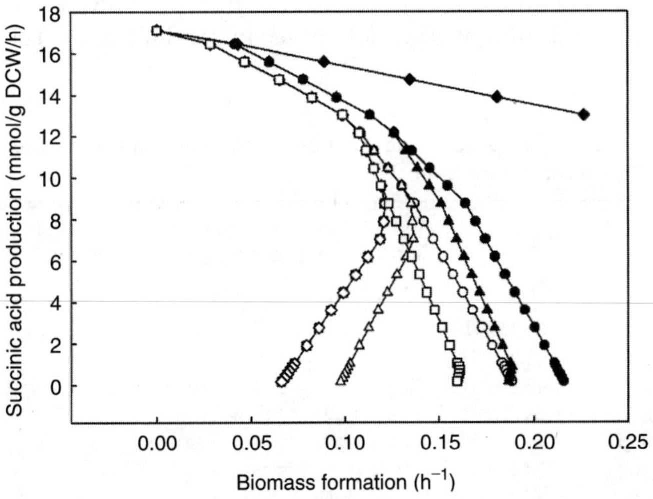

Figure 7.13 Computationally predicted tradeoff curves for several in silico mutants. Closed circles, wild-type; open circles, *ΔptsG*; closed triangles, *ΔpflB*; open triangles, *ΔptsG ΔpykFA*; close squares, *ΔptsG ΔpykFA ΔpflB*; open squares, *ΔptsG ΔpflB ΔldhA*; closed diamonds, *ΔptsG ΔpykFA ΔpflB ΔldhA* under aerobic condition; open diamonds, *ΔptsG ΔpykFA ΔpflB ΔldhA* under anaerobic condition. Abbreviations: *ptsG*, phosphotransferase systems; *pyk*, pyruvate kinase; *pflB*, pyruvate formate-lyase; *ldhA*, D-lactate dehydrogenase.

Table 7.4 ***E. coli*** **mutant strains and their genotypes**

Strain	Genotype or gene knockouts
E. coli W3110	Wild type
E. coli W3110A	*pykA*::Km^r
E. coli W3110G	*ptsG*::Sp^r
E. coli W3110GF	*ptsG*::Sp^r, *pykF*::Tc^r
E. coli W3110GFA	*ptsG*::Sp^r, *pykF*::Tc^r, *pykA*::Km^r
E. coli W3110GFAP	*ptsG*::Sp^r, *pykF*::Tc^r, *pykA*::Km^r, *pflB*::Cm^r
E. coli W3110GFAPL	*ptsG*::Sp^r, *pykF*::Tc^r, *pykA*::Km^r, *pflB*::Cm^r, *ldhA*::Pm^r
E. coli W3110GFO	*ptsG*::Sp^r, *pykF*::Tc^r, *mqo*::Cm^r
E. coli W3110GFH	*ptsG*::Sp^r, *pykF*::Tc^r, *sdh*::Km^r
E. coli W3110GFHO	*ptsG*::Sp^r, *pykF*::Tc^r, *sdh*::Km^r, *mqo*::Cm^r
E. coli W3110GFHOE	*ptsG*::Sp^r, *pykF*::Tc^r, *sdh*::Km^r, *mqo*::Cm^r, *aceBA*::Pm^r

Km^r, kanamycin resistance; Sp^r, spectinomycin resistance; Tc^r, tetracycline resistance; Cm^r, chloramphenicol resistance; Pm^r, phleomycin resistance.

system, and eventually to discover a true knowledge map for the functions of the living system. Although the metabolic system was the main focus of this chapter, other processes such as signaling and regulatory networks can be combined to predict more accurate behavior of the cell under the particular condition.

Systems biotechnology is a discipline that has only recently been introduced and presents a variety of technical challenges. The central

Table 7.5 Cultivation products formed by ***E. coli*** **W3110 mutants under anaerobic condition**

Strain	OD_{600}	Concentration of cultivation substrate or products (mM)[a]		Succinic acid ratio[c]	Fold
		Glucose[b]	Succinic aicd		
W3110	1.79 ± 0.11	5.07 ± 0.45	2.43 ± 0.03	0.017	1.00
W3110A	1.62 ± 0.04	5.05 ± 0.56	1.93 ± 0.02	0.012	0.75
W3110G	1.47 ± 0.01	5.66 ± 1.77	2.16 ± 0.08	0.014	0.86
W3110GF	1.46 ± 0.04	4.28 ± 1.42	2.83 ± 0.07	0.019	1.15
W3110GFO	1.49 ± 0.07	4.68 ± 0.35	2.67 ± 0.33	0.018	1.07
W3110GFH	1.35 ± 0.01	4.70 ± 0.39	2.51 ± 0.02	0.017	1.02
W3110GFA	0.73 ± 0.06	20.57 ± 3.02	8.16 ± 0.01	0.137	8.29
W3110GFOH	1.28 ± 0.11	4.82 ± 0.48	2.58 ± 0.03	0.016	1.00
W3110GFOHE	1.27 ± 0.05	4.25 ± 0.27	2.49 ± 0.18	0.017	1.03
W3110GFAP	0.682 ± 0.16	12.2 ± 0.04	8.45 ± 0.05	0.214	12.60
W3110GFAPL	0.05 ± 0.02	41.6 ± 0.8	0.39 ± 0.01	0.19	11.53

[a] 24 h anaerobic vial cultivation.
[b]Residual glucose concentration measured (50 mM initial glucose).
[c]Calculated as succinic aicd/(succinic acid + lactate + formate + acetate + ethanol).

task of systems biotechnology is to comprehensively collect the global cellular information and omics data at different scales ranging from DNA and RNA to proteins, metabolites, and fluxes, and to combine such data in order to generate predictive computational models of the biological system. In this context, bioinformatic data analysis and interpretations of x-omic data needs to be improved as more and more data are collected. Each x-ome alone is not enough for understanding cellular physiology and regulatory mechanisms due to the existence of an information gap among the x-omes and the behavioral phenotype [22]. As mentioned earlier, the transcriptome cannot explain translation, posttranslational regulations, and protein–protein interactions. Furthermore, the amount of protein is not always proportional to the activity of the protein, which in turn is not always proportional to the metabolic flux. Thus, it is indeed required to carry out combined analysis of all x-omes to better understand the cellular physiology and metabolism at the systems level. By doing so, strategies for metabolic and cellular engineering of organisms at the true whole-cell level can be developed.

In silico modeling and simulation of the complex biological system is invaluable to organize and integrate the available metabolic knowledge and to design the right experiments. Most importantly, the number of real wet experiments can be minimized by carrying out in silico experiments using the computational model. In silico perturbation of the metabolic system can provide crucial information on cellular behavior under varying genetic and environmental perturbations, thereby suggesting a multitude of strategies for the development of efficient biotechnology processes. The current predicting power of biological simulation, however, is limited by insufficient knowledge of global regulation and kinetic information, and thus in silico design-based process development is still far from perfection. Will we ever be able to have a whole-cell model that truly resembles the real cell in every aspect? We will reserve the answer, hopefully a positive one, to the near future since much effort is being devoted to whole-cell modeling and simulation, thus giving optimistic expectation to develop a system that allows more accurate simulation of metabolic and regulatory behaviors [41]. In the meantime, we can enjoy the great advantage of having large-scale in silico models and global-scale omics data for strain improvement as described in this chapter.

Systemic integration of heterogeneous sectors is an ongoing trend in every field. In industrial biotechnology, this trend is also being driven by combining omics together with in silico modeling and simulation. Although much remains to be done to integrate all these systems of biotechnological components, the future seems to be bright toward the goal of whole-cell level understanding and engineering.

ACKNOWLEDGMENTS

Our work described in this chapter was supported by the Korean Systems Biology Research Program (M10309020000-03B5002-00000) of the Ministry of Science and Technology and by the BK21 project. Further support from LG Chem Chair Professorship, KOSEF, through the Center for Ultramicrochemical Process Systems, and the IBM-SUR program, is greatly appreciated.

REFERENCES

1. Lee, S. Y. High cell density cultivation of *Escherichia coli*. *Trends in Biotechnology*, 14(3):98–105, 1996.
2. Makrides, S. C. Strategies for achieving high-level expression of genes in *Escherichia coli*. *Microbiological Reviews*, 60(3):512–38, 1996.
3. Baneyx, F. Recombinant protein expression in *Escherichia coli*. *Current Opinion in Biotechnology*, 10(5):411–21, 1999.
4. Zhou, S., L. P. Yomano, A. Z. Saleh, et al. Enhancement of expression and apparent secretion of *Erwinia chrysanthemi* endoglucanase (encoded by celZ) in *Escherichia coli* B. *Applied and Environmental Microbiology* 65(6):2439–45, 1999.
5. Anderson, A. J. and E. A. Dawes. Occurrence, metabolism, metabolic role, and industrial uses of bacterial polyhydroxyalkanoates. *Microbiological Reviews*, 54(4):450–72, 1990.
6. Khosla, C. and J. D. Keasling. Metabolic engineering for drug discovery and development. *Nature Reviews Drug Discovery*, 2(12):1019–25, 2003.
7. Martin, V. J., D. J. Pitera, S. T. Withers, et al. Engineering a mevalonate pathway in *Escherichia coli* for production of terpenoids. *Nature Biotechnology*, 21(7):796–802, 2003.
8. Lau, J., C. Tran, P. Licari, et al. Development of a high cell-density fed-batch bioprocess for the heterologous production of 6-deoxyerythronolide B in *Escherichia coli*. *Journal of Biotechnology*, 110(1):95–103, 2004.
9. Parekh, S., V. A. Vinci and R. J. Strobel. Improvement of microbial strains and fermentation processes. *Applied Microbiology and Biotechnology*, 54: 287–301, 2000.
10. Jeong, K. J. and S. Y. Lee. Excretion of human beta-endorphin into culture medium by using outer membrane protein F as a fusion partner in recombinant *Escherichia coli*. *Applied and Environmental Microbiology*, 68(10):4979–85, 2002.
11. Stephanopoulos, G. and J. J. Vallino. Network rigidity and metabolic engineering in metabolite overproduction. *Science*, 252(5013):1675–81, 1991.
12. Williams, K. L. Genomes and proteomes: towards a multidimensional view of biology. *Electrophoresis*, 20(4-5):678–88, 1999.
13. Ryu, D. D. and D.-H. Nam. Recent progress in biomolecular engineering. *Biotechnology Progress*, 16(1):2–16, 2000.
14. Ideker, T., V. Thorsson, J. A. Ranish, et al. Integrated genomic and proteomic analyses of a systematically perturbed metabolic network. *Science*, 292(5518):929–34, 2001.
15. Lee, K. H. Proteomics: a technology-driven and technology-limited discovery science. *Trends in Biotechnology*, 19(6):217–22, 2001.

16. Askenazi, M., E. M. Driggers, D. A. Holtzman, et al. Integrating transcriptional and metabolite profiles to direct the engineering of lovastatin-producing fungal strains. *Nature Biotechnology*, 21(2):150–6, 2003.
17. Stephanopoulos, G., H. Alper and J. Moxley. Exploiting biological complexity for strain improvement through systems biology. *Nature Biotechnology*, 22(10):1261–67, 2004.
18. Phair, R. D. and T. Misteli. Kinetic modelling approaches to in vivo imaging. *Nature Reviews Molecular Cell Biology*, 2(12):898–907, 2001.
19. Endy, D. and R. Brent. Modelling cellular behaviour. *Nature*, 18(409): 391–395, 2001.
20. Wiechert, W. Modeling and simulation: tools for metabolic engineering. *Journal of Biotechnology*, 94(1):37–63, 2002.
21. Blattner, F. R., G. Plunkett, C. A. Bloch, et al. The complete genome sequence of *Escherichia coli* K-12. *Science*, 277(5331):1453–62, 1997.
22. Hermann, T. Using functional genomics to improve productivity in the manufacture of industrial biochemicals. *Current Opinion in Biotechnology*, 15:444–8, 2004.
23. Patil, K. R., M. Åkesson and J. Nielsen. Use of genome-scale microbial models for metabolic engineering. *Current Opinion in Biotechnology*, 15(1): 64–6, 2004.
24. Kolisnychenko, V., G. Plunkett III, C. D. Gerring, et al. Engineering a reduced *Escherichia coli* genome. *Genome Research*, 12(4):640–7, 2002.
25. Ohnishi, J., S. Mitsuhashi., M. Hayashi, et al. A novel methodology employing *Corynebacterium glutamicum* genome information to generate a new L-lysine-producing mutant. *Applied Microbiology and Biotechnology*, 58(2):217–23, 2002.
26. Gonzalez, R., H. Tao, K. T. Shanmugam, et al. Global gene expression differences associated with changes in glycolytic flux and growth rate in *Escherichia coli* during the fermentation of glucose and xylose. *Biotechnology Progress*, 18(1):6–20, 2002.
27. Lockhart, D. J. and E. A. Winzeler. Genomics, gene expression and DNA arrays. *Nature*, 405(6788):827–36, 2000.
28. Mills, J. C., K. A. Roth, R. L. Cagan, et al. DNA microarrays and beyond: completing the journey from tissue to cell. *Nature Cell Biology*, 3(10):943, 2001
29. Choi, J. H., S. J. Lee, and S. Y. Lee. Enhanced production of insulin-like growth factor I fusion protein in *Escherichia coli* by coexpression of the down-regulated genes identified by transcriptome profiling. *Applied and Environmental Microbiology*, 69(8):4737–42, 2003.
30. Yoon, S. H., M. J. Han, S. Y. Lee, et al. Combined transcriptome and proteome analysis of *Escherichia coli* during the high cell density culture. *Biotechnology and Bioengineering*, 81(7):753–67, 2003.
31. Han, M. J, S. S. Yoon and S. Y. Lee. Proteome analysis of metabolically engineered *Escherichia coli* producing poly(3-hydroxybutyrate). *Journal of Bacteriology*, 183(1), 301–8, 2001.
32. Han, M. J., K. J. Jeong, J. S. Yoo, et al. Engineering *Escherichia coli* for increased production of serine-rich proteins based on proteome profiling. *Applied and Environmental Microbiology*, 69(10):5772–81, 2003.
33. Sauer, U. High-throughput phenomics: experimental methods for mapping fluxomes. *Current Opinion in Biotechnology*, 15(1):58–63, 2004.

34. Villas-Boas, S. G., J. F. Moxley, M. Akesson, et al. High-throughput metabolic state analysis: the missing link in integrated functional genomics of yeasts. *Biochemical Journal*, 388(Pt 2):669–77, 2005.
35. Sanford, K., G. Whited and G. Chotani. Genomics to fluxomics and physiomics—pathway engineering. *Current Opinion in Microbiology*, 5(3): 318–22, 2002.
36. Krömer, J. O., O. Sorgenfrei, K. Klopprogge, et al. In-depth profiling of lysine-producing *Corynebacterium glutamicum* by combined analysis of the transcriptome, metabolome, and fluxome. *Journal of Bacteriology*, 186(6):1769–84, 2004.
37. Lee, J. H., B. U. Lee and H. S. Kim. Global analyses of transcriptomes and proteomes of a parent strain and an L-threonine-overproducing mutant strain. *Journal of Bacteriology*, 185(18):5442–51, 2003.
38. Hong, S. H., J. S. Kim, S. Y. Lee, et al. The genome sequence of the capnophilic rumen bacterium *Mannheimia succiniciproducens*. *Nature Biotechnology*, 22(10):1275–81, 2004.
39. Lee, P.C., S. Y. Lee, S. H. Hong, et al. Isolation and characterization of a new succinic acid-producing bacterium, *Mannheimia succiniciproducens* MBEL55E, from bovine rumen. *Applied Microbiology and Biotechnology*, 58:663–8, 2002.
40. Lee, S. Y. and E. T. Papoutsakis. *Metabolic Engineering*. Marcel Dekker, New York, 1999.
41. Ishii, N., M. Robert, Y. Nakayama, et al. Toward large-scale modeling of the microbial cell for computer simulation. *Journal of Biotechnology*, 113:281–94, 2004.
42. Schomburg, I., A. Chang and D. Schomburg. BRENDA, enzyme data and metabolic information. *Nucleic Acids Research*, 30(1):47–9, 2002.
43. Sivakumaran, S., S. Hariharaputran, J. Mishra, et al. The Database of Quantitative Cellular Signaling: management and analysis of chemical kinetic models of signaling networks. *Bioinformatics*, 19(3):408–15, 2003.
44. Mendes, P. GEPASI: a software package for modeling the dynamics, steady states and control of biochemical and other systems. *Computer Applications in the Biosciences*, 9(5), 563–71, 1993.
45. Goryanin, I., T. C. Hodgman and E. Selkov. Mathematical simulation and analysis of cellular metabolism and regulation. *Bioinformatics*, 15(9): 749–58, 1999.
46. Tomita, M., K. Hashimoto, K. Takahashi, et al. E-Cell: software environment for whole-cell simulation. *Bioinformatics*, 15(1):72-84, 1999.
47. Sauro, H. M. SCAMP: a general-purpose simulator and metabolic control analysis program. *Computer Applications in the Biosciences*, 9(4):441–50, 1993.
48. Schaff, J., C. C. Fink, B. Slepchenko, et al. A general computational framework for modeling cellular structure and function. *Biophysical Journal*, 73(3): 1135–46, 1997.
49. Morton-Firth, C. J. and D. Bray. Predicting temporal fluctuations in an intracellular signalling pathway. *Journal of Theoretical Biology*, 192(1):117–28, 1998.
50. Kierzek, A. M. STOCKS: STOChastic Kinetic Simulations of biochemical systems with Gillespie algorithm. *Bioinformatics*, 18(3):470–81, 2002.
51. You, L, A. Hoonlor and J. Yin. Modeling biological systems using Dynetica-a simulator of dynamic networks. *Bioinformatics*, 19(3):435–6, 2003.

52. Teusink, B., J. Passarge, C. A. Reijenga, et al. Can yeast glycolysis be understood in terms of in vitro kinetics of the constituent enzymes? Testing biochemistry. *European Journal of Biochemistry*, 267(17):5313–29, 2000.
53. Clarke, B. L. Stoichiometric network analysis. *Cell Biophysics*, 12:237–53, 1988.
54. Seressiotis, A. and J. E. Bailey. MPS: an artificially intelligent software system for the analysis and synthesis of metabolic pathways. *Biotechnology and Bioengineering*, 31(6):587–602, 1988.
55. Mavrovouniotis, M. L., G. Stephanopoulos. and G. Stephanopoulos. Computer aided synthesis of biochemical pathways. *Biotechnology and Bioengineering*, 36(11):1119–32, 1990.
56. Liao, J. C., S. Y. Hou, and Y. P. Chao. Pathway analysis, engineering, and physiological considerations for redirecting central metabolism. *Biotechnology and Bioengineering*, 52(1):129–40, 1996.
57. Schilling, C.H., J. S. Edwards and B. Ø. Palsson. Toward metabolic phenomics: analysis of genomic data using flux balances. *Biotechnology Progress*, 15(3):288–95, 1999.
58. Schuster, S., D. A. Fell and T. Dandekar. A general definition of metabolic pathways useful for systematic organization and analysis of complex metabolic networks. *Nature Biotechnology*, 18(3):326–32, 2000.
59. Varma, A. and B. Ø. Palsson. Stoichiometric flux balance models quantitatively predict growth and metabolic by-product secretion in wild-type *Escherichia coli* W3110. *Applied and Environmental Microbiology*, 60(10):3 724–31, 1994.
60. Pramanik, J. and J. D. Keasling. Effect of *Escherichia coli* biomass composition on central metabolic fluxes predicted by a stoichiometric model. *Biotechnology and Bioengineering*, 60(2):230–8, 1998.
61. Wiback, S. J., R. Mahadevan and B. Ø. Palsson. Using metabolic flux data to further constrain the metabolic solution space and predict internal flux patterns: the *Escherichia coli* spectrum. *Biotechnology and Bioengineering*, 86(3):317–31, 2004.
62. Fischer, E. and U. Sauer. Metabolic flux profiling of *Escherichia coli* mutants in central carbon metabolism using GC-MS. *European Journal of Biochemistry*, 270(5):880–91, 2003.
63. Wittmann, C. and E. Heinzle. Modeling and experimental design for metabolic flux analysis of lysine-producing *Corynebacteria* by mass spectrometry. *Metabolic Engineering*, 3(2):173–91, 2001.
64. Schmidt, K., M. Carlsen, J. Nielsen, et al. Modeling isotopomer distributions in biochemical networks using isotopomer mapping matrices. *Biotechnology and Bioengineering*, 55(6):831–40, 1997.
65. Kelleher, J. K. Flux estimation using isotopic tracers. Common ground for metabolic physiology. *Metabolic Engineering*, 3(2):100–10, 2001.
66. Yang, C., Q. Hua and K. Shimizu. Metabolic flux analysis in *Synechocystis* using isotope distribution from ^{13}C-labeled glucose. *Metabolic Engineering*, 4(3):202–16, 2002.
67. Wiechert, W and A. A. de Graaf. In vivo stationary flux analysis by ^{13}C labeling experiments. *Advances in Biochemical Engineering/Biotechnology*, 54:109–54, 1996.
68. Wiechert, W., M. Mollney, N. Isermann, et al. Bidirectional reaction steps in metabolic networks. III. Explicit solution and analysis of isotopomer labeling systems. *Biotechnology and Bioengineering*, 66(2):69–85, 1999.

69. Zupke, C. and G. Stephanopoulos. Modeling of isotope distribution and intracellular fluxes in metabolic networks using atom mapping matrices. *Biotechnology Progress*, 10:489–98, 1994.
70. Christensen, B. and J. Nielsen. Isotopomer analysis using GC-MS. *Metabolic Engineering*, 1(4): 282–90, 1999.
71. Baxevanis, A. D. Using genomic databases for sequence-based biological discovery. *Molecular Medicine*, 9(9-12):185–92, 2003.
72. Stein, L. D. Integrating biological databases. *Nature Reviews Genetics*, 4:337–45, 2003.
73. Lemer, C., E. Antezana, F. Couche, et al. The aMAZE LightgBench: a web interface to a relational database of cellular processes. *Nucleic Acids Research*, 32:D443–8, 2004.
74. Philippi, S. Light-weight integration of molecular biological databases. *Bioinformatics*, 20(1):51–7, 2004.
75. Hou, B. K., J. S. Kim, J. H. Jun, et al. BioSilico: an integrated metabolic database system. *Bioinformatics*, 20(17):3270–2, 2004.
76. Lee, D.-Y., H. S. Yun, S. Y. Lee, et al. MetaFluxNet: the management of metabolic reaction information and quantitative metabolic flux analysis. *Bioinformatics*, 19(16):2144–6, 2003.
77. Klamt, S., J. Stelling, M. Ginkel, et al. FluxAnalyzer: exploring structure, pathways, and flux distributions in metabolic networks on interactive flux maps. *Bioinformatics*, 19(2):261–9, 2003.
78. Hucka, M., A. Finney, H. M. Sauro, et al. The systems biology markup language (SBML): a medium for representation and exchange of biochemical network models. *Bioinformatics*, 19(4):524–31, 2003.
79. Kitano, H. Computational systems biology. *Nature*, 420(6912):206–10, 2002.
80. Kitano, H. Systems biology: a brief overview. *Science*, 295(5560):1662–4, 2002.
81. Segre, D., J. Zucker, J. Katz, et al. From annotated genomes to metabolic flux models and kinetic parameter fitting. *OMICS*, 7(3):301–16, 2003.
82. Burgard, A.P. and C. D. Maranas. Probing the performance limits of the *Escherichia coli* metabolic network subject to gene additions or deletions. *Biotechnology and Bioengineering*, 74:364–75, 2001.
83. Segre, D., D. Vitkup and G. M. Church. Analysis of optimality in natural and perturbed metabolic networks. *Proceedings of the National Academy of Sciences USA*, 99(23):15112–17, 2002.
84. Burgard, A.P., P. Pharkya and C. D. Maranas. Optknock: a bilevel programming framework for identifying gene knockout strategies for microbial strain optimization. *Biotechnology and Bioengineering*, 84:647–57, 2003.
85. Edwards, J. S., R. U. Ibarra and B. Ø. Palsson. In silico predictions of *Escherichia coli* metabolic capabilities are consistent with experimental data. *Nature Biotechnology*, 19(2):125-30, 2001.
86. Levchenko, A. Computational cell biology in the post-genomic era. *Molecular Biology Reports*, 28(2):83–9, 2001.
87. Lee, S. Y. Plastic bacteria? Progress and prospects for polyhydroxyalkanoate production in bacteria. *Trends in Biotechnology*, 14: 431–8, 1996.
88. Madison, L. L. and G. W. Huisman. Metabolic engineering of poly (3-hydroxyalkanoates): from DNA to plastic. *Microbiology and Molecular Biology Reviews*, 63(1):21–53, 1999.

89. Choi, J., S. Y. Lee and K. Han. Cloning of the *Alcaligenes latus* polyhydroxyalkanoate biosynthesis genes and use of these genes for enhanced roduction of poly(3-hydroxybutyrate) in *Escherichia coli. Applied and Environmental Microbiology*, 64(12):4897–903, 1998.
90. Steinbüchel, A. and B. Fuchtenbusch. Bacterial and other biological systems for polyester production. *Trends in Biotechnology*, 16(10): 419–27, 1998.
91. Peoples, O. P. and A. J. Sinskey. Poly-beta-hydroxybutyrate biosynthesis in *Alcaligenes eutrophus* H16. Characterization of the genes encoding beta-ketothiolase and acetoacetyl-CoA reductase. *Journal of Biological Chemistry*, 264(26):15293–7, 1989.
92. Schubert, P., A. Steinb¸chel and H. G. Schlegel. Cloning of the *Alcaligenes eutrophus* genes for synthesis of poly-beta-hydroxybutyric acid (PHB) and synthesis of PHB in *Escherichia coli. Journal of Bacteriology*, 170(12): 5837–47, 1988.
93. Lee, S.Y., K. M. Lee, H. N. Chang and A. Steinbüchel. Comparison of *Escherichia coli* strains for synthesis and accumulation of poly(3-hydroxybutyric acid), and morphological changes. *Biotechnology and Bioengineering*, 44:1337–47, 1994.
94. Edwards, J. S. and B. Ø. Palsson. The *Escherichia coli* MG1655 in silico metabolic genotype: its definition, characteristics, and capabilities. *Proceedings of the National Academy of Sciences USA*, 97:5528–33, 2000.
95. Hong, S. H., S. J. Park, S. Y. Moon, et al. In silico prediction and validation of the importance of Entner-Doudoroff pathway in poly(3-hydroxybutyrate) production by metabolically engineered *Escherichia coli. Biotechnology and Bioengineering*, 83(7): 854–63, 2003.
96. van Wegen, R. J., S. Y. Lee and A. P. J. Middelberg. Metabolic and kinetic analysis of poly(3-hydroxybutyrate) production by recombinant *Escherichia coli. Biotechnology and Bioengineering*, 74(1):70-81, 2001.
97. Wong, H. H., R. J. van Wegen, J. Choi, et al. Metabolic analysis of poly(3-hydroxybutyrate) production by recombinant *Escherichia coli. Journal of Microbiology and Biotechnology*, 9(5):593–603, 1999.
98. Zeikus, J. G., M. K. Jain and P. Elankovan. Biotechnology of succinic acid production and markets for derived industrial products. *Applied Microbiology and Biotechnology*, 51:545–52, 1999.
99. Stols, L. and M. I. Donnelly. Production of succinic acid through overexpression of NAD(+)-dependent malic enzyme in an *Escherichia coli* mutant. *Applied and Environmental Microbiology*, 63(7):2695–701, 1997.
100. Jeong, H., B. Tombor, R. Albert, et al. The large-scale organization of metabolic networks. *Nature*, 407:651–4, 2000.
101. Sambrook, J., E. F. Fritsch, and T. Maniatis. *Molecular Cloning: A Laboratory Manual*, 2nd ed. Cold Spring Harbor Laboratory Press, Cold Spring Harbor, N.Y., 1989.
102. Park, S. J., W. S. Ahn, P. Green, et al. Biosynthesis of poly(3-hydroxybutyrate-co-3-hydroxyvalerate-co-3-hydroxyhexanoate) by metabolically engineered *Escherichia coli* strains. *Biotechnology and Bioengineering*, 74(1):81–6, 2001.

8

Genome-Scale Models of Metabolic and Regulatory Networks

Markus J. Herrgård & Bernhard Ø. Palsson

Mathematical models of biochemical networks have been built from the very early days of biochemistry and molecular biology [1]. However, these models have been focused on small subsystems that are thought to represent an independent module in the whole biochemical network of the cell. This subsystem-based approach was necessitated by the limited information available on the entire set of proteins that are present in a particular cell. The availability of genome sequences has allowed the development of systematic approaches to establish functions for unannotated genes and regulatory regions [2–5]. The resulting list of well-characterized coding and regulatory regions in a genome has further enabled the reconstruction of the structure of metabolic and regulatory networks at the genome scale. In this chapter we describe the process of reconstructing genome-scale metabolic and regulatory networks, provide an overview of methods that have been developed to study these reconstructions in silico, and describe applications of existing network reconstructions. Network-level steady-state assumption has been used to derive a number of powerful mathematical analysis tools that can be used to elucidate network function in silico. These tools are based on imposing a succession of constraints on the cellular function giving rise to the term "constraint-based analysis" to describe the overall framework [6–8].

While the basic chemistry underlying metabolic and regulatory networks is the same, the different nature of the molecular components makes the reconstruction and analysis of these two network types quite different in practice. We will first discuss the fairly mature field of genome-scale metabolic network reconstruction and analysis. Since metabolic network reconstruction can be done routinely based on database and literature information, the focus of this section is on describing existing genome-scale models and their uses in biological discovery. The other major aim of the metabolic network section is to give relevant background on the in silico constraint-based analysis methods that can be used to study the properties of metabolic network reconstructions and make experimentally testable predictions using network models. In the second part of this chapter the reconstruction and analysis of transcriptional regulatory networks is reviewed. The focus of this section will be on illuminating the differences between the reconstruction and analysis of metabolic and regulatory networks [9]. We will conclude with discussion of methods that have been developed to analyze the integrated function of metabolic and regulatory networks at the genome scale.

METABOLIC NETWORKS

Reconstruction of Metabolic Networks

The publication of the first prokaryotic genomic sequence [2] signaled the beginning of a new era, not only in experimental biology, but also in modeling and analysis biochemical reaction networks. For the first time it was possible to define exactly the genetic makeup of an organism and hence, in principle, all the biochemical reactions that can take place in a particular cell. Metabolic physiology has been extensively studied and the reaction stoichiometry of most commonly occurring metabolic reactions has been well established [10]. All these factors together allow using comparative genomics to define potential enzymatic functions for genes and build pathways and networks by connecting individual enzymatic conversions [11]. In order to assign enzymatic functions by comparative approaches, a similar function has to have been characterized for a gene product in another organism. In the case of prokaryotes, the most common model organism for metabolic physiology is the gram-negative bacterium *Escherichia coli* whereas for eukaryotes the equivalent position is held by the yeast *Saccharomyces cerevisiae*.

Metabolic network reconstruction benefits enormously from the availability of well-curated and comprehensive databases such as KEGG [12] and EcoCyc [13] that store information about individual metabolic reactions and whole metabolic pathways. The information in these databases is usually derived from both manual curation based on primary literature and automated sequence-based annotation. These databases form the basis for reconstructing more detailed models of metabolic networks, by providing comprehensive summaries of the known and hypothetical metabolic reactions present in an organism. However, a network reconstructed on the basis of database information alone would not function as a predictive model for a number of reasons.

Despite the best efforts of database curators, many of the networks contain gaps due to lack of direct biochemical information for a particular organism and extra reactions due to erroneous sequence homology-based function assignments. These types of errors can often be partially fixed by analysis of the network structure and by using the network as a basis for simulations as described later in this chapter. For example, the known physiology of an organism may indicate that it is capable of producing a particular amino acid, but steps in the biosynthetic pathway for this amino acid are missing. In this case, reference pathways from other organisms can be used to establish potential missing reactions and directed homology searches can be used to find gene associations for these reactions [6,14]. Similarly, often the transport mechanisms that are used to transport metabolites in and out of the cell are not known, and, based on known physiology, the ability to transport metabolites may have to be assumed without known transport facilitators. This type of careful analysis relies largely on books and reviews written on the genetics, biochemistry, molecular biology, and physiology of particular organisms such as *E. coli* or yeast [15,16].

In order to enable mathematical analysis of network function, the reactions in the network also need to be mass balanced [14]. This requires estimating the ionization states of metabolites at a particular pH and including protons and water molecules as part of the reaction formulas. A further complication that arises in metabolic network reconstruction for eukaryotic organisms is assigning the correct intracellular compartments for metabolites and reactions in the network. This type of information is not typically available in metabolic network databases and has to be collected either from studies of individual gene product localizations published in the literature or from high-throughput protein localization screen data [17,18]. A genome-scale metabolic network reconstruction is basically defined by the components listed above: the correctly balanced stoichiometry of each individual metabolic reaction including transport reactions, and the subcellular localization assignment of each metabolic reaction in the network. While this information already allows analysis of network structure and capabilities, further details are needed to be defined in order to connect reactions in the network to the genes and proteins in an organism. These connections are established through gene–protein–reaction associations [14,19] that describe, for example, how two alternative isozymes can catalyze the same reaction or how two components of a complex come together to catalyze one reaction.

Genome-scale metabolic network reconstructions have been developed for a number of organisms, including *H. influenzae* [20], *H. pylori* [21], *S. aureus* [22], *E. coli* [14], *S. cerevisiae* [19,23], *S. coelicolor* [125], and the human mitochondrion [24]. Table 8.1 summarizes the properties of these reconstructions including the numbers of genes, metabolites, and reactions. In the following we will focus primarily on the work done with the *E. coli* [14,25] and *S. cerevisiae* metabolic network models [19,23], since these models have been most extensively tested experimentally.

Table 8.1 Examples of existing genome-scale metabolic network reconstructions

Organism	Number of genes	Number of metabolites	Number of reactions	Reference
Escherichia coli	904	625	931	[14]
Saccharomyces cerevisiae	750	646	1149	[19]
Staphylococcus aureus	619	571	640	[22]
Haemophilus influenzae	362	343	488	[20]
Helicobacter pylori	268	340	444	[21]
Human mitochondria	N/A	226	186	[24]
Streptomyces coelicolor	711	500	971	[125]

Constraints on Metabolic Network Function

In order to study the properties of reconstructed genome-scale metabolic networks, mathematical modeling tools have to be applied to the reconstructions. The constraint-based analysis framework approaches this modeling task by imposing a series of constraints restricting allowable metabolic flux space (figure 8.1). For the purpose of constraint-based analysis of metabolic networks, the reconstructed network structure is represented in the form of a stoichiometric matrix S. This matrix has M rows with each row corresponding to a metabolite in the network and N columns with each column corresponding to a reaction in the network. The nonzero elements in each column are the stoichiometric coefficients of the metabolites participating in a particular reaction with negative elements corresponding to substrates of the reaction and positive elements to the products of the reaction. In addition to representing all the reactions internal to the metabolic network, including transport reactions, the stoichiometric matrix also

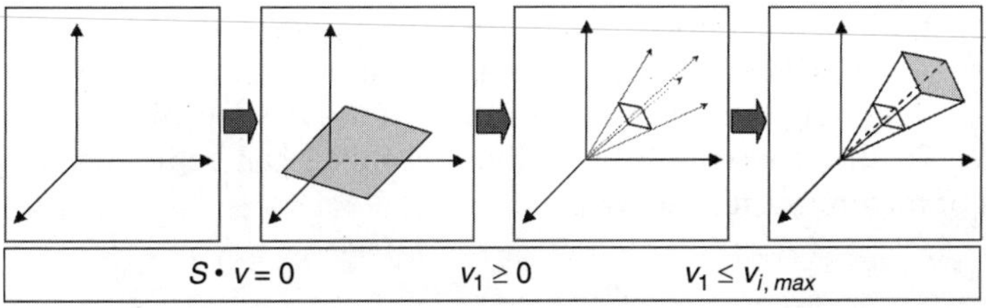

Figure 8.1 Constraint-based analysis framework applied to metabolic network reconstructions. Imposing a succession of physicochemical constraints on the metabolic network limits the allowable phenotypic space. Here only stoichiometric, reaction directionality, and maximum reaction rate constraints are illustrated. Further constraints due to, for example, transcriptional regulation can also be imposed on the network to further limit the allowable phenotypic space.

includes exchange reactions that allow the exchange of metabolites (either substrates or byproducts) with the environment.

Given the stoichiometric matrix S for a metabolic network, one could write a system of ordinary differential equations to describe the time evolution of the metabolite concentrations X of the system:

$$\frac{dX}{dt} = Sv(X) \tag{1}$$

where $v = v(X)$ are the metabolic fluxes through each reaction. Note that eq. (1) assumes that all the metabolites are in a single well-mixed compartment in the cell. Since metabolic reactions have relatively short characteristic time scales, for most purposes the equation above can be simplified by assuming that the system is in a steady state, that is,

$$Sv = 0 \tag{2}$$

This equation represents the conservation of mass in the metabolic network under steady-state conditions, that is, the production rate of each metabolite equals its consumption rate. Equation (2) defines all the flux distributions allowed by the stoichiometry of the entire reaction network under steady-state conditions and hence represents a constraint on the allowable range of functionalities of the metabolic network.

Reaction stoichiometry is the most fundamental of the constraints acting on the overall functionality of the metabolic network, but other constraints are also commonly imposed on the system in constraint-based analysis. By separating reversible reactions into separate forward and backward reactions, all fluxes are restricted to be nonnegative:

$$v_i \geq 0 \tag{3}$$

Directionality constraints that restrict the flux through a reaction to proceed in one direction due to thermodynamic reasons can be enforced by only retaining the reaction operating in the desired direction in the model. Further constraints on individual fluxes can be imposed in the form of maximum reaction rates:

$$v_i \leq v_i^{\max} \tag{4}$$

Typically maximal reaction rates are not known for intracellular reactions, but upper limits for these rates can be estimated based on enzyme concentrations in a cell. For exchange reactions, on the other hand, the maximal rates are usually at least partially known, since these rates can be experimentally measured as a function of substrate concentrations [26]. Alternatively, one can measure condition-dependent substrate uptake rates and byproduct secretion rates and use these values as the maximum allowed rates in constraint-based analysis [27].

Further constraints, including restrictions on allowed flux distributions due to free energy balancing in loops in metabolic networks, can be applied to further reduce the size of the solution space [28–30]. Additionally, regulatory constraints due to transcriptional regulation acting on metabolic enzyme expression can be imposed on the network resulting in time- and condition-dependent changes in the solution space. These regulatory constraints will be discussed in more detail in the section focusing on regulatory networks below.

Constraints (2)–(4) together with any additional constraints form the basis for the analysis of allowable flux distributions in a reconstructed metabolic network. A number of different mathematical methods described below can be applied to analyzing these allowable flux distributions (figure 8.2). The methods can be roughly classified into two categories: (1) methods that assume an objective for the network (e.g., biomass production) and result in prediction of a single optimal flux distribution or a set of optimal distributions; and (2) methods that do not assume a particular objective and aim to characterize all the allowed flux distributions. Methods in the former class are computationally more efficient, and result in experimentally verifiable predictions, whereas the methods in the latter class tend to be more computationally demanding, and can be used to analyze general network properties.

Finding Optimal Network States

Flux balance analysis. The most commonly used constraint-based analysis method is flux balance analysis (FBA), which allows identifying a particular point (corresponding to a particular flux distribution in the network) in the subspace defined by constraints (2)–(4), based on optimizing a specific objective function. The objective usually assumed for metabolic physiology of microbial cells is cellular growth, which can be mathematically represented by the biomass composition c of the cell. In vector c the production flux for each biomass component is represented by a coefficient equivalent to the experimentally measured fraction of the cell biomass. While the biomass composition varies depending on growth conditions, for practical purposes a constant biomass composition is usually assumed. In addition to the biomass components, this objective function typically also includes the ATP maintenance requirements of the cell for cellular processes outside metabolism. Given a biomass composition represented by the nonzero components of the vector c, we can predict flux distributions using FBA as the solution of the following linear program:

$$\max c^T v \qquad \text{subject to } Sv = 0,\ 0 \le v_i \le v_i^{\max} \tag{5}$$

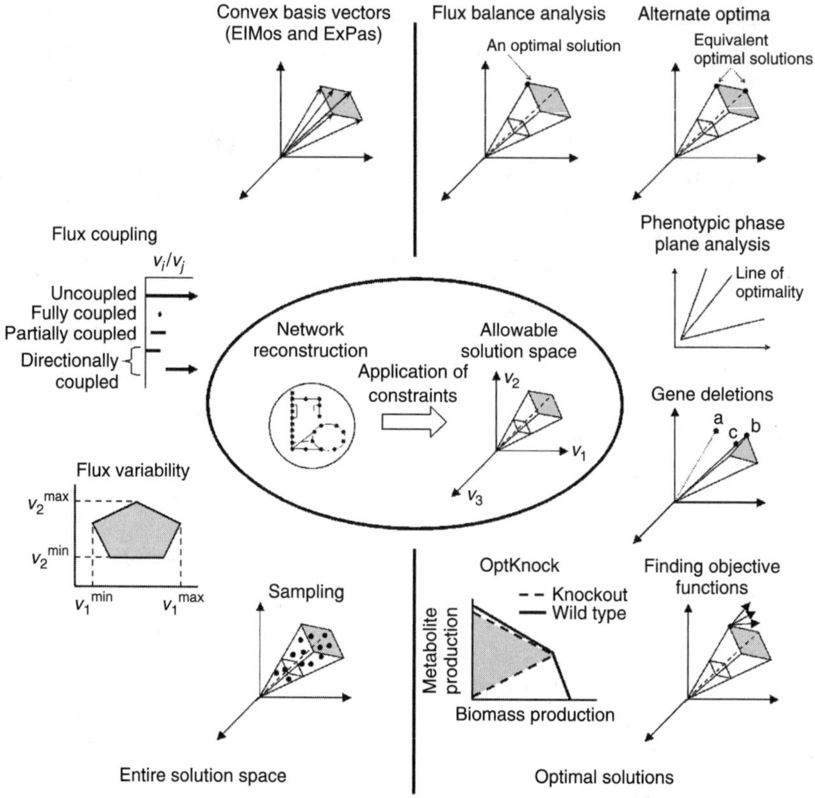

Figure 8.2 Commonly utilized constraint-based analysis methods. A number of constraint-based analysis methods can be applied to studying the solution spaces defined by imposing physicochemical and other constraints on metabolic network function. The methods illustrated here can be roughly classified to ones that study optimal solutions within the solution space and those that analyze the entire solution space. See the main text for a more detailed description of the different methods.

where the environmental conditions chosen for a particular simulation are set by the maximal flux constraints for exchange fluxes. The maximal value of the objective function $\mu = \max c^T v$ corresponds to the growth rate predicted by the FBA model under particular environmental conditions. The comparison of these growth rates and other predictions by FBA models to experimental data will be discussed in greater detail later in this chapter.

Alternative optimal solutions. Because of redundancies in the metabolic reaction network there are usually multiple alternative optimal solutions to the optimization problem (4), so that FBA does not necessarily predict a single optimal solution. Methods for enumerating alternative optimal solutions, and for determining the allowed ranges of fluxes for optimal solutions, have been developed and applied to genome-scale models [31–33]. These approaches are especially useful when flux predictions are compared to experimental data on intracellular or exchange fluxes.

Sensitivity analysis. In addition to obtaining an optimal flux distributions or a set of optimal distributions, one may also be interested in analyzing the effects of changes in parameters of the model on the optimal solutions (figure 8.3). The simplest question of this type is how changes in a single maximal reaction rate affect the predicted growth rate, for example in the case of carbon source uptake [34]. Because of the linear programming structure of problem (5), the growth rate μ is a piecewise linear function of each of the $v_i^{\max}$ parameters, and the slope of each of the linear regions corresponds to a

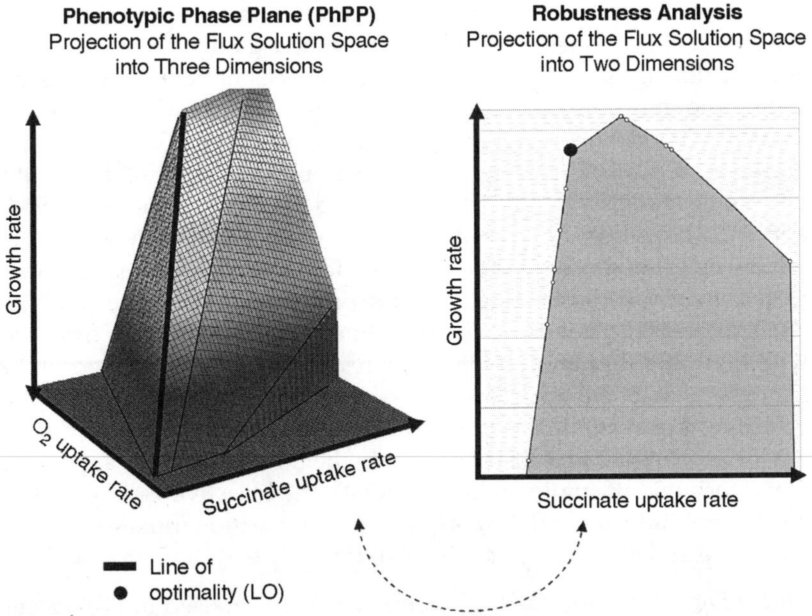

Figure 8.3 Sensitivity analysis of optimal solutions. Phenotype phase plane analysis allows studying the effects of varying two parameters (here oxygen and succinate uptake rates) on the predicted optimal growth (or biomass production) rate. The line of optimality shows the conditions that result in the optimal biomass yield (biomass production rate/substrate uptake rate). Robustness analysis of the optimal growth rate to changes in a single parameter is shown on the right. The point corresponding to the optimal yield is shown in the graph.

reduced cost of the linear programming problem [35]. Similar sensitivity analysis can also be performed using two variable parameters simultaneously, for example, the maximal glucose and oxygen uptake rates. This approach, called phenotype phase plane analysis (PhPP), allows subdividing the two-dimensional space spanned by the two parameters into linearly constrained regions, based on the shadow price structure of the corresponding linear programming problem. Each of these regions or phases corresponds to a particular mode of operation of the metabolic network with qualitatively similar flux patterns within the phases. In the PhPP analysis there is typically a particular line separating two phases (line of optimality) that corresponds to the maximum biomass yield at all parameter values. As described below, the PhPP approach has proven to be useful for analyzing experimental data as well as designing experiments [36,37].

Assessing effects of genetic changes. Growth phenotyping studies of gene deletion strains are a prominent source of data for metabolic reconstruction validation and expansion, and hence the ability to predict phenotypes of genetically modified strains is crucial to the utilization of genome-scale models. The standard FBA approach can be applied directly to predicting the consequences of gene deletions by setting the maximum and minimum fluxes through the reactions catalyzed by the deleted gene to zero and repeating the FBA calculation. However, it is unclear whether the assumption of optimality should hold for genetically modified strains, and variants of the basic FBA approach have been proposed to predict deletion phenotypes. In the minimization of metabolic adjustment (MoMA) approach it is assumed that, instead of maximizing biomass production, the gene deletion strain attempts to minimize the adjustment of cellular fluxes from the wild-type flux distribution [38]. This results in an optimization problem where the linear objective in problem (5) is replaced by a quadratic objective representing the Euclidean distance between the wild-type and deletion strain flux distributions. In the third alternative approach, regulatory on/off minimization (ROOM), the objective is assumed to be the minimization of on/off changes of reaction fluxes between wild-type and genetically modified strains [39].

Evaluating cellular objectives. The objective function used in FBA calculations is an approximation for the true cellular objective. It is likely that biological systems are simultaneously trying to satisfy multiple objectives, and that the objective coefficients c used in FBA do not represent all these condition-dependent objectives. However, we can investigate how good a particular objective function is by asking what objective functions could give rise to an experimentally measured flux distribution. This problem is essentially the inverse of the FBA problem (5), since we are now given a flux distribution v and try to find an objective function c that would result in v as the optimal solution of (5). An optimization approach developed for this inverse task [40] has been applied to analyze objective functions based on experimentally measured aerobic and anaerobic flux distributions in *E. coli*. The objective functions defined through this approach were found to be similar to the assumed biomass composition and were not found to vary significantly between the aerobic and anaerobic conditions. These results indicate that, at least for the *E. coli* model, the objective function assumed previously was consistent with the flux measurements.

Analyzing Allowable Phenotypes

While FBA and its variations described above are the most common approaches utilized in constraint-based analysis of metabolic networks, there are many other mathematical methods that can be applied to studying the properties of genome-scale stoichiometric matrices and making experimentally verifiable predictions based on these matrices [8,41]. The linear constraints (2)–(4) define a closed convex subspace representing all the allowed flux distributions, optimal or not, under particular conditions, and a number of methods have been developed to analyze this whole space of flux distributions. It should be emphasized that, in contrast to the optimization-based approaches described above, the methods discussed here do not require specifying a particular objective function and do not predict a particular flux distribution. In this sense the methods described here are more objective than the optimization-based approaches and can be used to analyze the relationship of metabolic network topology to its function.

Network-based pathway analysis. Extreme pathway and elementary mode analysis are based on enumerating basis vectors for the convex subspace defined by constraints (2) and (3) [42–44]. Biologically, these basis vectors correspond to network-based definitions of pathways whose combinations can be used to represent an arbitrary flux distribution within the allowed space of flux distributions. Extreme pathways correspond directly to the edges of the open convex solution space defined by constraints (2) and (3), and hence are the smallest set of pathways that represent the overall functionality of the metabolic network. Elementary flux modes are a superset of extreme pathways, where each elementary mode corresponds to a flux distribution through a minimal set of reactions that can operate at steady state. The pathway enumeration methods are computationally extremely intensive because the number of basis vectors scales exponentially with the number of reactions in the network. For this reason, these enumeration methods have been limited to smaller metabolic networks or to cases where the numbers of inputs and outputs are limited.

Flux coupling analysis. Recently, other more scalable methods have been developed to analyze flux subspaces defined by constraints (2) and (3) based on optimization techniques [45] or random sampling of the solution space [46–48]. The flux coupling finder (FCF) [45] is an optimization-based approach for finding general dependences between pairs of fluxes in the metabolic network under steady-state conditions without assuming a particular objective for the network. The FCF approach enumerates coupled reaction subsets and directional couplings between reactions. Coupled subsets are sets of reactions that always have to be co-utilized because of stoichiometric constraints, and in most cases the fluxes through the reactions are also constrained to fixed ratios with respect to each other. Directional couplings are unidirectional relationships between pairs of fluxes in the network that represent the requirement for a nonzero flux through one reaction for the functioning of another reaction, but not vice versa. The FCF method is scalable to genome-scale metabolic networks, because it only requires solving a fixed number of linear optimization problems.

Random sampling. Random sampling approaches represent perhaps the most unbiased way to analyze the overall properties of the flux spaces defined by constraints (2)–(4) as well as the effect of imposing further constraints on the network [46–48]. While uniform sampling of points (each corresponding to an allowed flux distribution) within the closed convex set defined by (2)–(4) is in principle straightforward, in practice efficient algorithms are required to ensure that the sampling converges to a uniform distribution in a reasonable time. The standard approach for sampling convex spaces, the hit-and-run algorithm [49], is based on uniformly randomly selecting a direction from a random starting point within the space and moving a random distance (drawn from a uniform distribution) along this direction so that the resulting point is still within the constraints. This process is then repeated until the sampling distribution has converged to a stationary uniform distribution. Extensions to this basic method have been proposed that potentially allow faster convergence of the sampler [50]. The output of the sampling methods is a large set of flux distributions whose characteristics can be analyzed using statistical multivariate analysis tools.

The methods described above are all based on the stoichiometric matrix as opposed to using a graph representation of the metabolic network. It should be emphasized that representing the connectivity of the metabolic network in the form of a stoichiometric matrix allows automatically accounting for the mass balancing of the entire metabolic reaction network. This is in contrast to most graph-based methods [51,52], which usually fail to account for the correct cofactor connectivity and balancing in these networks. Since metabolism is fundamentally driven by these cofactors such as ATP, NADH, and NADPH, the failure to account for the balancing of these factors severely limits the power of graph theoretic analysis of metabolic networks [53].

Applications of Genome-Scale Metabolic Models

The methods described above can be applied to genome-scale metabolic network reconstructions to obtain experimentally verifiable predictions and to analyze general network properties. In the following we will describe five major application areas for constraint-based analysis of genome-scale metabolic network reconstructions.

OPTIMALITY AND ADAPTABILITY OF METABOLIC NETWORKS

Constraint-based metabolic models can predict quantitatively growth rates, substrate uptake rates, and byproduct secretion rates under defined environmental conditions. Using an *E. coli* genome-scale metabolic model [25], it was shown that for some carbon sources including acetate and succinate the model predictions of these physiological parameters agreed very closely with experimental measurements of the same parameters for wild-type *E. coli* strain MG1655 grown in batch cultures [27]. By varying the sugar and oxygen uptake rates experimentally it was shown that wild-type *E. coli* in general grows on the line of optimality in the phenotype phase planes predicted by the model. These results show that the usage of the *E. coli* metabolic network is optimized in vivo to maximize biomass production under the experimental conditions considered.

For other carbon sources such as glycerol and lactate the model significantly overpredicted growth rates compared to the experimental data [36]. Since the FBA predictions are based on the assumption of optimality under stoichiometric and maximum uptake constraints, it seemed likely that for some reason the experimental system was not operating optimally. Indeed, it was shown [36] that by experimentally evolving wild-type *E. coli* strains on glycerol for 700 generations, it was possible to reach the experimentally predicted growth rates. Furthermore, after reaching the line of optimality the evolved strain continued growing on this line. These results show that experimental evolution can result in in vivo optimization of cellular physiology, and that the endpoint of evolution can be predicted based on fundamental physicochemical constraints acting on the metabolic network. Similar results have in general been obtained for other carbon sources including lactate [54].

In addition to studying wild-type strains, the adaptive evolution of *E. coli* knockout strains has been investigated in the context of genome-scale metabolic models. Six different metabolic gene knockout strains were evolved on four different carbon sources each for 25–50 days until stable growth rates were established [55]. For the starting point strains, the in silico model generally predicted up to three times higher growth rates than were experimentally observed. After experimental evolution, which resulted in growth rate increase for all strains, 39 of the 50 cases tested (78%) reached the in silico predicted growth rates (figure 8.4). For the majority of the remaining cases the experimental evolution reduced the discrepancy between experimental data and in silico predictions, but the in silico predictions were still in most cases higher than the experimentally measured growth rates.

Based on the results discussed above, it appears that the FBA predictions of at least exchange fluxes match those obtained for experimentally evolved strains, and in some cases even for nonevolved strains. This indicates that, in general, experimental evolution under constant evolutionary pressure results in removal of any regulatory and other constraints so that only the basic stoichiometric and maximum nutrient uptake constraints remain. However, even after experimental evolution there are cases where the model either over- or underpredicts compared to the experimental data. Such discrepancies between model predictions and in vivo behavior may indicate either missing reactions or additional constraints that are still active in evolved strains. These constraints may, for example, be due to kinetic or transcriptional regulation and identifying them will allow developing improved constraint-based modeling strategies.

GENOME-SCALE ANALYSIS OF DELETION PHENOTYPES

In addition to analyzing in detail the physiology of evolved and nonevolved strains, as described above, genome-scale metabolic models have been widely used for qualitative prediction of growth phenotypes for large nonevolved knockout strain collections. Although, based on the data for evolved *E. coli* knockout strains, it is known that FBA generally overpredicts growth rates for nonevolved knockout strains, it should still be able to predict qualitative growth phenotypes (e.g., slow/normal growth) correctly. Large-scale knockout studies have been performed in *E. coli* [56,57], *S. cerevisiae* [19,58], *H. pylori* [21], and *H. influenzae* [59]. Overall the FBA predictions of gene deletion phenotypes have been accurate in 60–90% of the cases studied, depending on the completeness of the metabolic network for each organism. While the correct prediction rates have been quite high for all the organisms studied so far, the interesting part of the deletion

---

Figure 8.4 Adaptive evolution of *E. coli* metabolic gene knockout strains on specific carbon sources. The *x*-axis corresponds to days of experimental evolution (up to 50 days). The *y*-axis shows the strain growth rate measured as a function of days of evolution together with the in silico predicted growth rate (horizontal line). Data for at least two independently evolved strains is shown in each panel. For more details see ref. [55].

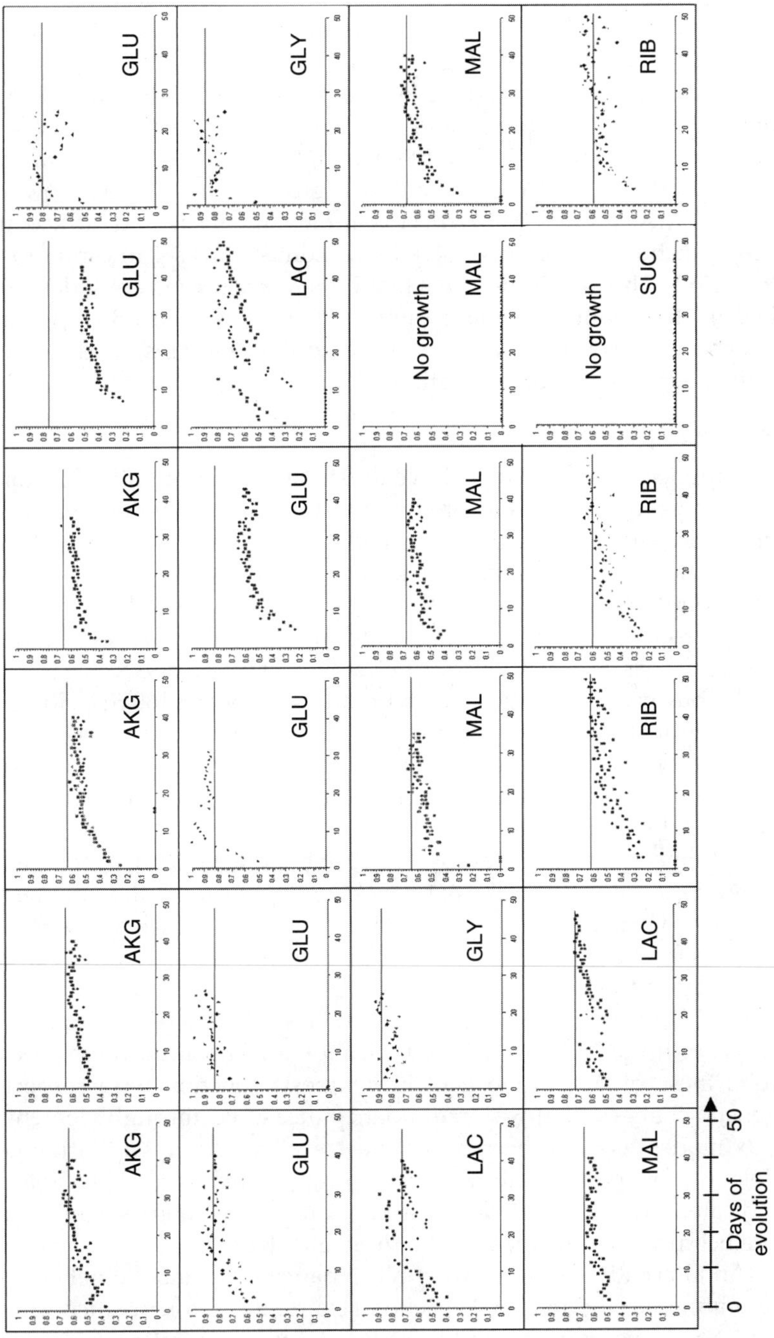

ack
frd
pck
ppc
tpi
zwf
AKG
GLU
LAC
MAL
GLY
RIB
SUC
No growth
Days of evolution
0
50

studies described above are the incorrect predictions. Detailed analysis of incorrect predictions allows identifying potential changes to the in silico models and in general increasing our understanding of the metabolic physiology of the organism [19].

In addition to the standard FBA approach, the MoMA and ROOM methods described above have also been used to perform large-scale gene deletion studies [38,39]. In general, these methods perform similarly to FBA in predicting qualitative growth phenotypes. However, both MoMA and ROOM predict experimentally measured intracellular flux distributions for nonevolved knockout strains significantly better than FBA. This result indicates that in vivo organisms are indeed limited in their ability to adapt to genetic changes. The FBA approach, which assumes full optimality for the knockout strains, tends to overestimate the degree of adaptation allowed.

ROBUSTNESS AND REDUNDANCY

The applications described above illustrate only the ability of genome-scale metabolic models to predict directly experimentally measurable quantities such as growth rates. In addition to these direct predictions, metabolic reconstructions can also be studied using, for example, flux coupling or extreme pathway analysis. The results from these analyses are primarily intended to address global network-level questions such as the relationship between metabolic network structure and robustness to genetic and environmental changes [60,61]. However, the insights obtained in this way on the global organization of metabolic networks can be used to interpret indirectly experimental data sets such as synthetic lethality data [62,63] and to derive experimentally testable hypotheses.

Extreme pathway analysis has been used to study the built-in redundancy in metabolic networks in the form of the number of distinct routes a metabolic network can use to produce a given set of outputs from a given set of inputs. For example, comparison of extreme pathway redundancy for amino acid biosynthesis between *H. pylori* and *H. influenzae* indicated a much smaller degree of redundancy in *H. pylori* [61,64]. This decreased redundancy has implications for the robustness of the amino acid biosynthesis function with respect to genetic changes. Pathway redundancy measured through elementary flux modes has also been shown to correlate with quantitative gene deletion phenotypes in *E. coli* central metabolism. This analysis indicated that the pathway-based redundancy metrics can indeed predict in vivo robustness and redundancy [60].

An alternative way to study redundancy in metabolic networks is to enumerate the minimal reaction sets required for in silico growth under various growth conditions [65]. This enumeration task can be reformulated as a mixed-integer optimization problem where the number of reactions that are allowed to carry flux is minimized while still requiring that a reasonable level of biomass production is sustained. Upon application of this method to a genome-scale *E. coli* metabolic network it was found that only 31% and 17% of the reactions in the model were required to support growth on glucose minimal and rich medium, respectively. This high degree of redundancy implies robustness of the metabolic network function against both environmental and genetic changes.

Genome-scale models can also be used to obtain insights into the experimentally observed apparent nonessentiality of a large fraction of genes in many organisms under standard laboratory conditions [66,67]. In a recent study, FBA predictions of metabolic gene deletion strain growth rates under multiple media conditions in yeast were used to identify reasons for the observed in vivo dispensability [68]. The analysis indicated that the dominant explanation for apparent dispensability were condition-specific roles for metabolic genes. Much fewer dispensable genes were buffered by a duplicate gene capable of acting as an isozyme and only a small fraction of genes were actually buffered by metabolic network flux reorganization. These results indicate that while the metabolic network functionality is in general robust to most gene deletions, the majority of metabolic genes are still necessary to allow growth in a wide range of nutritional environments.

GLOBAL ORGANIZATION OF METABOLIC NETWORKS

In addition to the robustness and redundancy questions discussed above, genome-scale models have also been used to analyze more generally the functional organization of metabolic networks. The most common type of analysis of this type is the identification of functional modules in biological networks. The modularization of biochemical networks has been identified as one of the central goals for analyzing the behavior and function of these networks [69]. Correlated reaction sets (Co-Sets) contain reactions that are functionally connected in all possible functional states of the network under a particular environmental condition [70]. The difference between modules defined using analysis of the stoichiometric matrix and modules defined, for example, using graph theoretic methods is that the former definition accounts correctly for all the functional connections implicit in the network whereas the latter definitions may ignore important connections due to, for instance, cofactor balancing.

Pathway-based methods can be used to enumerate Co-Sets by identifying reactions that are co-utilized in all extreme pathways or elementary flux modes. Co-Sets were calculated with extreme pathway analysis of the metabolic networks of *H. pylori* and *H. influenzae* [61,64]. Nonobvious sets were identified, including reactions involved in amino acid and phospholipid metabolism coupled with reactions involved in nucleotide metabolism. Thus, Co-Sets in metabolic networks identify functional and not necessarily intuitive groups of reactions. Co-Sets in *E. coli* for growth on a variety of different carbon sources were also enumerated from alternate optimal solution calculations [32].

The flux coupling finder method can be directly used to define Co-Sets that constitute functional modules in the network [45]. The percentage of reactions in such Co-Sets under aerobic glucose minimal medium growth conditions was found to be ~60% for *H. pylori*, ~30% for *E. coli*, and ~20% for *S. cerevisiae*. The high fraction of coupled reactions in *H. pylori* provides another indication of the low degree of flexibility that the metabolic network of this pathogen has. The directional couplings obtained from flux coupling analysis can be used to establish the core set of reactions required for biomass formation in all possible flux distributions within the allowed flux space for a particular media condition. In *H. pylori*, 59% of the reactions in the model were required for biomass formation whereas for *E. coli* and yeast these percentages were 28% and 14% respectively.

The flux distributions obtained by random sampling methods described above can also be used to establish functional modules in metabolic networks [47,48]. By calculating pairwise correlation coefficients between all reaction fluxes based on the sampled uniform flux distributions, the degree of co-utilization of fluxes can be quantified. Sampling can also be used to establish other types of general features of metabolic network function such as the distribution of flux magnitudes allowed through each reaction in the metabolic network. This type of analysis identified a high-flux reaction backbone in the genome-scale *E. coli* metabolic network that carries the vast majority of the metabolic flux through the network [46]. This backbone remains relatively intact under a variety of environmental conditions, whereas the remaining low-flux pathways are used differently in different conditions.

STRAIN DESIGN AND METABOLIC ENGINEERING

The high degree of functional connectivity in metabolic networks implied by the results described in the previous sections also poses a challenge for designing strains for overproduction of desired metabolic byproducts. Accounting for all the cofactor balancing requirements, as well as cellular growth requirements, is essential for determining the minimal genetic modifications needed to increase desired byproduct secretion compared to a wild-type strain. Genome-scale models provide an attractive starting point for strain design, because they automatically include all the necessary requirements as comprehensively as is possible based on our current understanding of the metabolic physiology of a given organism.

A number of different constraint-based analysis approaches have been used for strain engineering. Elementary flux mode calculations were applied to a stoichiometric model of a recombinant yeast strain [71] in order to analyze effects of knockouts and altered culturing conditions on poly-β-hydroxybutyrate (PBH) yield. However, the application of the pathway-based approach is hindered by the computational complexity of enumerating elementary flux modes for genome-scale models as well as the need to guess ahead of time which genetic modifications should be tested. The MoMA approach for simulating flux changes in response to gene deletions in a genome-scale metabolic model [38] was used to identify single and multiple gene deletions that would improve lycopene yield in *E. coli* [72]. When the model-based strain design strategy was combined with a transposon insertion-based combinatorial design approach [73], significant further improvements in lycopene production were achieved.

To complement the model-based metabolic engineering approaches described above, an efficient in silico method for automatically designing genetically modified strains for metabolite overproduction has been introduced [74,75]. This method, named OptKnock, attempts to identify the best possible gene deletions to couple the production of a desired metabolite to biomass production. The idea is that when a strain designed by OptKnock is evolved experimentally under a suitable selection pressure, increased metabolite production would result as a side effect of the growth rate increase. The OptKnock approach uses bilevel optimization where the inner problem is the standard FBA problem (5) and the outer level is a combinatorial optimization problem over all possible metabolic network structures with the objective of maximizing the secretion of a particular metabolite. OptKnock has been used to identify nonobvious multiple gene knockout strategies for the production of intermediate metabolites (succinate and lactate) as well as downstream metabolites (1,3-propanediol, chorismate, alanine, serine, aspartate, and glutamate) [74,76]. The same basic approach used in OptKnock has also been extended to designing optimal gene addition strategies and media compositions for metabolite overproduction [75]. Initial experimental validation of designed lactate production strains has shown that the OptKnock approach has great promise in designing strains for metabolic engineering [126].

TRANSCRIPTIONAL REGULATORY NETWORKS

Reconstruction of Regulatory Networks

Transcriptional regulatory networks play a key role in cellular response to different environments. Recent developments in experimental techniques have allowed the generation of vast amounts of gene expression and transcription factor binding data that hold the promise to speed up network reconstruction significantly. Large-scale regulatory network reconstructions can currently be converted to qualitative in silico models that allow systematic analysis of network behavior in response to changes in environmental conditions. These models can be combined with genome-scale metabolic models to build integrated models of cellular function including both metabolism and its regulation. Since metabolic network reconstruction can be routinely done for any organism with a sequenced genome, the challenges and opportunities associated with regulatory network reconstruction are best understood through a comparison between the two network types. Some of the key differences between regulatory and metabolic networks and their respective reconstruction processes are summarized in table 8.2.

Table 8.2 Some key differences between regulatory and metabolic networks that affect the network reconstruction process

	Metabolic networks	Regulatory networks
Structure	Fixed reaction stoichiometry	Qualitative statements representing combinatorial action of different transcription factors
Evolutionary conservation	Enzyme sequences highly conserved across species	Limited conservation of *cis*-regulatory sites between closely related species
Malleability	Fixed structure in terms of the substrates that a particular enzyme can process	Adjustable structure, because of the possiblity that mutations in the *cis*-regulatory sites change binding specificity
Level of biochemical characterization	Fairly complete understanding of most subsystems	Most subnetworks have not been wellcharacterized even in model organisms
Modeling approaches	Quantitative constraint-based models can be constructed at the genome scale	Quantitative models can be currently constructed only on a small scale; qualitative discrete network models can be used to study large networks
Role of noise	Relatively small, because of high concentrations of metabolites involved in most reactions	Possibly significant in determining both structural features of the network and the overall response of the network to a simulus

As described above, well-curated metabolic network databases [12,13] form the basis for any more detailed metabolic network reconstruction or modeling task. For regulatory networks, similar comprehensive databases covering genome-scale regulatory networks in multiple organisms do not currently exist. The lack of databases can be primarily attributed to the low degree of direct molecular level conservation of regulatory network components (transcription factors and especially binding sites) between distantly related species [77]. However, for individual organisms such network databases containing experimentally verified regulatory interactions have been established, the most prominent one being RegulonDB for *E. coli* [78]. In addition to databases describing regulatory network structures, there are also comprehensive databases specializing in describing transcription factor (TF) binding sites such as TRANSFAC [79]. Although these databases contain valuable information for regulatory network reconstruction, they are not complete and for most part lack information about the synergistic effects between TFs acting on one gene. Nevertheless, these databases and primary research literature can be used as a starting point to reconstruct regulatory networks for well-characterized organisms such as *E. coli* based on genetic and biochemical information reported in the primary literature [57,80].

The major advantage that regulatory network reconstruction has over metabolic network reconstruction is the availability of high-throughput experimental data directly relevant to the network structure. Data on both network structure and network states in the form of genome-wide promoter occupancy and mRNA expression data are widely available for many commonly studied model organisms. Gene expression data can be readily generated for well-studied organisms using a number of standard technologies [81]. Advances in statistical data analysis allow both establishing significant changes in gene expression between conditions [82,83] and derivation of hypotheses about regulatory interactions or coregulated gene modules directly from the data [84-86]. In particular, gene expression changes in response to deletions or overexpression of regulatory genes have been productively used to obtain sets of potential target genes for many regulatory proteins, for example, in yeast and *E. coli* [57,87–90].

The development of the ChIP-chip (chromatin immunoprecipitation followed by microarray analysis) method [91,92], which allows directly detecting genomic target sites for DNA-binding proteins such as TFs, has resulted in even more significant improvement in our ability to reconstruct regulatory network structures. So far ChIP-chip has been most extensively applied in yeast, where it has been used to map the target genes of 203 TFs under a limited number of environmental conditions [93,94], but in principle the technique can be readily extended to other organisms [95,96]. ChIP-chip has also been applied to study the stimulus-dependent binding of TFs [97], opening up the possibility of using this technique to map combinatorial interactions between TFs on a genome-wide scale.

The combination of expression profiling with ChIP-chip as well as promoter sequence motif analysis has been shown to allow generation of hypothetical regulatory network structures using a variety of data integration methods [85,93,98–101]. However, deriving full regulatory network structures based solely on experimental data appears to be challenging due to the large quantities of high-quality data that would be required for such a task. One alternative to this purely data-driven approach would be to utilize regulatory network structures derived from databases and primary literature as a starting point for expanding the network based on high-throughput data. Such a combination of knowledge-driven and data-driven regulatory network reconstruction strategies has already been shown to accelerate network reconstruction in *E. coli* [57] and yeast [102].

Both knowledge- and data-driven network reconstruction strategies have so far been primarily applied to the two best characterized microbial organisms, *E. coli* and *S. cerevisiae*. Existing genome-scale regulatory network reconstructions and data collections for these organisms are summarized in table 8.3.

Integration of Regulation and Metabolism

Regulatory network reconstructions can be represented at different levels of detail, ranging from connectivity diagrams to kinetic descriptions, depending on the intended application of the resulting network model. Different in silico modeling approaches have been extensively reviewed elsewhere [103] and will not be discussed in detail here.

Table 8.3 Examples of reconstructed transcriptional regulatory network structures and data sets for reconstruction in *E. coli* and *S. cerevisiae*

	E. coli full metabolic [57]	*E. coli* database [80]	*S. cerevisiae* metabolic [102]	*S. cerevisiae* database [124]	*S. cerevisiae* ChIP-chip [94]
Regulatory genes	104	123	55	109	203
Target genes	451	762[a]	348	418	1296[b]
Regulatory interactions	—	1468[a]	775	945	3353[b]
Regulated reactions	555	—	—	—	

[a]Counting each gene in operon separately.
[b]Includes only high confidence interactions.

The representation of network structure is another major difference between regulatory and metabolic network reconstruction as the latter are naturally described through the reaction stoichiometry, whereas for the former there is no single widely accepted description. Clearly, more detailed descriptions of regulatory networks require increasingly large amounts of parameters [104].

Because the role of transcriptional regulation is to modulate other cellular processes, integrating the reconstructed regulatory networks with models of these other processes is central to understanding regulatory network function in the context of the whole organism. Currently, there are major challenges for achieving this integration relating both to obtaining the relevant data, and to the modeling frameworks that are able to support the required large-scale integration. Possibilities for a suitable modeling framework include discrete network models, such as Boolean and Bayesian networks, which allow representing key combinatorial interactions between regulators acting on the same gene either deterministically or probabilistically [105,106]. These qualitative network models, unlike graph-based models, allow simulating the network behavior but require significantly fewer parameters than linear or nonlinear kinetic models.

Boolean regulatory network models can be readily integrated with genome-scale metabolic models to formulate integrated models of cellular function [107]. In the regulated flux balance analysis (rFBA) approach [107], each of the N reactions in the metabolic model is given a binary variable y_i describing whether the reaction is active or not. The variables y_i are evaluated through a series of Boolean rules that encode how the activity of a reaction depends on the expression of the genes related to this reaction, and how the expression of all the metabolic genes in turn depends on the activities of transcription factors. Finally, the activities of transcription factors are determined from extracellular metabolite concentrations X^{ext} so that the reactions states y_i are given by

$$y_i = f_i(X^{\text{ext}}) \tag{6}$$

where f_i are the composite Boolean functions describing the overall relationship between the reaction states y_i and the concentration vector X^{ext}.

The reaction states y_i are then used to determine the constraints for an FBA calculation in the following form:

$$\max c^T v \qquad \text{subject to } Sv = 0,\ 0 \le v_i \le v_i^{\max} y_i \tag{7}$$

where $y_i = \{0,1\}$ as determined by eq. (6). In order to then calculate extracellular concentrations X^{ext}, it is assumed that the production or consumption of metabolites and the production of biomass X^{BM} is determined by the following set of differential equations derived from eq. (1):

$$\frac{dX^{BM}}{dt} = \mu X^{BM}$$
$$\frac{dX^{ext}}{dt} = v^{ext} X^{BM} \tag{8}$$

where μ is the growth rate obtained from the FBA calculation (7) and v^{ext} are the exchange fluxes corresponding to all extracellular metabolites that trigger regulatory effects (negative fluxes correspond to uptake and positive fluxes to secretion of metabolites). Equations (6)–(8) can be solved iteratively starting from a particular initial set of extracellular metabolite concentrations to obtain a time course of biomass and extracellular metabolite concentrations. Overall the rFBA approach allows accounting for environment- and time-dependent constraints acting on the metabolic network to limit its phenotypic capabilities (figure 8.5). Similarly to the rFBA method, other constraint-based methods, such as extreme pathway analysis can be extended to allow accounting for regulatory effects [108].

Applications of Integrated Models

The kind of integrated models that combine genome-scale metabolic and regulatory networks discussed above have so far been formulated for *E. coli* [57,109] and yeast [102]. The dynamic time-course analysis discussed above was used to predict metabolite concentration profiles for a number of metabolic shifts using a regulated *E. coli* core metabolic model [109]. Building on this model, a genome-scale regulated metabolic model of *E. coli* was established [57]. This model, which accounts for a total of 1010 genes, represents the first genome-scale integrated model of multiple cellular functions in a microbial organism (table 8.3). The major advantage of such integrated models is that even when the modeling of the regulatory network function is done at the qualitative level, the integrated regulatory/metabolic model can be used to quantitatively predict phenotypes such as growth rates. These predictions can then be compared to experimentally measured regulatory or metabolic gene knockout strain growth rates.

In addition to the growth rate predictions, the qualitative gene expression change predictions made by the integrated model can be compared with experimentally measured expression profiles [57]. This type of systematic comparison allowed iterative improvement of the regulatory and metabolic network models to establish models with improved predictive capability for both growth rates and expression changes (figure 8.5). The integrated models discussed here are a

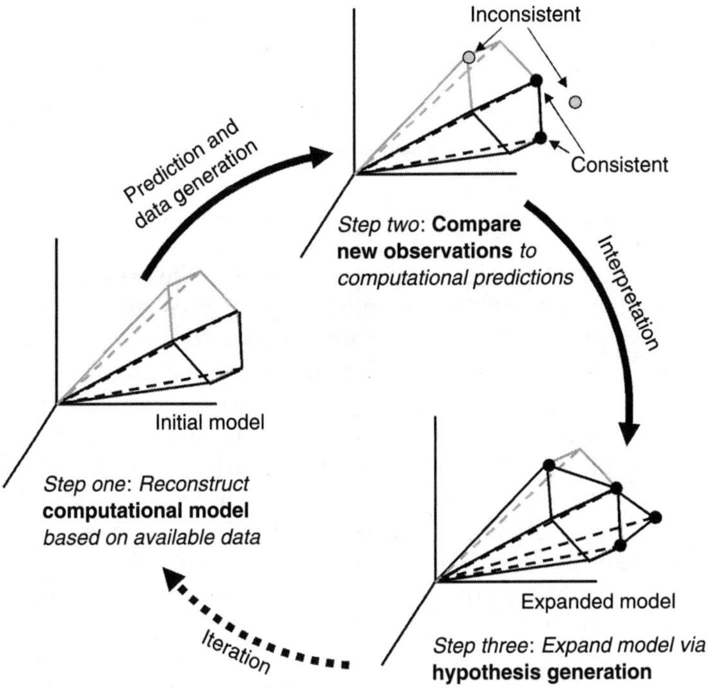

Figure 8.5 Iterative model-building process for regulated metabolic models. An intial model (dark lines including regulation, lighter lines corresponding to unregulated model) is used to define an allowable solution space. Experimental data is compared to model predictions and discrepancies are identified. The model is then expanded to improve agreement with data and tested against an independent data set to avoid overfitting. The process is then repeated by designing a new set of experiments to further expand the model.

powerful way to bring together multiple types of high-throughput data (e.g., gene expression and phenotyping) and to interpret these data sets. Discrepancies between model predictions and experimental data can be used to iteratively improve the regulatory network reconstruction.

The approaches described above aim to construct a model of the regulatory network based on, for example, ChIP-chip and gene expression data and then combine this model with the genome-scale metabolic model. An alternative to this approach is to use gene expression data directly as an additional constraint in the analysis of metabolic networks on a condition-by-condition basis [110]. Constraining fluxes through downregulated reactions in a genome-scale model of yeast was shown to result in improved prediction of exchange and intracellular fluxes [110]. Since the relationship between metabolic flux and gene expression patterns appears to be quite complicated [111], further advances in integrated metabolic/regulatory network modeling methods will be needed [112–115]. Constraint-based modeling provides a good framework for developing such methods through successive application of more complex regulatory constraints.

CONCLUSIONS AND FUTURE DIRECTIONS

Reconstruction of genome-scale biochemical networks is necessary for understanding integrated network properties and for building quantitative predictive models of these networks. Genome sequences as well as extensive biochemical literature have allowed reconstruction of genome-scale metabolic reaction networks for many microbial organisms. Methods have been developed to allow constraint-based analysis of the resulting stoichiometric matrices in order to predict experimentally measurable physiological parameters. These predictions have been shown to agree quantitatively with experimental measurements for evolved strains and qualitatively with growth phenotyping data for nonevolved knockout strains. In addition, constraint-based analysis of genome-scale metabolic networks has allowed studying in silico global properties such as redundancy and robustness of these networks. Recent developments in in silico methods have also allowed utilizing genome-scale models in metabolic engineering [72,74,76] and other practical applications in, for example, bioremediation are already emerging [116].

With the increased availability of high-throughput experimental data, the reconstruction of transcriptional regulatory networks is becoming feasible at the genome scale, for any organisms that has been sufficiently well characterized and for which suitable data is available. Specific properties of regulatory networks, such as lack of evolutionary conservation of binding sites, make both reconstruction and modeling of these networks more challenging than the corresponding tasks for metabolic networks. New developments in experimental techniques, data-based reconstruction approaches, and in silico modeling methods are still needed to improve our ability to build quantitative models of regulatory networks. Nevertheless, the first large-scale regulatory network models that have been reconstructed show that we are already in the position to build integrated models that incorporate both regulatory and metabolic networks [57,102]. These models allow simultaneous analysis of multiple data types as well as the studying the effects of the interplay between metabolic and regulatory networks on overall cellular function.

Genome-scale models of cellular processes, other than metabolism or transcriptional regulation, are also beginning to be developed. In particular, sufficient amounts of information on signaling networks in many cell types have accumulated to allow reconstruction of these networks in a stoichiometric formalism [117,118]. These reconstructions can then be subjected to the same constraint-based analysis methods used to study metabolic networks, to investigate network-level characteristics such as cross-talk or redundancy. The basic processes involved in protein synthesis, transcription and translation, are also amenable to genome-scale constraint-based modeling [119,120]. Nonmetabolic processes such as signaling or transcription have traditionally been modeled on a relatively small scale using either stochastic or deterministic kinetic approaches [121]. Large-scale constraint-based models can complement these models by providing a more complete picture of the interconnectivity between different modules in, for example, signaling pathways.

One of the future challenges will be integration of the individual models of these different cellular processes to allow systematic analysis of whole cell function. Achieving this integration will require further developments in constraint-based analysis, to represent solution spaces for protein and metabolite concentrations, in addition to metabolic fluxes [122,123]. Extension of genome-scale constraint-based models to additional types of cellular processes, and continued development of in silico methods, will enable more accurate prediction of cellular behavior. Combination of in silico analysis of genome-scale models and experimentation will also allow systematic study of the underlying operational principles of biochemical networks.

ACKNOWLEDGMENTS

We thank Nathan Price, Markus Covert, Jennifer Reed, and Stephen Fong for assistance in preparing figures. We acknowledge the support of the National Science Foundation and the National Institutes of Health.

REFERENCES

1. Novick, A. and M. Weiner. Enzyme induction as an all-or-none phenomenon. *Proceedings of the National Academy of Sciences USA*, 43:553–67, 1957.
2. Fleischmann, R. D., et al. Whole-genome random sequencing and assembly of *Haemophilus influenzae* Rd. *Science*, 269(5223):496–8, 507–12, 1995.
3. Goffeau, A., et al. Life with 6000 genes. *Science*, 274(5287):546, 563–7, 1996.
4. Venter, J. C., et al. The sequence of the human genome. *Science*, 291(5507): 1304–51, 2001.
5. Lander, E. S., et al. Initial sequencing and analysis of the human genome. *Nature*, 409(6822):860–921, 2001.
6. Covert, M. W., et al. Metabolic modeling of microbial strains in silico. *Trends in Biochemical Sciences*, 26:179–86, 2001.
7. Papin, J. A., N. D. Price and B. O. Palsson. In silico cells: studying genotype-phenotype relationships with constraints-based models. In B. Kholodenko

and H. V. Westerhoff (Eds.), *Metabolic Engineering in the Post-Genomic Era* (pp. 209–36). Horizon Bioscience, Norwich, U.K., 2004.
8. Price, N. D., et al. Genome-scale microbial in silico models: the constraints-based approach. *Trends in Biotechnology*, 21(4):162–9, 2003.
9. Herrgård, M. J., M. W. Covert and B.O. Palsson. Reconstruction of microbial transcriptional regulatory networks. *Current Opinion in Biotechnology*, 15(1):70–7, 2004.
10. Michal, G. *Biochemical Pathways: An Atlas of Biochemistry and Molecular Biology*. Wiley-Spektrum, New York, 1999.
11. Osterman, A. and R. Overbeek. Missing genes in metabolic pathways: a comparative genomics approach. *Current Opinion in Chemical Biology*, 7(2): 238–51, 2003.
12. Kanehisa, M., et al. The KEGG resource for deciphering the genome. *Nucleic Acids Research*, 32 (Database issue):D277–80, 2004.
13. Keseler, I. M., et al. EcoCyc: a comprehensive database resource for *Escherichia coli*. *Nucleic Acids Research*, 33 (Database Issue):D334–7, 2005.
14. Reed, J. L., et al. An expanded genome-scale model of *Escherichia coli* K-12 (*i*JR904 GSM/GPR). *Genome Biology*, 4(9):R54.1–12, 2003.
15. Neidhardt, F. C. *Escherichia coli and Salmonella: Cellular and Molecular Biology*. ASM Press, Washington, D.C., 1996.
16. Strathern, J. N., E. W. Jones and J.R. Broach. *The Molecular Biology of the Yeast Saccharomyces: Metabolism and Gene Expression*. Cold Spring Harbor Laboratory, Cold Spring Harbor, N.Y., 1982.
17. Huh, W. K., et al. Global analysis of protein localization in budding yeast. *Nature*, 425(6959):686–91, 2003.
18. Prokisch, H., et al. Integrative analysis of the mitochondrial proteome in yeast. *PLoS Biology*, 2(6):e160, 2004.
19. Duarte, N. C., M. J. Herrgård and B. Palsson. Reconstruction and validation of *Saccharomyces cerevisiae* *i*ND750, a fully compartmentalized genome-scale metabolic model. *Genome Research*, 14(7):1298–309, 2004.
20. Schilling, C. H. and B. O. Palsson. Assessment of the metabolic capabilities of *Haemophilus influenzae* Rd through a genome-scale pathway analysis. *Journal of Theoretical Biology*, 203(3):249–83, 2000.
21. Schilling, C. H., et al. Genome-scale metabolic model of *Helicobacter pylori* 26695. *Journal of Bacteriology*, 184(16):4582–93, 2002.
22. Becker, S. A. and B. O. Palsson. Genome-scale reconstruction of the metabolic network in *Staphylococcus aureus* N315: an initial draft to the two-dimensional annotation. *BMC Microbiology*, 5(1):8, 2005.
23. Forster, J., et al. Genome-scale reconstruction of the *Saccharomyces cerevisiae* metabolic network. *Genome Research*, 13(2):244–53, 2003.
24. Vo, T. D., H. J. Greenberg and B.O. Palsson. Reconstruction and functional characterization of the human mitochondrial metabolic network based on proteomic and biochemical data. *Journal of Biological Chemistry*, 279(38):39532–40, 2004.
25. Edwards, J. S. and B. O. Palsson. The *Escherichia coli* MG1655 in silico metabolic genotype: its definition, characteristics, and capabilities. *Proceedings of the National Academy of Sciences USA*, 97(10):5528–33, 2000.
26. Sainz, J., et al. Modeling of yeast metabolism and process dynamics in batch fermentation. *Biotechnology and Bioengineering*, 81(7):818–28, 2003.

27. Edwards, J. S., R. U. Ibarra and B.O. Palsson. In silico predictions of *Escherichia coli* metabolic capabilities are consistent with experimental data. *Nature Biotechnology*, 19:125–30, 2001.
28. Beard, D. A., S. D. Liang and H. Qian. Energy balance for analysis of complex metabolic networks. *Biophysical Journal*, 83(1):79–86, 2002.
29. Price, N. D., et al. Extreme pathways and Kirchhoff's second law. *Biophysical Journal*, 83(5):2879–82, 2002.
30. Qian, H., D. A. Beard and S. D. Liang. Stoichiometric network theory for nonequilibrium biochemical systems. *European Journal of Biochemistry*, 270(3):415–21, 2003.
31. Lee, S., et al. Recursive MILP model for finding all the alternate optima in LP models for metabolic networks. *Computers in Chemical Engineering*, 24:711–16, 2000.
32. Reed, J. L. and B. O. Palsson. Genome-scale in silico models of *E. coli* have multiple equivalent phenotypic states: assessment of correlated reaction subsets that comprise network states. *Genome Research*, 14(9):1797–805, 2004.
33. Mahadevan, R. and C. H. Schilling. The effects of alternate optimal solutions in constraint-based genome-scale metabolic models. *Metabolic Engineering*, 5(4):264–76, 2003.
34. Edwards, J. S. and B. O. Palsson. Robustness analysis of the *Escherichia coli* metabolic network. *Biotechnology Progress*, 16(6):927–39, 2000.
35. Chvatal, V. *Linear Programming*. W. H. Freeman, New York, 1983.
36. Ibarra, R. U., J. S. Edwards and B. O. Palsson. *Escherichia coli* K-12 undergoes adaptive evolution to achieve in silico predicted optimal growth. *Nature*, 420(6912):186–9, 2002.
37. Ibarra, R. U., et al. Quantitative analysis of *Escherichia coli* metabolic phenotypes within the context of phenotypic phase planes. *Journal of Molecular Microbiology and Biotechnology*, 6(2):101–8, 2004.
38. Segre, D., D. Vitkup and G. M. Church. Analysis of optimality in natural and perturbed metabolic networks. *Proceedings of the National Academy of Sciences USA*, 99(23):15112–17, 2002.
39. Shlomi, T., O. Berkman and E. Ruppin. Regulatory on/off minimization of metabolic flux changes after genetic perturbations. *Proceedings of the National Academy of Sciences USA*, 102(21):7695–700, 2005.
40. Burgard, A. P. and C. D. Maranas. Optimization-based framework for inferring and testing hypothesized metabolic objective functions. *Biotechnology and Bioengineering*, 82(6):670–7, 2003.
41. Price, N. D., J. L. Reed and B. O. Palsson. Genome-scale models of microbial cells: evaluating the consequences of constraints. *Nature Reviews in Microbiology*, 2(11):886–97, 2004.
42. Schilling, C. H., D. Letscher and B. O. Palsson. Theory for the systemic definition of metabolic pathways and their use in interpreting metabolic function from a pathway-oriented perspective. *Journal of Theoretical Biology*, 203(3):229–48, 2000.
43. Schuster, S., T. Dandekar and D. A. Fell. Detection of elementary flux modes in biochemical networks: a promising tool for pathway analysis and metabolic engineering. *Trends in Biotechnology*, 17(2):53–60, 1999.
44. Papin, J. A., et al. Comparison of network-based pathway analysis methods. *Trends in Biotechnology*, 22(8):400–5, 2004.

45. Burgard, A. P., et al. Flux coupling analysis of genome-scale metabolic network reconstructions. *Genome Research*, 14(2):301–12, 2004.
46. Almaas, E., et al. Global organization of metabolic fluxes in the bacterium *Escherichia coli. Nature*, 427(6977):839–43, 2004.
47. Price, N. D., J. Schellenberger and B. O. Palsson. Uniform sampling of steady state flux spaces: means to design experiments and to interpret enzymopathies. *Biophysical Journal*, 87(4):2172–86, 2004.
48. Thiele, I., et al. Candidate metabolic network states in human mitochondria: impact of diabetes, ischemia, and diet. *Journal of Biological Chemistry*, 280(12):11683–95, 2005.
49. Lovasz, L. Hit-and-run mixes fast. *Mathematical Programming*, 86(3):443–61, 1999.
50. Kaufman, D. E. and R. L. Smith. Direction choice for accelerated convergence in hit-and-run sampling. *Operations Research*, 46(1):84–95, 1998.
51. Jeong, H., et al. The large-scale organization of metabolic networks. *Nature*, 407(6804):651–4, 2000.
52. Ravasz, E. et al. Hierarchical organization of modularity in metabolic networks. *Science*, 297(5586):1551–5, 2002.
53. Mahadevan, R. and B. O. Palsson. Properties of metabolic networks: structure versus function. *Biophysical Journal*, 88(1):L07–9, 2005.
54. Fong, S. S. et al. In silico design and adaptive evolution of *Escherichia coli* for production of lactic acid. *Biotechnology and Bioengineering*, 91(5):643–8, 2005.
55. Fong, S. S. and B. O. Palsson. Metabolic gene deletion strains of *Escherichia coli* evolve to computationally predicted growth phenotypes. *Nature Genetics*, 36(10):1056–8, 2004.
56. Edwards, J. S. and Palsson, B. O. Metabolic flux balance analysis and the in silico analysis of *Escherichia coli* K-12 gene deletions. *BMC Bioinformatics*, 1:1, 2000.
57. Covert, M. W., et al. Integrating high-throughput and computational data elucidates bacterial networks. *Nature*, 429(6987):92–6, 2004.
58. Forster, J., et al. Large-scale evaluation of in silico gene knockouts in *Saccharomyces cerevisiae. OMICS*, 7(2):193–202, 2003.
59. Edwards, J. S. and B. O. Palsson. Systems properties of the *Haemophilus influenzae* Rd metabolic genotype. *Journal of Biological Chemistry*, 274(25): 17410–16, 1999.
60. Stelling, J., et al. Metabolic network structure determines key aspects of functionality and regulation. *Nature*, 420:190–3, 2002.
61. Price, N. D., J. A. Papin and B. O. Palsson. Determination of redundancy and systems properties of *Helicobacter pylori*'s metabolic network using genome-scale extreme pathway analysis. *Genome Research*, 12(5):760–9, 2002.
62. Segre, D., et al. Modular epistasis in yeast metabolism. *Nature Genetics*, 37(1):77–83, 2005.
63. Tong, A. H., et al. Systematic genetic analysis with ordered arrays of yeast deletion mutants. *Science*, 294(5550):2364–8, 2001.
64. Papin, J. A., et al. The genome-scale metabolic extreme pathway structure in *Haemophilus influenzae* shows significant network redundancy. *Journal of Theoretical Biology*, 215(1):67–82, 2002.

65. Burgard, A. P., S. Vaidyaraman and C. D. Maranas. Minimal reaction sets for *Escherichia coli* metabolism under different growth requirements and uptake environments. *Biotechnology Progress*, 17(5):791–7, 2001.
66. Giaever, G., et al. Functional profiling of the *Saccharomyces cerevisiae* genome. *Nature*, 418(6896):387–91, 2002.
67. Kamath, R. S., et al. Systematic functional analysis of the *Caenorhabditis elegans* genome using RNAi. *Nature*, 421(6920):231–7, 2003.
68. Papp, B., C. Pal and L. D. Hurst. Metabolic network analysis of the causes and evolution of enzyme dispensability in yeast. *Nature*, 429(6992):661–4, 2004.
69. Hartwell, L. H., et al. From molecular to modular cell biology. *Nature*, 402(6761 Suppl):C47–52, 1999.
70. Papin, J. A., et al. Metabolic pathways in the post-genome era. *Trends in Biochemical Sciences*, 28(5):250–8, 2003.
71. Carlson, R., D. Fell and F. Srienc. Metabolic pathway analysis of a recombinant yeast for rational strain development. *Biotechnology and Bioengineering*, 79(2):121–34, 2002.
72. Alper, H., et al. Identifying gene targets for the metabolic engineering of lycopene biosynthesis in *Escherichia coli. Metabolic Engineering*, 7(3):155–64, 2005.
73. Alper, H., K. Miyaoku and G. Stephanopoulos. Construction of lycopene-overproducing *E. coli* strains by combining systematic and combinatorial gene knockout targets. *Nature Biotechnology*, 23(5):612–16, 2005.
74. Burgard, A. P., P. Pharkya and C. D. Maranas. OptKnock: a bilevel programming framework for identifying gene knockout strategies for microbial strain optimization. *Biotechnology and Bioengineering*, 84(6): 647–57, 2003.
75. Pharkya, P., A. P. Burgard and C. D. Maranas. OptStrain: a computational framework for redesign of microbial production systems. *Genome Research*, 14(11):2367–76, 2004.
76. Pharkya, P., A. P. Burgard and C. D. Maranas. Exploring the overproduction of amino acids using the bilevel optimization framework OptKnock. *Biotechnology and Bioengineering*, 84(7):887–99, 2003.
77. Babu, M. M., et al. Structure and evolution of transcriptional regulatory networks. *Current Opinion in Structural Biology*, 14(3):283–91, 2004.
78. Salgado, H., et al. RegulonDB (version 3.2): transcriptional regulation and operon organization in *Escherichia coli* K-12. *Nucleic Acids Research*, 29(1):72–4, 2001.
79. Matys, V., et al. TRANSFAC: transcriptional regulation, from patterns to profiles. *Nucleic Acids Research*, 31(1):374–8, 2003.
80. Shen-Orr, S. S., et al. Network motifs in the transcriptional regulation network of *Escherichia coli. Nature Genetics*, 31(1):64–8, 2002.
81. Holloway, A. J., et al. Options available—from start to finish—for obtaining data from DNA microarrays II. *Nature Genetics*, 32(Suppl):481–9, 2002.
82. Quackenbush, J. Microarray data normalization and transformation. *Nature Genetics*, 32(Suppl):496–501, 2002.
83. Churchill, G. A. Fundamentals of experimental design for cDNA microarrays. *Nature Genetics*, 32(Suppl):490–5, 2002.
84. Ihmels, J., et al. Revealing modular organization in the yeast transcriptional network. *Nature Genetics*, 22:22, 2002.

85. Segal, E., et al. Module networks: identifying regulatory modules and their condition-specific regulators from gene expression data. *Nature Genetics*, 34(2):166–76, 2003.
86. Wang, W., et al. A systematic approach to reconstructing transcription networks in *Saccharomyces cerevisiae. Proceedings of the National Academy of Sciences USA*, 99(26):16893–8, 2002.
87. Mnaimneh, S., et al. Exploration of essential gene functions via titratable promoter alleles. *Cell*, 118(1):31–44, 2004.
88. Hughes, T. R., et al. Functional discovery via a compendium of expression profiles. *Cell*, 102(1):109–26, 2000.
89. Ideker, T., et al. Integrated genomic and proteomic analyses of a systematically perturbed metabolic network. *Science*, 292(May 4):929–34, 2001.
90. Oshima, T., et al. Transcriptome analysis of all two-component regulatory system mutants of *Escherichia coli* K-12. *Molecular Microbiology*, 46(1):281–91, 2002.
91. Ren, B., et al. Genome-wide location and function of DNA binding proteins. *Science*, 290(5500):2306–9, 2000.
92. Iyer, V. R., et al. Genomic binding sites of the yeast cell-cycle transcription factors SBF and MBF. *Nature*, 409(6819):533–8, 2001.
93. Lee, T. I., et al. Transcriptional regulatory networks in *Saccharomyces cerevisiae. Science*, 298(5594):799–804, 2002.
94. Harbison, C. T., et al. Transcriptional regulatory code of a eukaryotic genome. *Nature*, 431(7004):99–104, 2004.
95. Laub, M. T., et al. Genes directly controlled by CtrA, a master regulator of the *Caulobacter* cell cycle. *Proceedings of the National Academy of Sciences USA*, 99(7):4632–7, 2002.
96. Odom, D. T., et al. Control of pancreas and liver gene expression by HNF transcription factors. *Science*, 303(5662):1378–81, 2004.
97. Zeitlinger, J., et al. Program-specific distribution of a transcription factor dependent on partner transcription factor and MAPK signaling. *Cell*, 113(3):395–404, 2003.
98. Hartemink, A. J., et al. Combining location and expression data for principled discovery of genetic regulatory network models. *Pacific Symposium in Biocomputing*, 437–49, 2002.
99. Liu, X. S., D. L. Brutlag and J. S. Liu. An algorithm for finding protein DNA binding sites with applications to chromatin-immunoprecipitation microarray experiments. *Nature Biotechnology*, 8:8, 2002.
100. Bar-Joseph, Z., et al. Computational discovery of gene modules and regulatory networks. *Nature Biotechnology*, 21(11):1337–42, 2003.
101. Yeang, C. H., T. Ideker and T. Jaakkola. Physical network models. *Journal of Computational Biology*, 11(2-3):243–62, 2004.
102. Herrgård, M. J., B.-S. Lee and B. O. Palsson. Integrated analysis of regulatory and metabolic networks reveals novel regulatory mechanisms in *Saccharomyces cerevisiae. Genome Research*, 16(5):627–35, 2006.
103. de Jong, H. Modeling and simulation of genetic regulatory systems: a literature review. *Journal of Computational Biology*, 9(1):67–103, 2002.
104. Bolouri, H. and E. H. Davidson. Modeling transcriptional regulatory networks. *Bioessays*, 24(12):1118–29, 2002.

105. Davidson, E. H., et al. A provisional regulatory gene network for specification of endomesoderm in the sea urchin embryo. *Developmental Biology*, 246(1):162–90, 2002.
106. Pe'er, D., et al. Inferring subnetworks from perturbed expression profiles. *Bioinformatics*, 17(1):S215–24, 2001.
107. Covert, M. W., C. H. Schilling and B. Palsson. Regulation of gene expression in flux balance models of metabolism. *Journal of Theoretical Biology*, 213(1):73–88, 2001.
108. Covert, M. and B. O. Palsson. Constraints-based models: regulation of gene expression reduces the steady-state solution space. *Journal of Theoretical Biology*, 221(3):309–25, 2003.
109. Covert, M. W. and B. O. Palsson. Transcriptional regulation in constraints-based metabolic models of *Escherichia coli*. *Journal of Biological Chemistry*, 277(31):28058–64, 2002.
110. Akesson, M., J. Forster, and J. Nielsen. Integration of gene expression data into genome-scale metabolic models. *Metabolic Engineering*, 6(4):285–93, 2004.
111. Daran-Lapujade, P., et al. Role of transcriptional regulation in controlling fluxes in central carbon metabolism of *Saccharomyces cerevisiae*. A chemostat culture study. *Journal of Biological Chemistry*, 279(10):9125–38, 2004.
112. Patil, K. R. and J. Nielsen. Uncovering transcriptional regulation of metabolism by using metabolic network topology. *Proceedings of the National Academy of Sciences USA*, 102(8):2685–9, 2005.
113. Ihmels, J., R. Levy, and N. Barkai. Principles of transcriptional control in the metabolic network of *Saccharomyces cerevisiae*. *Nature Biotechnology*, 22(1):86–92, 2004.
114. Haugen, A. C., et al. Integrating phenotypic and expression profiles to map arsenic-response networks. *Genome Biology*, 5(12):R95, 2004.
115. Kharchenko, P., D. Vitkup and G. M. Church. Filling gaps in a metabolic network using expression information. *Bioinformatics*, 20(Suppl 1): I178–85, 2004.
116. Lovley, D. R. Cleaning up with genomics: applying molecular biology to bioremediation. *Nature Reviews in Microbiology*, 1(1):35–44, 2003.
117. Papin, J. A. and B. O. Palsson. The JAK-STAT signaling network in the human B-cell: an extreme signaling pathway analysis. *Biophysical Journal*, 87(1):37–46, 2004.
118. Papin, J. A. and B. O. Palsson. Topological analysis of mass-balanced signaling networks: a framework to obtain network properties including crosstalk. *Journal of Theoretical Biology*, 227(2):283–97, 2004.
119. Allen, T. E., et al. Genome-scale analysis of the uses of the *Escherichia coli* genome: model-driven analysis of heterogeneous data sets. *Journal of Bacteriology*, 185(21):6392–9, 2003.
120. Allen, T. E. and B. O. Palsson. Sequenced-based analysis of metabolic demands for protein synthesis in prokaryotes. *Journal of Theoretical Biology*, 220(1):1–18, 2003.
121. Tyson, J. J., K. C. Chen and B. Novak. Sniffers, buzzers, toggles and blinkers: dynamics of regulatory and signaling pathways in the cell. *Current Opinion in Cell Biology*, 15(2):221–31, 2003.

122. Famili, I. and B. O. Palsson. The convex basis of the left null space of the stoichiometric matrix leads to the definition of metabolically meaningful pools. *Biophysical Journal*, 85(1):16–26, 2003.
123. Famili, I., R. Mahadevan and B. O. Palsson. k-Cone analysis: determining all candidate values for kinetic parameters on a network scale. *Biophysical Journal*, 88(3):1616–25, 2005.
124. Guelzim, N., et al. Topological and causal structure of the yeast transcriptional regulatory network. *Nature Genetics*, 31(1):60–3, 2002.
125. Borodina, A., P. Krabben, and J. Nielsen. Genome-scale analysis of *Streptomyces coelicolor* A3(2) metabolism. *Genome Research*, 15(6):820-9, 2005.
126. Fong S. S., et al. In silico design and adaptive evolution of *Escherichia coli* for production of lactic acid. *Biotechnology and Bioengineering*, 91(5):643–8, 2005.

9

Biophysical Models of the Cardiovascular System

Raimond L. Winslow, Joseph. L. Greenstein, & Patrick A. Helm

Cardiac electrophysiology is a field with an extensive history of integrative modeling that is closely coupled with both the design and interpretation of experiments. The first models of the cardiac action potential (AP) were developed shortly after the Hodgkin–Huxley model of the squid AP, and were formulated in order to explain the experimental observation that, unlike the neuronal AP, cardiac APs exhibit a long duration plateau phase. It was not long after the formulation of these early myocyte models that initial models of electrical conduction in cardiac tissue were formulated and applied to yield clinically useful insights into mechanisms of arrhythmia. This close interplay between experiment and integrative modeling continues today, with new model components and applications being developed in close coordination with the emergence of new subcellular, cellular, and whole-heart data describing cardiac function in health and disease.

This chapter reviews the current state of integrative modeling of the heart, focusing on three topics: (a) review of the anatomical and biophysical properties of cardiac muscle cells and how key processes may be modeled; (b) use of diffusion tensor magnetic resonance imaging for rapid acquisition of geometric and microanatomic data on cardiac ventricular structure at high spatial resolution; and (c) the integration of cellular models with this imaging to formulate computational models of cardiac ventricular electrical conduction. These approaches will facilitate a new interplay between experiment and modeling of electrical conduction in the heart, thus contributing to our understanding of the systems biology of the heart in both health and disease.

STRUCTURE AND BIOPHYSICAL PROPERTIES OF THE CARDIAC MYOCYTE

There are a diversity of cell types within the heart. This includes both cardiac myocytes as well as smooth muscle cells, fibroblasts, neurons, and endothelial cells. Distinct types of cardiac mycoytes include those of the sinoatrial node, atrium, atrioventricular node, the cardiac conduction system, and the cardiac ventricles. Each of these cells differs in its structure as well as in the type and expression of membrane currents and transporters shaping its electrical excitability. A review of each of these cell types is beyond the scope of this book (see ref. [1]). Instead, we focus on a review of properties of the cardiac ventricular myoycte, as these myocytes are the most abundant in the heart and their properties are of fundamental importance to electrical conduction during both health and disease.

Structure of the Ventricular Myocyte

Structural organization of the ventricular myocyte is shown in figure 9.1. Ventricular myocytes are long (~120 μm), thin (diameter ~25 μm) muscle cells bounded by a cell membrane known as the sarcolemma. The basic unit of contraction in the ventricular myocyte is the sarcomere (figure 9.1a). Individual sarcomeres are approximately 2.0 μm in length and are bounded on both ends by the Z-disks. The H-zone is the region of the thick filament that contains only myosin tails and not the head regions. While the cardiac sarcomere is similar in many respects to that of skeletal muscle, some important differences exist. Specifically, cardiac muscle cannot be extended to lengths above 2.3 μm and hence always operates on the ascending limb of the force–sarcomere length relationship. The A-band is the region spanned by the length of the thick filaments. The shaded region in figure 9.1a represents the region of overlap of thick and thin filaments. Muscle contraction is accomplished by the sliding motion of the thick and thin filaments relative to one another in this region in response to elevated levels of intracellular Ca^{2+} and adenosine triphosphate (ATP) hydrolysis.

Figure 9.1a also shows that sarcomeres are bounded on each end by the T-tubules [2]. T-tubules are cylindrical invaginations of the sarcolemma that extend into the myocyte (figure 9.1b), approaching an organelle known as the sarcoplasmic reticulum (SR). The SR is composed of two components known as junctional SR (JSR) and network SR (NSR). The NSR is a lumenal organelle extending throughout the myocyte. NSR membrane contains a high concentration of the SR Ca^{2+}-ATPase (SERCA2α) pump, which transports Ca^{2+} from the cytosol into the lumen of the NSR. The JSR is that portion of the SR most closely approximating the T-tubules. The close proximity of these two structures (they are separated by a distance of ~10 nm) forms a restricted region commonly referred to as the dyad with an approximate diameter of 400 nm. Ca^{2+}-sensitive Ca^{2+}-release channels (known as ryanodine receptors or RyR) are located in the dyadic region of the JSR membrane. In addition, sarcolemmal L-type Ca^{2+} channels (LCCs) are located preferentially within the dyadic region of the T-tubules, where they are in close apposition to the RyR. It has been estimated

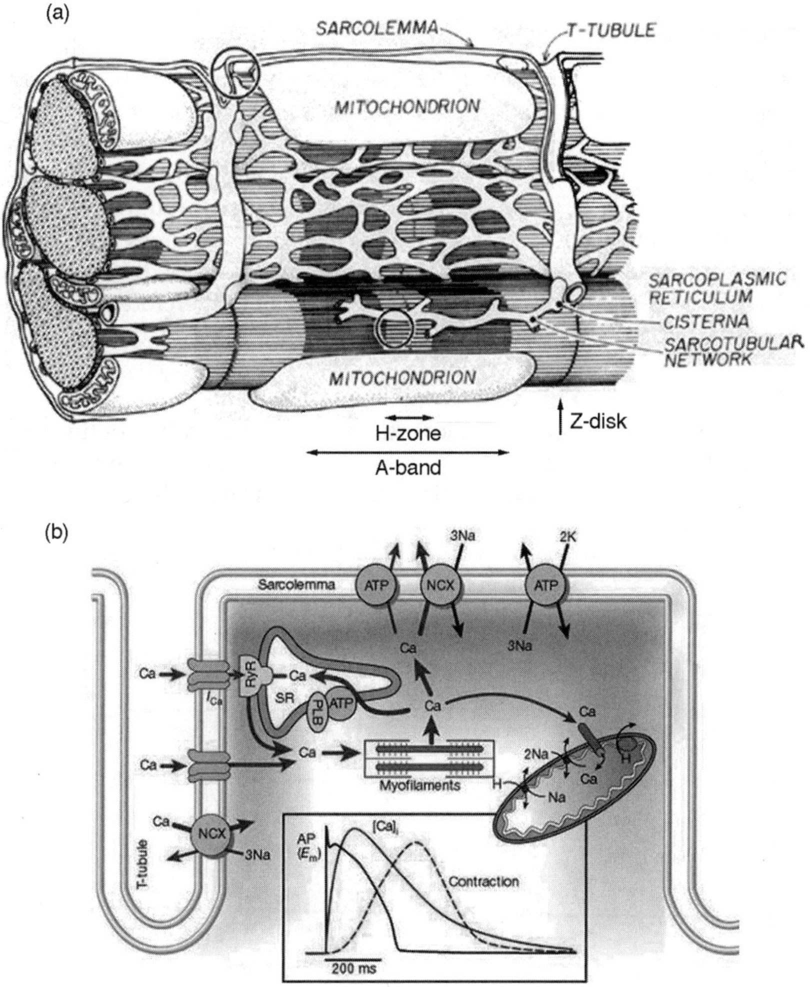

Figure 9.1 (a) Ultrastructure of the cardiac ventricular myocyte, illustrating the organization of sarcomeres and the T-tubules. (Reproduced from fig. 1.14 of Katz [92] with permission). (b) Illustration of the physical organization and channel localization in the cardiac dyad. (Reproduced from fig. 1 of Bers [33] with permission).

that there are approximately 10 LCCs per dyad [3], ~50,000 active LCCs per myocyte, and therefore ~5000 dyads per myocyte. As will be described in a subsequent section, the detailed microstructure of the dyad has a profound influence on the macroscopic function of the myocyte.

Contraction of muscle cells occurs when LCCs located in the cell membrane open in response to membrane depolarization, thereby producing a Ca^{2+} flux into the cell. The resulting increase of intracellular Ca^{2+} leads to Ca^{2+} binding to the RyR and an increase of their open probability. RyR opening in turn produces a large flux of Ca^{2+} from the JSR into the dyadic space. These two sources of Ca^{2+} flux (a small "trigger" flux through LCCs and a larger "release" flux through the RyRs) produce the intracellular Ca^{2+} transient leading ultimately to muscle contraction. Understanding of the molecular basis of this so-called calcium-induced calcium-release (CICR) process is therefore of fundamental importance to understanding cardiac muscle function.

The Cardiac Action Potential

The center panel of figure 9.2 shows a schematic illustration of the large mammalian cardiac AP as well as the membrane currents giving rise to this AP. The currents mediating the AP upstroke (phase 0) are the

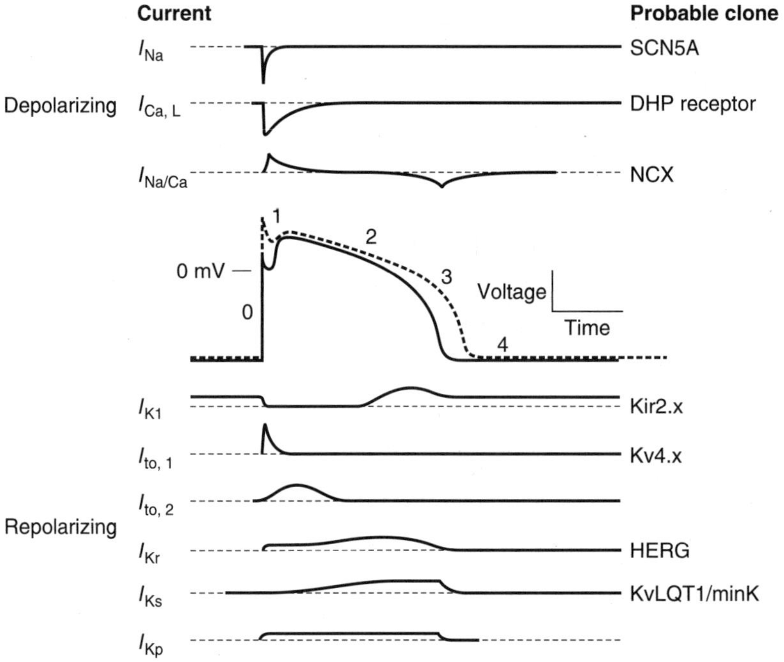

Figure 9.2 Schematic illustration of the large mammalian cardiac ventricular myocyte action potential (membrane potential in mV as a function of time) illustrating depolarizing and repolarizing current (left) and alias gene names encoding each of these currents. (Reprinted from Tomaselli and Marban [93] by permission of the American Heart Association).

fast inward sodium (Na^+) current (I_{Na}; for review see ref. [4]), and to a lesser extent, the L-type Ca^{2+} current ($I_{Ca,L}$; for review see ref. [5]). The phase 1 notch, which is apparent in ventricular myocytes isolated from epi- and mid-myocardial regions, but which is largely absent in those isolated from the endocardium, is produced by activation of the voltage-dependent transient outward potassium (K^+) current ($I_{to,1}$). In some species, a transient voltage-independent Ca^{2+}-modulated Cl^- current contributes to the phase 1 notch ($I_{to,2}$); however, this current is not known to be expressed in human. The phase 2 plateau is a time during which membrane conductance is very low, with potential being determined by a delicate balance between small inward and outward currents. The major inward plateau current is $I_{Ca,L}$, and major outward plateau currents are generated by the rapid and slow-activating delayed outward rectifier K^+ currents I_{Kr} and I_{Ks}, respectively, and the plateau K^+ current I_{Kp}. Repolarization phase 3 is produced by the hyperpolarizing activated inward rectifier K^+ current I_{K1}.

Three major ion transporters and exchangers shape properties of the cardiac AP, Ca^{2+} transient and influence long-term regulation of intracellular ion concentrations. These are the sarcolemmal Na^+-K^+ pump, the sarcolemmal Na^+-Ca^{2+} exchanger, and the SR Ca^{2+}-ATPase. The sarcolemmal Na^+-K^+ pump extrudes three Na^+ ions while importing two K^+ ions on each cycle. This pump functions to keep intracellular Na^+ low, thereby maintaining the external versus internal gradient of Na^+, by extruding Na^+ that enters during each AP. Cycling of this pump requires hydrolysis of one ATP molecule, and generates a net outward movement of one positive charge, thus contributing to outward membrane current (this current is not shown in figure 9.2) and influencing resting membrane potential.

The sarcolemmal Na^+-Ca^{2+} exchanger imports three Na^+ ions for every Ca^{2+} ion extruded, yielding a net charge movement. It is driven by both transmembrane voltage and intra- and extracellular Na^+ and Ca^{2+} ion concentrations. It functions in forward mode during diastole (the time interval between successive cardiac action potentials), in which case it extrudes Ca^{2+} and imports Na^+, thus generating a net inward current (labeled $I_{Na/Ca}$ in figure 9.2). It is the principal means by which Ca^{2+} is extruded from the myocyte following each AP, particularly during the diastolic interval. Because of the voltage- and Ca^{2+}-sensitivity of the exchanger, experimental evidence indicates that it can function in reverse mode during the plateau phase of the AP, in which case it extrudes Na^+ and imports Ca^{2+}, thus generating a net outward current.

A second major cytoplasmic Ca^{2+} extrusion mechanism is the SR Ca^{2+}-ATPase (figure 9.1b). Rather than extruding Ca^{2+} from the cell, this ATPase pumps Ca^{2+} from the cytosol into the NSR. The SR Ca^{2+}-ATPase has both forward and reverse components [6], with the reverse component serving to prevent overloading of the SR with Ca^{2+} at rest. An additional Ca^{2+} extrusion mechanism is the sarcolemmal Ca^{2+}-ATPase. This Ca^{2+} pump hydrolyzes ATP to transport Ca^{2+} out of the cell. However, it contributes a sarcolemmal current that is small relative to that of the Na^{+}-Ca^{2+} exchanger, with estimates indicating that perhaps as little as 3% of Ca^{2+} extrusion from the myocyte is mediated by this pump.

COMPUTATIONAL MODELS OF THE VENTRICULAR MYOCYTE

Development of myocyte models began in the early 1960s with publication of Purkinje fiber AP models. Subsequent elaboration of these models led to development of the first biophysically based cell model describing interactions between voltage-gated membrane currents, membrane pumps, and exchangers that regulate Ca^{2+}, Na^{+}, and K^{+} levels, and additional intracellular Ca^{2+} cycling processes in the cardiac myocyte: the DiFrancesco–Noble model of the Purkinje fiber [7]. This important model established the conceptual framework from which all subsequent models of the myocyte have been derived (ventricular myocytes [8–11], SA node cells [12–15], and atrial myocytes [16,17]). These models have proven reproductive and predictive properties and have been applied to advance our understanding of myocyte function in both health and disease.

The chief distinction between models of the ventricular myocyte in use today concerns the representation of the CICR process. In all present-day models, CICR is described either phenomenologically or through use of a formulation known as the "common pool." Common pool models are ones in which Ca^{2+} flux through all LCCs is lumped into a single trigger flux, Ca^{2+} flux through all RyRs is lumped into a single release flux, and both the trigger and release flux are directed into a common Ca^{2+} compartment (the "subspace") with volume equal to the sum of the volume of all dyads in the cell. In such models, activation of the JSR release mechanism is controlled by Ca^{2+} concentration in this common pool.

A consequence of this physical arrangement is that once RyR Ca^{2+} release is initiated, the resulting increase of Ca^{2+} concentration in the common pool stimulates regenerative, all-or-none rather than graded Ca^{2+} release from the JSR. Graded release refers to a distinctive feature of cardiac CICR, originally observed by Fabiato [18–20], that Ca^{2+} release from JSR is a smooth and continuous function of trigger Ca^{2+} entering the cell via LCCs. Common pool models therefore fail to capture one of the most important properties of CICR observed experimentally. This "latch up" of Ca^{2+} release can be avoided and graded JSR release can be achieved using phenomenological models in which Ca^{2+} release flux is an explicit function of only Ca^{2+} flux through LCCs, rather than as a function of Ca^{2+} concentration in the common pool. Models of this type can exhibit graded JSR Ca^{2+} release; however, such phenomenological formulations lack mechanistic descriptions of the processes that are the underlying basis of CICR. Accordingly, in the following presentation, we focus on properties of common pool models and their extensions.

Modeling Ion Channels and Currents in Cardiac Myocytes

For many years, Hodgkin–Huxley models have been the standard for describing membrane current kinetics [21,22]. However, data obtained using new experimental approaches, in particular those for producing recombinant channels by coexpression of genes encoding pore-forming and accessory channel subunits in host cells, have shown that these models have significant limitations. First, while these models can be expanded to an equivalent Markov chain representation having multiple closed and inactivated states [23], many single-channel behaviors such as mean open time, first latency, and a broad range of other kinetics behaviors cannot be described using these equivalent Markov models [24,25]. Second, where it has been studied in detail, Hodgkin–Huxley models are insufficient for reproducing behaviors that may be critically state-dependent, such as how ionic channels interact with drugs and toxins [26,27]. Accordingly, much recent effort in modeling of cardiac ionic currents has focused on development of biophysically detailed Markov chain models of channel gating. We will therefore illustrate fundamental concepts involved in modeling of ion channel function and membrane current properties using the example of the cardiac LCC. Examples of the application of these modeling methods to other sarcolemmal membrane currents abound (see I_{Na} [28], I_{to1} [29], and I_{Kr} [30]).

The structure of the cardiac LCC model, adapted from Jafri et al. [11], is shown in figure 9.3. The channel is assumed to occupy any of 11 states. The top row of closed states corresponds to zero to four voltage sensors being activated (C_0 through C_4) plus an additional conformational change required for opening ($C_4 \rightarrow O$). The bottom row of closed-inactivated states corresponds to channel inactivation produced by local increases of Ca^{2+} concentration within the subspace—a mode of inactivation known as Ca^{2+}-dependent inactivation. Horizontal state transitions are assumed to be voltage-dependent (with the exception of the $C_4 \rightarrow O$ transition) and closed to closed-inactivated transitions (vertical transitions) are voltage-independent, with rate dependent on Ca^{2+} concentration in the subspace. Voltage-dependence of transition rates is given by Eyring rate theory [23]:

$$\lambda = \frac{kT}{h} \exp\left(\frac{-\Delta H_\lambda}{RT} + \frac{\Delta S_\lambda}{R} + \frac{z_\lambda F V}{RT} \right) \tag{1}$$

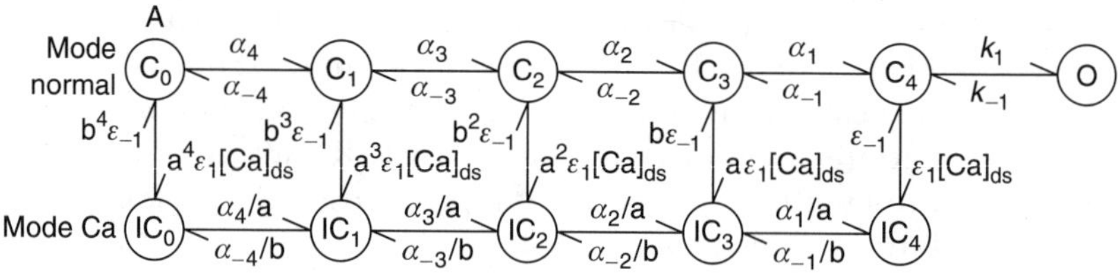

Figure 9.3 The 11-state LCC model developed by Jafri et al. [11].

where k is the Boltzmann constant, T is the absolute temperature, h is the Planck constant, R is the gas constant, F is Faraday's constant, ΔH_λ is the change in enthalpy, ΔS_λ is the change in entropy, z_λ is the effective valence (i.e., the charge moved times the fractional distance the charge is moved through the membrane), and V is the membrane potential in volts.

The probability of occupying any particular channel state, including the open state, is described by a set of ordinary differential equations, written in matrix notation as

$$\frac{d\boldsymbol{P}(t)}{dt} = \mathbf{W}\boldsymbol{P}(t) \tag{2}$$

where $\boldsymbol{P}(t)$ is a vector of state occupancy probabilities and $\mathbf{W}$ is the state transition matrix. $\mathbf{W}$ is in general a function of voltage and time, since state transition rates are also in general functions of voltage and time. However, for the voltage-clamp conditions generally used to constrain ion current models, $\mathbf{W}$ is piecewise time-independent, thus eq. (2) has the analytic solution

$$\boldsymbol{P}(t) = \exp(\mathbf{W}t)\boldsymbol{P}(0) \tag{3}$$

Current through an ensemble of Ca^{2+} channels, denoted $I_{Ca,L}$, is calculated as

$$I_{Ca,L}(t) = NG_{Ca}P_{open}(t)(V(t) - E_{Ca}(t)) \tag{4}$$

where N is the number of Ca^{2+} channels, G_{Ca} is single-channel conductance, $P_{open}(t)$ is the probability of occupying the open state O, $V(t)$ is membrane potential, and $E_{Ca}(t)$ is the reversal potential for Ca^{2+} given by the Nernst equation.

The number of coupled differential equations, and hence the number of parameters that need to be constrained for the model, may be reduced through application of the fundamental principle that the state occupancy probabilities for a Markov chain model must sum to one. Thus, an N-state Markov model may be reduced to an N–1 state model through application of this constraint. Second, closed loops in a state model such as that of figure 9.3 must satisfy the principle of microscopic reversibility. Microscopic reversibility is derived from the law of conservation of energy and states that the product of rate constants when traversing a loop clockwise must be equal to the product when traversing the same loop counterclockwise. Application of this constraint further limits the number of free parameters (transition rates) that must be determined.

Modeling Intracellular Ion Concentration Changes

In the common pool model formulation, there are four distinct Ca^{2+} compartments (the cytosol, subspace, NSR, and JSR compartments) and one Na^+ and potassium (K^+) compartment (the cytosol). Concentration is assumed to be uniform within each of these compartments. The time rate of change of concentration C_i of the ith ionic species in a given compartment is given by

$$\frac{dC_i(t)}{dt} = \frac{I_i(t)}{z_i FV} \tag{5}$$

where $C_i(t)$ is concentration (typically mM) of species i, t is time (typically ms), $I_i(t)$ is net current into the compartment carried by species i (typically in pA), z_i is valence of the ith species, F is Faraday's constant, and V is compartment volume (typically in units of pl). One such equation may be defined for the concentration of each ionic species in each model compartment. Ion flux between compartments, related to the term $I_i(t)$ in eq. (5), is produced by: (a) diffusion due to differences in ion species concentration between adjacent compartments (for example, flux produced by Ca^{2+} diffusion from the subspace to the cytosol); (b) gating of ion channels in the sarcolemmal or JSR membrane (for example, Ca^{2+} flux from the extracellular space into the subspace through sarcolemmal LCCs); or (c) the action of membrane transporters and exchangers (for example, Ca^{2+} flux from the cytosol into the NSR through the SR Ca^{2+}-ATPase). The form of the algebraic equations describing function of membrane transporters and exchangers, including their concentration-, voltage-, and in some instances ATP-dependence, may be found in the published equations for a number of myocyte models. In addition, buffering of Ca^{2+} by negatively charged phospholipid head groups in the sarcolemmal and JSR subspace membrane, by cytosolic calmodulin and myofilaments (troponin) and calsequestrin in the JSR, is modeled.

Composite Equations for Common Pool Models

The equations defining the common pool model of the cardiac myocyte specify an initial value problem composed of an initial condition and a set of coupled nonlinear ordinary differential algebraic equations describing the time rate of change of model state variables. These state variables are: (a) probability of occupancy of ion channel states such as those of eq. (2) and current flux through open channels as in eq. (4); (b) concentrations of ion species in model compartments as in eq. (5); and (c) time evolution of membrane potential. Currently, all biophysically detailed models of the myocyte assume that since these cells are spatially compact, they are isopotential, with time rate of change of membrane potential given by:

$$\frac{dv(t)}{dt} = -\left\{ \sum_i I_i^{\text{ion}}[v(t)] + \sum_i I_i^{\text{pump}}[v(t), c(t)] \right\} \tag{6}$$

where $v(t)$ is membrane potential, $I_i^{\text{ion}}[v(t)]$ is current carried by the ith membrane current, and $I_i^{\text{pump}}[v(t), c(t)]$ is current through the ith membrane pump/exchanger, which can depend on both membrane potential $v(t)$ and the relevant ion concentration $c(t)$.

Successes and Failures of Common Pool Models

Common pool models have been used to explore diverse properties of the cardiac myocyte. In this section we focus on some major successes of these models, but more importantly, on a recently discovered, significant weakness of this modeling approach.

Figures 9.4a and 9.4b show examples of APs (figure 9.4a) and Ca^{2+} transients (figure 9.4b) measured from ventricular myocytes isolated from normal (solid line) and failing (dashed line) canine hearts. Figures 9.4c–d show corresponding APs (figure 9.4c) and Ca^{2+} transients (figure 9.4d) computed by numerical solution of the model initial value problem. These results were obtained using the canine ventricular myocyte model of Winslow et al. [31] for normal (solid line) and failing (dashed line) myocytes. These data demonstrate that common pool models have been quite successful in reconstruction of the AP and the time-varying waveform of the cytosolic Ca^{2+} transient.

Results of a more challenging simulation, obtained using the Jafri–Rice–Winslow model of the guinea pig ventricular myocyte [11], are shown in figure 9.5. In this simulation, the model was used to elucidate mechanisms of the interval-force relation, a classical characterization of cardiac muscle in which developed force is strongly determined by pacing history. In these simulations, the model cell was paced at an interstimulus interval of 1.5 s until a steady state was reached. Panels (a) and (b) of figure 9.5 show model Ca^{2+} (figure 9.5a) and isometric force (figure 9.5b) transients in response to the pacing stimuli. Isometric force was computed using a model developed by Rice et al. [11,32] that relates cytosolic Ca^{2+} level to isometric force. Ca^{2+} and force transients labeled SS in figures 9.5a–b are responses

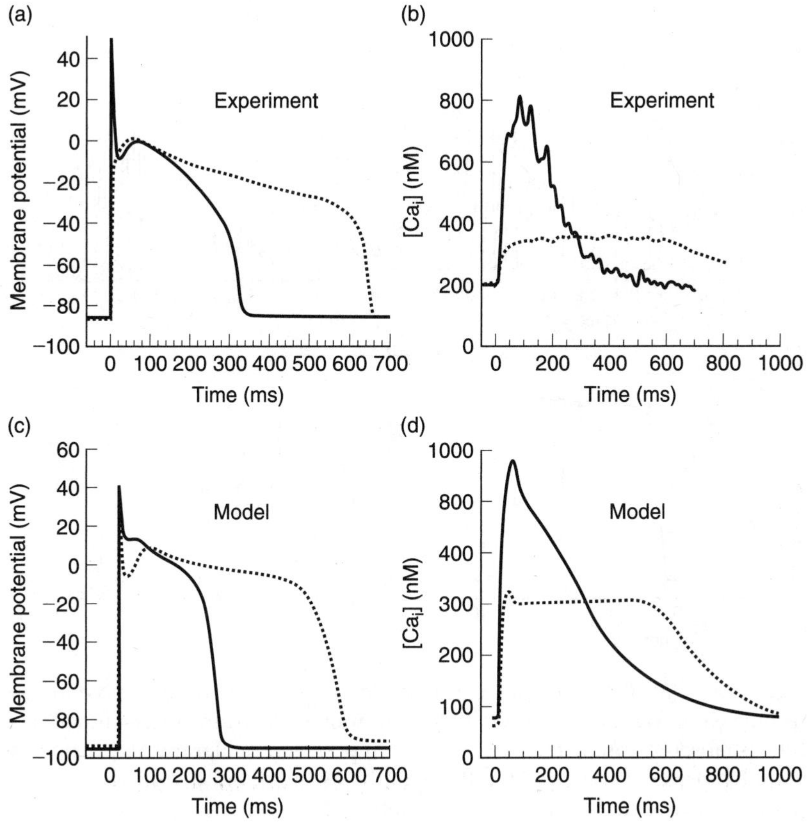

Figure 9.4 Model versus experimental action potentials and Ca^{2+} transients. Each action potential and Ca^{2+} transient is in response to a 1 Hz pulse train, with responses measured in the steady-state. (a) Experimentally measured membrane potential (mV, ordinate) as a function of time (ms, abscissa) in normal (solid) and failing (dashed) canine myocytes. (b) Experimentally measured cytosolic Ca^{2+} concentration (nmol/l, ordinate) as a function of time (ms, abscissa) for normal (solid) and failing (dashed) canine ventricular myocytes. (c) Membrane potential (mV, ordinate) as a function of time (ms, abscissa) simulated using the normal canine myocyte model (solid) and failing (dashed) canine myoycte model (end-stage heart failure modeled by downregulation of I_{to1} by 66%, downregulation of I_{K1} by 32%, downregulation of SERCA2 by 62%, and upregulation of NCX1 by 75%). (d) Cytosolic Ca^{2+} concentration (nmol/l, ordinate) as a function of time (ms, abscissa) simulated using the normal (solid) and heart failure (dashed) model.

to the last stimulus of this periodic pulse train. Following cessation of the periodic stimulus train, a second stimulus (denoted S2) is delivered at a variable interval. Responses labeled *a* and *b* in figures 9.5a–b are Ca^{2+} and force transients in response to short or long S2 intervals, respectively, following cessation of the periodic pulse train. A third

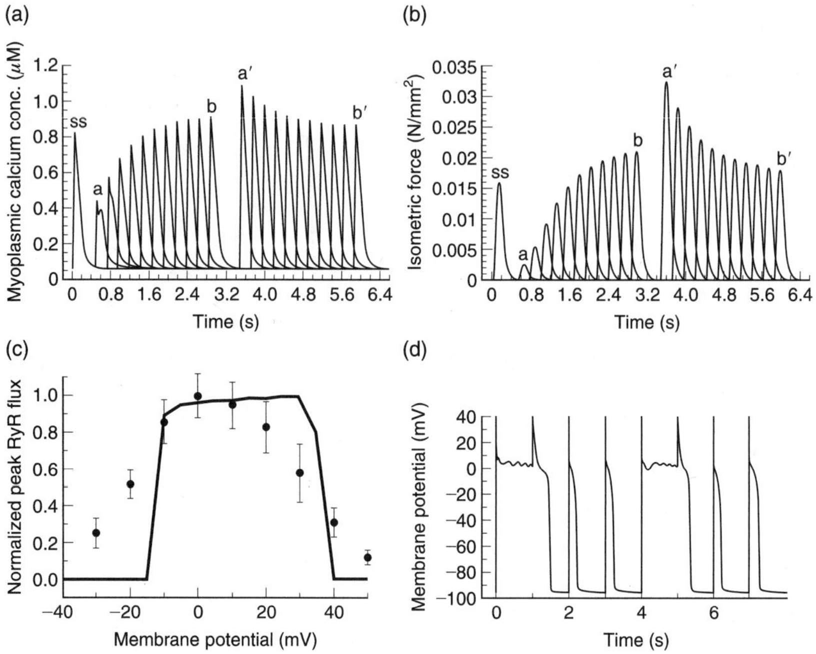

Figure 9.5 (a) Extrasystolic restitution and post-extrasystolic potentiation responses showing cytosolic Ca^{2+} concentration (μM, ordinate) as a function of time (s, abscissa) predicted by the common pool ventricular myocyte using the stimulus protocols described in the text. (b) As in (a), but with isometric force (N/mm^2) predicted from the Ca^{2+} transients in (a), force plotted on the ordinate. (c) Normalized peak RyR release flux (ordinate) as a function of membrane potential (mV, abscissa) for rat ventricular myocytes (symbols) and for the common pool model (solid line). (d) Behavior of the common pool ventricular myocyte model when Ca^{2+}-dependent inactivation of LCCs is adjusted to be stronger than voltage-dependent inactivation as per data of Linz and Meyer [38]. Note instability of action potentials.

stimulus (denoted S3) is delivered 3 s after the corresponding S2 stimulus. Responses to the S3 stimuli delivered 3 s after the corresponding S2 stimuli *a* and *b* are labeled in figures 9.5a–b as *a′* and *b′*, respectively. As seen experimentally, a small extrasystolic response to S2 leads to a greater response to S3, an effect known as post-extrasystolic potentiation [33].

S2 responses are seen to increase in amplitude as a function of the S2 interval. This property is known as extrasystolic restitution and, in the model, is determined primarily by the rate of recovery of the RyR from adaptation following opening and release of Ca^{2+} from the JSR [32]. S3 responses are seen to decline in amplitude depending on the corresponding S2 interval. The mechanism can be understood by contrasting the magnitude of Ca^{2+} release events for S2 responses *a* and *b* in figure 9.5a. Since the S2 interval is long for response *b*, this provides ample time for SR Ca^{2+} levels to be restored by the SR Ca^{2+}-ATPase, and Ca^{2+} release amplitude is large. This large release event depletes SR Ca^{2+} levels substantially so that the release event *b'* initiated 3 s later by the S3 stimulus is small. The converse is true for release events *a* and *a'*.

The responses shown in figures 9.5a–b agree very well with experimental data (contrast these model data with experimental findings in Wier et al., fig. 3 [34]), and show that common pool models are capable of describing complex intracellular Ca^{2+} cycling processes determined by the rate of SR Ca^{2+} loading by the SR Ca^{2+}-ATPase. However, common pool models, as noted previously, are not able to reproduce a very fundamental behavior of cardiac myocytes—SR Ca^{2+} release, which is a smooth and continuously graded function of trigger Ca^{2+} through sarcolemmal LCCs. This failure is demonstrated in figure 9.5c. This figure shows normalized peak Ca^{2+} flux through RyR channels (ordinate) as a function of membrane potential (mV, abscissa). Closed circles are experimental measurements from the work of Wier et al. [35], showing that release flux increases smoothly to a maximum flux at about 0 mV, and then decreases to near zero at more depolarized potentials. Release flux increases from –40 to 0 mV, since over this potential range the open probability of LCCs increases very steeply, reaching a maximum value. Release flux decreases over the potential range greater than 0 mV because electrical driving force on Ca^{2+} decreases monotonically. The solid line shows release flux for the Jafri–Rice–Winslow guinea pig ventricular myocyte model. Release is all-or-none, with regenerative release initiated at a membrane potential causing opening of a sufficient number of LCCs (~–15 mV), and release terminating at a potential for which electrical driving force is reduced to a critical level (~+40 mV).

This all-or-none behavior of Ca^{2+} release in common pool models has very important implications for the integrative function of the cell. As noted previously, LCCs not only undergo voltage- but also Ca^{2+}-dependent inactivation [36,37]. Inactivation depends on local subspace Ca^{2+} concentration, and occurs as Ca^{2+} binding to calmodulin [37], which is tethered to the LCC, induces the channel to switch from a normal mode of gating to a mode in which transitions to open states are extremely rare or nonexistent (these Ca^{2+}-inactivated states were labeled "Mode Ca" in the model of figure 9.3). Recent experimental data have demonstrated that voltage-dependent inactivation of LCCs is a slow and weak process, whereas Ca^{2+}-dependent inactivation is relatively fast and strong [37,38]. This in turn implies that there is a very strong coupling between Ca^{2+} release from JSR into the local subspace, and regulation of inactivation of the LCC. When this newly revealed balance between voltage- and Ca^{2+}-dependent inactivation is incorporated into common pool models, the models become unstable, exhibiting alternating short and long duration APs. An example of this behavior is shown in figure 9.5d. The reason for this is intuitively clear: since JSR Ca^{2+} release is all-or-none in these models, Ca^{2+}-dependent inactivation of LCCs is all-or-none, depending on whether release has or has not occurred. Since L-type Ca^{2+} current is a major contributor to inward current during the plateau phase of the AP (see figure 9.2), its biphasic inactivation leads to instability of AP duration. This, unfortunately, constitutes a fatal weakness of biophysically based common pool models.

Challenges for Single Myocyte Modeling

The fundamental failure of common pool models described above demonstrates that capturing the property of graded release is critically important for modeling dynamics of single cardiac myocytes. Understanding of the mechanisms by which Ca^{2+} influx via LCCs triggers Ca^{2+} release from the JSR has advanced tremendously with the development of experimental techniques for simultaneous measurement of LCC currents and Ca^{2+} transients and detection of local Ca^{2+} transients, and this has given rise to the local control hypothesis of CICR [35,39–41]. This hypothesis asserts that opening of an individual LCC in the T-tubular membrane triggers Ca^{2+} release from the small cluster of RyRs located in the closely apposed JSR membrane (see figure 9.1b). Thus, the local control hypothesis asserts that release is all-or-none at the level of these individual groupings of LCCs and RyRs. However, LCC:RyR clusters are physically separated at the ends of the sarcomeres [42]. These clusters therefore function in an approximately independent fashion. The local control hypothesis asserts that graded control of SR Ca^{2+} release is achieved by the statistical recruitment of elementary Ca^{2+} release events in these independent dyadic spaces.

We have recently implemented a local control model of myocyte function [43]. As a compromise between structural and biophysical detail versus tractability, a "minimal model" of local control of Ca^{2+} release, referred to as the Ca^{2+} release unit (CaRU) model, was developed. This model is intended to mimic the properties of Ca^{2+} sparks in the T-tubule/SR (T-SR) junction (Ca^{2+} sparks are elementary SR Ca^{2+} release events arising from opening of a cluster of RyRs [44]). Each CaRU in this model consists of a dyadic space, 4 LCCs and 20 RyRs in the JSR and sarcolemmal membranes, respectively. All 20 RyRs in the CaRU communicate via Ca^{2+} diffusion within the single local JSR (dyadic) volume. The 5:1 RyR to LCC stoichiometry is chosen to be consistent with recent estimates indicating that a single LCC typically triggers the opening of 4–6 RyRs [45]. Since LCC:RyR clusters are physically separated [42], each model CaRU is assumed to function independently of other CaRUs.

Simulation of the dynamics of the CaRU model requires both numerical integration of the ordinary differential equations (ODEs) describing time-varying global Ca^{2+}, Na^+, and K^+ concentrations and K^+ and Na^+ channel states in the myocyte model as well as Monte Carlo simulation of LCC and RyR channel gating in the approximately ~12,500 CaRUs in the cell (yielding ~50,000 LCCs per ventricular myocyte). The state of each LCC and RyR channel is described by a set of discrete valued random variables that evolve in time as described by Markov processes. Time steps for CaRU simulations are adaptive and are chosen to be sufficiently small based on channel transition rates. The CaRU simulations occur within the (larger) time step used for the numerical integration of the system of ordinary differential equations describing the time evolution of the global state variables. As a result of the embedded Monte Carlo simulation, all model state variables and ionic currents/fluxes will contain a component of stochastic noise. These fluctuations introduce a degree of variability to simulation output.

Figure 9.6 shows macroscopic properties of APs and SR Ca^{2+} release in this CaRU model. Figure 9.6a shows relative balance between the fraction of LCCs *not* voltage-inactivated (dotted line) and *not* Ca^{2+}-inactivated (dashed line) during the simulated AP (solid line). These fractions were designed to fit the experimental data of Linz and Meyer [38] showing that Ca^{2+}-dependent inactivation is stronger than is voltage-dependent inactivation. The solid line shows an AP predicted by the local control model. This AP should be contrasted with those produced by the common pool mode (figure 9.5d) when the same relationship between LCC voltage- and Ca^{2+}-dependent inactivation shown in figure 9.6a is used. Clearly, the local control model exhibits stable APs whereas the common pool model does not. Figure 9.6b shows the voltage-dependence of peak LCC Ca^{2+} influx ($F_{LCC(max)}$, closed circles, ordinate) and peak RyR Ca^{2+} release flux ($F_{RyR(max)}$, open circles, ordinate) in response to voltage-clamp steps to the indicated potentials (mV, abscissa). These data demonstrate graded release, as Ca^{2+} release flux is a smooth and continuous function of membrane potential and hence triggers Ca^{2+} flux through LCCs.

These simulations offer an intriguing glimpse of how the colocalization and stochastic gating of individual channel complexes can have a profound effect on the overall, integrative behavior of the cell. The results also point out that a key challenge remains. That is, computational models of the ventricular myocyte incorporating a phenomenological description of CICR have reduced predictive power as they do not describe the biophysical basis of the release process.

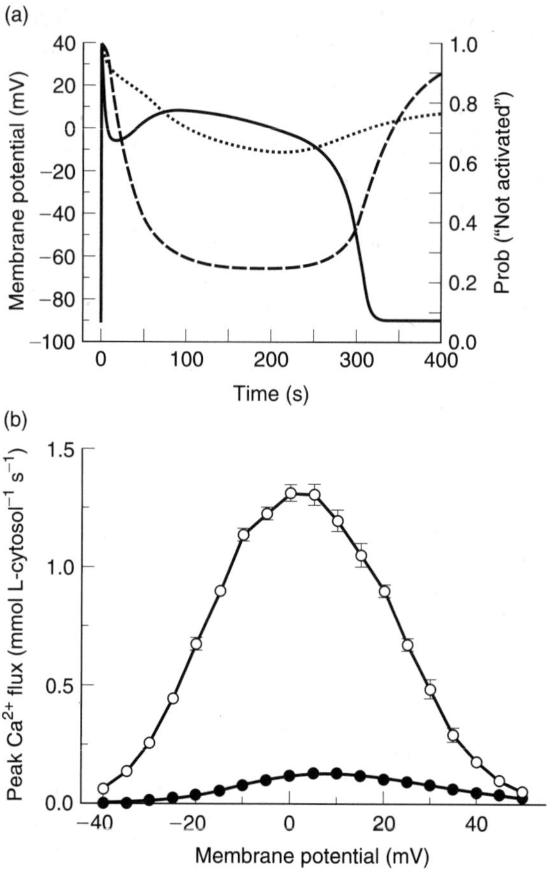

Figure 9.6 (a) Action potential (solid line) obtained in the CaRU model when the fraction of LCCs not voltage-inactivated (dotted line) and not Ca^{2+}-inactivated (dashed line) is adjusted to match the experimental data of Linz and Meyer [38]. Not stable AP, in contrast to those predicted when using the common pool model in figure 9.5d. (b) Mean peak Ca^{2+} flux amplitudes, $F_{\mathrm{LCC(max)}}$ (closed circles) and $F_{\mathrm{RyR(max)}}$ (open circles), as a function of membrane voltage for the CaRU model, $n = 5$ simulations at each voltage.

Common pool models do incorporate biophysical detail but cannot describe either graded CICR or predict stable APs. The CaRU model described above is able to capture key experimentally measured properties of CICR, but only at very high computational cost. A key challenge that remains is to find simpler approaches to the integrative modeling of CICR in cardiac myocytes that retain predictive power while achieving reduced computational cost.

INTEGRATIVE MODELING OF ELECTRICAL CONDUCTION IN THE CARDIAC VENTRICLES

Computational models of the cardiac myocyte have contributed greatly to our understanding of myocyte function. This is in large part due to a rich interplay between experiment and modeling—an interplay in which experiments inform modeling, and modeling suggests new experiments. Studies of cardiac ventricular conduction have to a large extent lacked this interplay. Stated more precisely, both two- and three-dimensional models of cardiac tissue as well as three-dimensional structurally detailed models of the cardiac ventricles have been used to simulate conduction in the heart. However, the extent to which these models can reproduce as well as predict experimental data has seldom been tested. "Closing the loop" between experimental and modeling studies of electrical conduction within the ventricles is a necessary step toward relating molecular and cellular events that are the basis of arrhythmia to their functional consequences at the level of whole-heart function.

The following sections describe one approach to this problem. In this approach, we (a) model ventricular geometry and fiber orientation in the same hearts that have been electrically mapped using data obtained from diffusion tensor magnetic resonance imaging (DTMRI); (b) construct computational models of the imaged hearts; and (c) compare simulated conduction properties with those measured experimentally in the same heart as a quantitative test of the models.

Measuring the Fiber Structure of the Cardiac Ventricles Using DTMRI

In addition to the biophysical properties of cells, the geometry and spatial orientation of ventricular fibers plays a critical role in shaping electrical propagation within the ventricles. Conduction is influenced by properties of tissue geometry [46–48] and is anisotropic, with current spread being most rapid in the direction of the fiber long axis [49–55]. The spatial rate of change of fiber orientation also governs conduction properties [56,57]. Remodeling of both ventricular geometry and fiber organization is a prominent feature of several cardiac pathologies [58–68], and these alterations may figure importantly in arrhythmogenesis [69]. A detailed knowledge of ventricular geometry and fiber orientation, how it may be remodeled in cardiac pathology, and the effects of this remodeling on ventricular conduction is therefore of fundamental importance to the understanding of cardiac ventricular electromechanics in health and disease.

Present understanding of ventricular fiber organization is based on fiber dissections of whole hearts [70–72] and histological measurement of fiber orientation in transmural plugs of ventricular tissue [73–77]. The principle conclusions of this work are that cardiac fibers are arranged as counter-wound helices encircling the ventricular cavities, and that fiber orientation depends on transmural location, with fiber direction being oriented predominantly in the base–apex direction on the epicardial and endocardial surfaces, and rotating to a circumferential direction in the midwall. These early studies of transmural variation of inclination angle were restricted to a small number of measurement sites. Nielsen et al. [78] overcame this limitation by using a custom-built histological apparatus to measure fiber orientation at up to ~14,000 points throughout the myocardium, with a resolution of ~500, 2500, and 5000 μm in the radial, longitudinal, and circumferential directions, respectively. This work led to the first complete reconstruction of ventricular geometry and fiber inclination angle—an achievement that has had significant impact on the cardiac mechanics and electrophysiology communities.

Despite this important advance, present histological methods suffer from the disadvantage that they require many weeks to even months for reconstruction of a single heart. As a consequence, only small numbers of canine and rabbit hearts have been reconstructed. The complete ventricular fiber organization of the human heart has never been reconstructed and modeled. Full reconstruction of ventricular fiber organization in diseased hearts has never been performed.

The use of DTMRI to measure cardiac geometry and fiber orientation solves many of these problems. DTMRI is based on the principle that proton diffusion in the presence of a magnetic field gradient causes signal attenuation, and that measurement of this attenuation in several different directions can be used to estimate a diffusion tensor at each image voxel [79,80]. Several studies have now confirmed that the principal eigenvector of the diffusion tensor measured at each imaging voxel is locally aligned with the long axis of cardiac fibers [81–83]. Use of DTMRI for reconstruction of cardiac fiber orientation provides several advantages over traditional histological methods:

1. DTMRI yields estimates of the absolute orientation of cardiac fibers, whereas histological methods yield estimates of only fiber inclination angle (defined as the angle formed by projecting fiber orientation onto the epicardial tangent plane, and computing the angle between this projection and a circumferentially oriented epicardial tangent plane vector).
2. DTMRI performed using formalin-fixed tissue (a) yields high-resolution images of the cardiac boundaries, thus enabling precise reconstruction of ventricular geometry using image segmentation software, and (b) eliminates flow artifacts present in perfused heart, enabling longer imaging times, increased signal-to-noise ratio, and improved spatial resolution.
3. DTMRI provides estimates of fiber orientation at more than several orders of magnitude more points than is possible with histological methods.

4. Reconstruction time is greatly reduced (tens of hours versus weeks to months) relative to that for histological methods.

Preparation of hearts for DTMR imaging is relatively straightforward, and proceeds as follows. Hearts are first excised and fixed in a 3% formalin mixture. Following fixation, excised hearts are placed in an acrylic container filled with Fomblin, a perfluoropolyether used to increase contrast and eliminate unwanted susceptibility effects near boundaries of the heart. Images presented here were acquired with a 3-D Fast Spin Echo Diffusion Tensor (3D FSE DT) sequence on a 1.5 T GE CV/I MRI Scanner (GE, Medical System, Wausheka, WI) using an enhanced gradient system with 40 mT/m maximum gradient amplitude and a 150 T/m/s slew rate. Depending on heart size, the MR parameters were varied to minimize the number of slices acquired. Generally, the FOV was 9 cm with an image size of 256 x 256 yielding an in-plane resolution of ~350 μm. The volume was imaged with 130–140 slices at ~800 μm thickness. Diffusion gradients were applied in sixteen noncollinear directions with a maximum *b* value of 900 s/mm^2. With an echo train length of 2 and a TR of ~700 ms, the total scan time was ~64 hours.

Recent advances in our lab have enabled automation of DTMRI data acquisition and analysis for ventricular reconstruction. Briefly, once image data are acquired, software written in the MatLab programming language is used to estimate epicardial and endocardial boundaries in each short-axis section of the image volume using either the method of region growing or the method of parametric active contours [84]. Diffusion tensor eigenvalues and eigenvectors are computed from the DTMRI data sets at those image voxels corresponding to myocardial points, and fiber orientation at each image voxel is computed as the primary eigenvector of the diffusion tensor.

Representative results from DTMR imaging are shown in figure 9.7. Figure 9.7a shows a finite-element model (see following section) describing the epicardial surface of the canine ventricles derived from DTMRI data. The line segments on the surface of this heart show epicardial fiber orientation measured using DTMRI. Figure 9.7b shows the volume-rendered DTRMI data, with voxels color-coded according to fiber inclination angle. Inclination angle (also known as fiber or helix angle) is known to vary continuously as a function of transmural depth. This variation is shown in figure 9.7b. Figure 9.7c shows data similar to that of figure 9.7b, the difference being that a single short-axis section showing fiber inclination angle estimated using the DTMRI data is rendered with an MR intensity image of this same heart. Inclination angle is again color-coded. In addition, diffusion tensors are visualized using a family of objects called glyphs [85]. This visualization technique facilitates voxel-based display of tensor data.

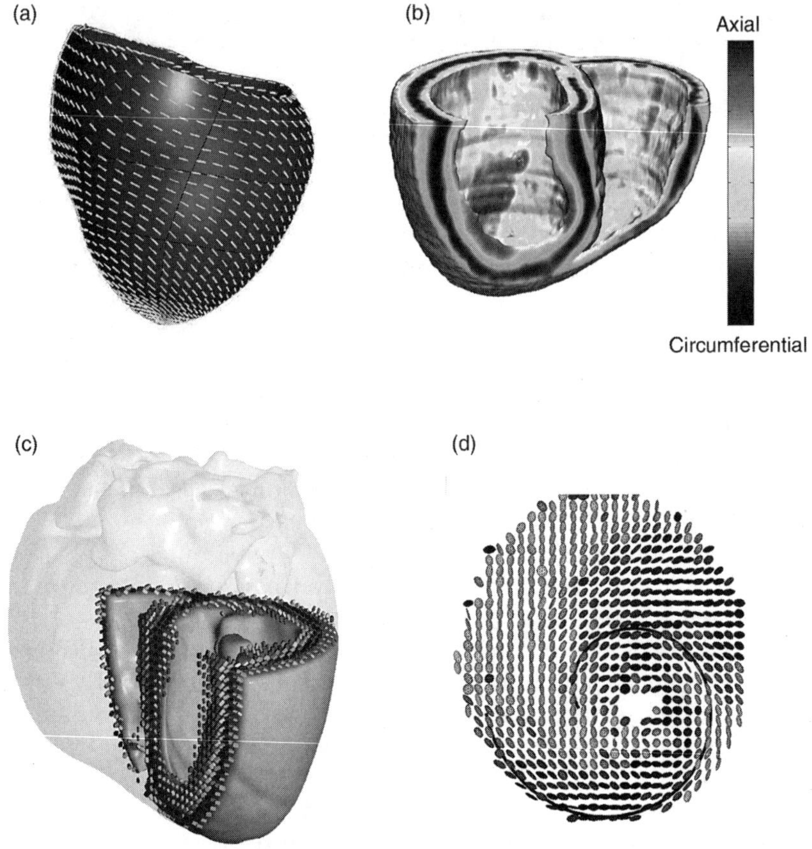

Figure 9.7 (a) Finite-element model of the canine ventricles with fiber orientation displayed on the epicardial surface using short line segments. (b) DTMRI-based reconstruction of fiber inclination angle (see color code) throughout the canine ventricles. (c) Transmural variation of fiber inclination angle (color coded as in (b)) with diffusion tensor data displayed using glyphs (see text). (d) Spatial rotation of epicardial fibers near the apex of the heart.

Eigenvectors of the diffusion tensor define glyph orientation and eigenvalues define glyph shape. In particular, cylindrical glyphs indicate a single preferred direction of diffusion with diffusion being isotropic in the transverse directions, and cuboid glyphs represent orthotropic diffusion (three distinct eigenvalues indicative of myocardial sheets or laminae). Figure 9.7d shows the spiral pattern that cardiac fibers take near the apex of the heart as they plunge into the infundibulum. Reconstruction of the high rate of twist of cardiac fibers in this region is a challenging test of the spatial resolution of modern DTMRI methods. These data demonstrate that DTMRI may be used to reconstruct ventricular anatomy at high spatial resolution.

Finite-Element Modeling of Cardiac Ventricular Anatomy

The structure of the cardiac ventricles can be modeled using finite-element modeling (FEM) methods developed by Nielsen et al. [78]. The geometry of the heart to be modeled is described initially using a predefined mesh with six circumferential elements and four axial elements. Elements use a cubic Hermite interpolation in the transmural coordinate (η), and bilinear interpolation in the longitudinal (μ) and circumferential (θ) coordinates. Voxels in the 3D DTMR images identified as being on the epicardial and endocardial surfaces by the semiautomated contouring algorithms described above are used to deform this initial FEM template. Deformation of the initial mesh is performed to minimize an objective function F(n):

$$F(\boldsymbol{n}) = \sum_{d=1}^{D} \gamma_d \left\| v(\boldsymbol{\varepsilon}_d) - v_d \right\|^2 + \int_{\Re^2} \{\alpha \nabla^2 \boldsymbol{n} + \beta (\nabla^2 \boldsymbol{n})^2\} \partial \boldsymbol{\varepsilon} \tag{7}$$

where $\boldsymbol{n}$ is a vector of mesh nodal values, v_d are the surface voxel data, $v(\varepsilon_d)$ are the projections of the surface voxel data on the mesh, and α and β are user-defined constants. This objective function consists of two terms. The first describes distance between each surface image voxel (v_d) and its projection onto the mesh $v(\varepsilon_d)$. The second, known as the weighed Sobelov norm, limits the stretching (first derivative terms) and the bending (second derivative terms) of the surface. The parameters α and β control the degree of deformation of each element. The weighted Sobelov norm is particularly useful in cases where there is an uneven distribution of surface voxels across the elements. A linear least-squares algorithm is used to minimize this objective function.

After the geometric mesh is fitted to DTMRI data, the fiber field is defined for the model. Principal eigenvectors lying within the boundaries of the mesh computed above are transformed into the local geometric coordinates of the model using the following transformation:

$$\boldsymbol{V}_G = [\boldsymbol{F}\ \boldsymbol{G}\ \boldsymbol{H}]^T [R] \boldsymbol{V}_S \tag{8}$$

where R is a rotation matrix that transforms a vector from scanner coordinates ($\boldsymbol{V}_S$) into the FEM model coordinates $\boldsymbol{V}_G$, and $\boldsymbol{F}, \boldsymbol{G}, \boldsymbol{H}$ are orthogonal geometric unit vectors computed from the ventricular geometry as described by LeGrice et al [86]. Once the fiber vectors are represented in geometric coordinates, DTMRI inclination and imbrication angles (α and ϕ) are fitted using a bilinear interpolation in the local ε_1 and ε_2 coordinates, and a cubic Hermite interpolation in the ε_3 coordinate. A graphical user interface for fitting FEMs to both the ventricular surfaces and fiber field data has been implemented using the MatLab programming language (available at www.ccbm.jhu.edu). An example of an FEM fitted to the epicardial surface of a reconstructed normal canine heart obtained using this software tool was shown in figure 9.7a.

Generation of Computational Models from DTMRI Data

Having developed a computational model describing both the geometry of the heart surfaces as well as fiber organization within the heart, we are ready to solve the equations governing electrical current conduction within the myocardium. These equations are known as the bidomain equations. Several excellent reviews have been published detailing the assumptions in, structure of, and solution methods for the bidomain equations [87–89]. The following is a brief review of the origins of these equations.

The bidomain equations are derived by applying conservation of current between the intra- and extracellular domains. The equations consist of parabolic [eq. (9)] and elliptic [eq. (10)] equations that must be satisfied within the myocardium (a region designated as H):

$$\frac{\partial v}{\partial t}(\boldsymbol{x},t)=\frac{1}{C_m}\left[-I_{\text{ion}}(\boldsymbol{x},t)-I_{\text{app}}(\boldsymbol{x},t)-\frac{1}{\beta}\left(\nabla\cdot M_e(\boldsymbol{x})\nabla\phi_e(\boldsymbol{x},t)\right)\right] \quad (9)$$

$$\nabla\cdot M_i(\boldsymbol{x})\nabla v(\boldsymbol{x},t)=-\nabla\cdot M(\boldsymbol{x})\nabla\phi_e(\boldsymbol{x},t), \qquad \forall\boldsymbol{x}\text{ in }H \quad (10)$$

and an additional elliptic equation [eq. (11)] that must be satisfied in the bath or tissue surrounding the heart (a region designated as B):

$$\nabla\cdot M_b(\boldsymbol{x})\nabla\phi_b(\boldsymbol{x},t)=0, \qquad \forall\boldsymbol{x}\text{ in }B \quad (11)$$

where $\boldsymbol{x}$ is spatial position; $\phi_i(\boldsymbol{x},t)$ and $\phi_e(\boldsymbol{x},t)$ are the transmembrane intra- and extracellular potentials, respectively; $v(\boldsymbol{x},t) = \phi_i(\boldsymbol{x},t) - \phi_e(\boldsymbol{x},t)$ is the transmembrane voltage; C_m is the membrane capacitance per unit area; $I_{\text{ion}}(\boldsymbol{x},t)$ is the sum of the ionic currents per unit area through the membrane (positive outward), as given by the single-cell models described in the previous section; $I_{\text{app}}(\boldsymbol{x},t)$ is an applied cathodal extracellular current per unit area; β is the ratio of membrane area to tissue volume; $M_e(\boldsymbol{x})$ and $M_i(\boldsymbol{x})$ are the extracellular and intracellular conductivity tensors, with $M(\boldsymbol{x}) = M_e(\boldsymbol{x}) + M_i(\boldsymbol{x})$; $\phi_b(\boldsymbol{x},t)$ is the bath potential; $M_b(\boldsymbol{x})$ is the bath conductivity tensor. These parameters may be set, in models of the normal heart, using values described by Pollard et al. [87] and Henriquez et al. [89]. Additionally, boundary conditions on the interface between the heart and the surrounding tissue, δH, and the body surface, δB, must be specified. The first boundary condition specifies continuity of potential:

$$\phi_e=\phi_b \text{ on } \delta H \quad (12)$$

The second specifies continuity of current at the interface:

$$\sigma_i\frac{\partial\phi_i}{\partial n}+\sigma_e\frac{\partial\phi_e}{\partial n}=\sigma_b\frac{\partial\phi_b}{\partial n} \text{ on } \delta H \quad (13)$$

where the σ's are the conductivities normal to the interface, and the $\partial/\partial n$ is the normal derivative operator.

In order for the problem to be well posed, a third boundary condition on δH is required. While the first two follow necessarily for all electrical phenomena, there are a number of ways to formulate the third boundary condition. Typically, we specify:

$$\sigma_i \frac{\partial \phi_i}{\partial n} = 0 \text{ on } \delta H \tag{14}$$

which has the physical interpretation that at the heart/body interface, all intracellular current must flow first through the extracellular space before it flows into the surrounding tissue. A boundary condition at δB for the Laplace equation in ϕ_b is also required. Given that air is a poor conductor, this is simply

$$\sigma_b \frac{\partial \phi_b}{\partial n} = 0 \text{ on } \delta B \tag{15}$$

Finally, an initial condition on the transmembrane voltage must be specified, $v(x,t=0) = V(x)$. Then from this, initial conditions on $\phi_e(x,t=0)$ and $\phi_b(x,t=0)$ can be found by solving the appropriate elliptic equation. Equations (9)–(15) specify the bidomain problem.

Under some restrictive assumptions, the bidomain equations can be simplified dramatically. If the surrounding tissue is taken to be a good insulator, then $\sigma_b = 0$ in B. Then we have

$$\begin{aligned} \phi_e &= 0 \\ \frac{\partial \phi_e}{\partial \eta} &= 0 \\ \frac{\partial \phi_i}{\partial n} &= 0 \text{ on } \delta H \end{aligned} \tag{16}$$

and the Laplace equation for ϕ_b need not be considered. Additionally, under the assumption of equal anisotropy, namely, that

$$M_i(x) = \frac{1}{k} M_e(x) \tag{17}$$

where k is called the anisotropy ratio, eqs. (9) and (10) uncouple, requiring then only solution of the parabolic equation,

$$\frac{\partial v}{\partial t}(x,t) = \frac{1}{C_m}\left[-I_{\text{ion}}(x,t) - I_{\text{app}}(x,t) + \frac{1}{\beta}\left(\frac{\kappa}{\kappa+1}\right)\nabla \cdot (M_i(x)\nabla v(x,t))\right] \tag{18}$$

Equation (18) is referred to as the monodomain equation.

Conductivity tensors at each point within the heart are specified by fiber orientation and by specific conductivities in each of the local coordinate directions. The conductivity tensor in the local coordinate system, $G_i(x)$, is defined as

$$G_i(\boldsymbol{x}) = \begin{bmatrix} \sigma_{1,i} & & \\ & \sigma_{2,i} & \\ & & \sigma_{3,i} \end{bmatrix} \tag{19}$$

where $\sigma_{1,i}$ is the longitudinal and $\sigma_{2,i}$ and $\sigma_{3,i}$ are the transverse intracellular conductivities, respectively. This local tensor may be expressed in global coordinates to give the conductivity tensor of eq. (18) using the transformation

$$M_i(x) = P(x)G(x)P^T(x) \tag{20}$$

where $P(x)$ is the coordinate transformation matrix from local to global coordinates. $P(x)$ is in turn determined by the underlying fiber organization of the heart, and is obtained using DTMRI as described previously.

Figure 9.8 shows the results of applying these methods to the analysis of conduction in a normal canine heart. Figure 9.8a shows activation time (color bar, in ms) measured experimentally in response to a stimulus pulse applied at the epicardial locations marked by the silver balls positioned on the left aspect of the heart. Epicardial conduction was measured using electrode arrays consisting of a nylon mesh with 256 electrodes and electrode spacing of ~5 mm sewn around its surface. Bipolar epicardial twisted-pair pacing electrodes were sewn onto the right atrium (RA) and the right ventricular (RV) free wall. Four to ten glass beads filled with gadolinium-DTPA (~5 mM) were attached to the sock as localization markers, and responses to different pacing protocols were recorded. After all electrical recordings were obtained, the animal was euthanized with a bolus of potassium chloride, and the heart was scanned with high-resolution T1-weighted imaging in order to locate the gadolinium-DTPA-filled beads in scanner coordinates. Following electrical mapping, this heart was excised, imaged using DTMRI, and an FEM was then fitted to the resulting geometry and fiber orientation data sets. Figure 9.8a shows experimentally measured activation times displayed on this FEM. The stimulus wave front can be seen to follow the orientation of the epicardial fibers, which is indicated by the dark line segments in figure 9.8a. Figure 9.8b shows results of simulating conduction using a computational model of the very same heart that was mapped electrically in figure 9.8a. Results can be seen to agree qualitatively; however, model conduction is more rapid in the region where the right and left ventricles join.

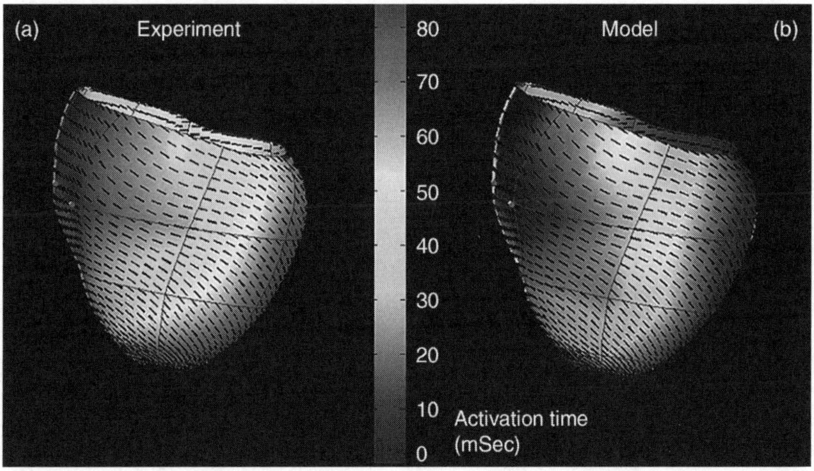

Figure 9.8 (a) Electrical activation times (indicated by color bar) in response to right RV pacing as recorded using electrode arrays. Data were obtained from a normal canine heart that was subsequently reconstructed using DTMRI. Activation times are displayed on the epicardial surface of a finite-element model fitted to the DTMRI reconstruction data. Fiber orientation on the epicardial surface, as fitted to the DTMRI data by the FEM model, is shown by the short line segments. (b) Activation times predicted using a computational model of the heart mapped in (a). To view this figure in color, see the companion web site for *Systems Biology*, http://www.oup.com/us/sysbio.

Nonetheless, these results demonstrate the feasibility of combined experimentation and modeling of electrical conduction in specific imaged and reconstructed hearts.

CONCLUSION

This chapter has reviewed modeling research in three broad areas: (1) models of single ventricular myocytes; (2) methods for the reconstruction and modeling of ventricular geometry and microanatomy; and (3) integrative modeling of the cardiac ventricles. We have seen that the level of biophysical detail, and hence the accuracy and predictability of current ventricular myocyte models, is considerable. Nonetheless, much remains to be done.

One emerging area of research is modeling of mitochondrial energy production. Approximately 2% of cellular ATP is consumed on each heartbeat. The major processes consuming ATP in the myocyte are muscle contraction, activity of the SR Ca^{2+}-ATPase, and Na-K pumping. Cellular ATP levels also influence ion channel function including the sarcolemmal ATP-modulated K channel [90]. Recently, an integrated thermokinetic model of cardiac mitochondrial energetics comprising the tricarboxylic acid (TCA) cycle, oxidative phosphorylation, and mitochondrial Ca^{2+} handling has been formulated [91]. The model is able to reproduce experimental data concerning mitochondrial bioenergetics, Ca^{2+} dynamics, and respiratory control, relying only on the fundamental properties of the system. The time-dependent behavior of the model, under conditions simulating an increase in workload, closely reproduce the experimentally observed mitochondrial NADH dynamics in heart trabeculae subjected to changes in pacing frequency. The steady-state and time-dependent behavior of the model support the role of mitochondrial matrix Ca^{2+} in matching energy supply with demand in cardiac cells. Further development and testing of this model, its integration into models of the myocyte, and the use of these models to investigate myocyte responses to ischemia, are required.

In real cardiac myocytes, there exist a diversity of mechanisms that act to modulate cellular excitability. This includes α- and β-adrenergic signaling pathways acting through G-protein-coupled membrane receptors to modulate properties of LCCs, various K^+ channels, and Ca^{2+} transporters such as the SR Ca^{2+}-ATPase. The addition of these modulatory mechanisms to the cell models remains an important goal for the future.

As demonstrated, magnetic resonance imaging now offers a relatively rapid way to measure ventricular fiber structure at high spatial resolution. The ability to rapidly acquire fiber orientation data throughout the ventricles in large populations of normal and diseased hearts will enable quantitative statistical comparison of normal and abnormal cardiac structure, and will provide insights into the possible structural basis of arrhythmia in heart disease. Unfortunately, a detailed understanding of the spatial heterogeneities within the heart, such as variation of intercellular coupling, regional expression of ionic currents, and Ca^{2+}-handling proteins, is still unavailable, although significant progress has certainly been made. Understanding and modeling of these spatial heterogeneities remains a challenge for the future.

These are exciting times for cardiovascular biology. A national infrastructure supporting the acquisition, distribution, and analysis of cardiovascular genomic and proteomic data is now in place (in particular, the Programs for Genomic Applications and Innovative Proteomics Centers supported by the National Heart, Lung, and Blood Institute of the National Institutes of Health). The data and models produced from these efforts will without question enhance our understanding of the systems biology of the heart in both health and disease. Major challenges in data collection, representation, storage, dissemination, and modeling remain. Meeting these challenges will contribute to the creation of a truly integrated cardiovascular research community, the whole of which is far greater than the sum of its parts.

ACKNOWLEDGMENTS

This work was supported by NIH HL61711, HL60133, HL70894, HL72488, HL52307, the Falk Medical Trust, the Whitaker Foundation, and IBM Corporation. Source code for models used in simulations described in this chapter is available at www.ccbm.jhu.edu.

REFERENCES

1. Fozzard, H. A., et al. (Eds.). *The Heart and Cardiovascular System: Scientific Foundations,* 2nd ed. Raven Press, New York, 1991.
2. Brette, F. and C. Orchard. T-tubule function in mammalian cardiac myocytes. *Circulation Research,* 92(11):1182–92, 2003.
3. Langer, G. A. and A. Peskoff. Calcium concentration and movement in the diadic cleft space of the cardiac ventricular cell. *Biophysical Journal,* 70: 1169–82, 1996.
4. Marban, E., T. Yamagishi and G. F. Tomaselli. Structure and function of voltage-gated sodium channels. *Journal of Physiology (London),* 508(Pt 3): 647–57, 1998.
5. Kemp, T. J. and J. W. Hell. Regulation of cardiac L-type calcium channels by protein kinase A and protein kinase C. *Circulation Research,* 87:1095, 2000.
6. Shannon, T. R., K. S. Ginsburg and D. M. Bers. Reverse mode of the SR Ca pump limits SR Ca uptake in permeabilized and voltage clamped myocytes. In R. G. Johnson, E. G. Kranias and W. Hasselbach (Eds.), *Cardiac Sarcoplasmic Reticulum Function and Regulation of Contractility.* New York Academy of Sciences, New York, 1997.
7. DiFrancesco, D. and D. Noble. A model of cardiac electrical activity incorporating ionic pumps and concentration changes. *Philosophical Transactions of the Royal Society B: Biological Sciences,* 307:353–98, 1985.
8. Luo, C. and Y. Rudy. A model of the ventricular cardiac action potential: depolarization, repolarization and their interaction. *Circulation Research,* 68:1501–26, 1991.
9. Luo, C. H. and Y. Rudy. A dynamic model of the cardiac ventricular action potential. I. Simulations of ionic currents and concentration changes. *Circulation Research,* 74:1071–96, 1994.
10. Noble, D. S., et al. The role of sodium-calcium exchange during the cardiac action potential. *Annals of the New York Academy of Science,* 639:334–54, 1991.
11. Jafri, S., J. J. Rice and R. L. Winslow. Cardiac Ca^{2+} dynamics: the roles of ryanodine receptor adaptation and sarcoplasmic reticulum load. *Biophysical Journal,* 74:1149–68, 1998.
12. Wagner, M. B., et al. Modulation of propagation from an ectopic focus by electrical load and by extracellular potassium. *American Journal of Physiology,* 272(4 Pt 2):H1759–69, 1997.
13. Noble, D. and S. J. Noble. A model of sino-atrial node electrical activity based on a modification of the DiFrancesco-Noble (1984) equations. *Proceedings of the Royal Society of London, B: Biological Sciences,* 222(1228):295–304, 1984.
14. Demir, S. S., et al. A mathematical model of a rabbit sinoatrial node cell. *American Journal of Physiology,* 266(3 Pt 1):C832–52, 1994.

15. Dokos, S., B. Celler and N. Lovell. Ion currents underlying sinoatrial node pacemaker activity: a new single cell mathematical model. *Journal of Theoretical Biology*, 181(3):245–72, 1996.
16. Courtemanche, M., R. J. Ramirez and S. Nattel. Ionic mechanisms underlying human atrial action potential properties: insights from a mathematical model. *American Journal of Physiology*, 275(1):H301–21, 1998.
17. Nygren, A., et al. Mathematical model of an adult human atrial cell: the role of K^+ currents in repolarization. *Circulation Research*, 82(1):63–81, 1998.
18. Fabiato, A. Time and calcium dependence of activation and inactivation of calcium-induced release of calcium from the sarcoplasmic reticulum of a skinned canine cardiac Purkinje cell. *Journal of General Physiology*, 85:247–89, 1985.
19. Fabiato, A. Rapid ionic modifications during the aequorin-detected calcium transient in a skinned canine cardiac Purkinje cell. *Journal of General Physiology*, 85:189–246, 1985.
20. Fabiato, A. Simulated calcium current can both cause calcium loading in and trigger calcium release from the sarcoplasmic reticulum of a skinned canine cardiac Purkinje cell. *Journal of General Physiology*, 85:291–320, 1985.
21. Hodgkin, A. L. and A. F. Huxley. A quantitative description of membrane current and its application to conduction and excitation in nerve. *Journal of Physiology*, 117:500–44, 1952.
22. Hodgkin, A. L. and A. F. Huxley. Currents carried by sodium and potassium ions through the membrane of the giant axon of *Loligo*. *Journal of Physiology*, 116:449–72, 1952.
23. Hille, B. *Ionic Channels of Excitable Membranes* (pp. 141–5). Sinauer, Sunderland, Mass., 1992.
24. Chay, T. R. The Hodgkin-Huxley Na^+ channel model versus the five-state Markovian model. *Biopolymers*, 31(13):1483-502, 1991.
25. Horn, R. and C. A. Vandenberg. Statistical properties of single sodium channels. *Journal of General Physiology*, 84(4):505–34, 1984.
26. Liu, S. and R. L. Rasmusson. Hodgkin-Huxley and partially coupled inactivation models yield different voltage dependence of block. *American Journal of Physiology*, 272(4 Pt 2):H2013–22, 1997.
27. Irvine, L. Models of the cardiac Na channel and the action of lidocaine. PhD thesis, Department of Biomedical Engineering, The Johns Hopkins University School of Medicine, Baltimore, Md., 1998.
28. Irvine, L., M. S. Jafri and R. L. Winslow. Cardiac sodium channel Markov model with temperature dependence and recovery from inactivation. *Biophysical Journal*, 76:1868–85, 1999.
29. Greenstein, J., et al. Role of the calcium-independent transient outward current Ito1 in action potential morphology and duration. *Circulation Research*, 87:1026, 2000.
30. Mazhari, R., et al. Molecular interactions between two long-QT syndrome gene products, HERG and KCNE2, rationalized by in vitro and in silico analysis. *Circulation Research*, 89(1):33–8, 2001.
31. Winslow, R. L., et al. Mechanisms of altered excitation-contraction coupling in canine tachycardia-induced heart failure. II. Model studies. *Circulation Research*, 84(5):571–86, 1999.

32. Rice, J. J., M. S. Jafri and R. L. Winslow. Modeling short-term interval-force relations in cardiac muscle. *American Journal of Physiology*, 278:H913, 2000.
33. Bers, D. M. Cardiac excitation-contraction coupling. *Nature*, 415(6868): 198–205, 2002.
34. Wier, W. G. and D. T. Yue. Intracellular calcium transients underlying the short-term force-interval relationship in ferret ventricular myocardium. *Journal of Physiology*, 376:507–30, 1986.
35. Wier, W. G., et al. Local control of excitation-contraction coupling in rat heart cells. *Journal of Physiology*, 474(3):463–71, 1994.
36. Bers, D. M. and E. Perez-Reyes. Ca channels in cardiac myocytes: structure and function in Ca influx and intracellular Ca release. *Cardiovascular Research*, 42(2):339–60, 1999.
37. Peterson, B., et al. Calmodulin is the Ca^{2+} sensor for Ca^{2+}-dependent inactivation of L-type calcium channels. *Neuron*, 22:549–58, 1999.
38. Linz, K. W. and R. Meyer. Control of L-type calcium current during the action potential of guinea-pig ventricular myocytes. *Journal of Physiology (London)*, 513(Pt 2):425–42, 1998.
39. Stern, M. Theory of excitation-contraction coupling in cardiac muscle. *Biophysical Journal*, 63:497–517, 1992.
40. Bers, D. M. *Excitation Contraction Coupling and Cardiac Contractile Force*. Series in Cardiovascular Medicine, Vol. 122. Kluwer Academic Press, Boston, 1993.
41. Sham, J. S. K. Ca^{2+} release-induced inactivation of Ca^{2+} current in rat ventricular myocytes: evidence for local Ca^{2+} signalling. *Journal of Physiology*, 500(2):285–95, 1997.
42. Franzini-Armstrong, C., F. Protasi and V. Ramesh. Shape, size, and distribution of Ca^{2+} release units and couplons in skeletal and cardiac muscles. *Biophysical Journal*, 77(3):1528–39, 1999.
43. Greenstein, J. and R. L. Winslow. An integrative model of the cardiac ventricular myocyte incorporating local control of Ca^{2+} release. *Biophysical Journal*, 83:2918–45, 2002.
44. Cheng, H., W. J. Lederer and M. B. Cannell. Calcium sparks: elementary events underlying excitation-contraction coupling in heart muscle. *Science*, 262:740–44, 1993.
45. Wang, S. Q., et al. Ca^{2+} signalling between single L-type Ca^{2+} channels and ryanodine receptors in heart cells. *Nature*, 410(6828):592–6, 2001.
46. Rohr, S. and B. M. Salzberg. Characterization of impulse propagation at the microscopic level across geometrically defined expansions of excitable tissue: multiple site optical recording of transmembrane voltage (MSORTV) in patterned growth heart cell cultures. *Journal of General Physiology*, 104(2): 287-309, 1994.
47. Rohr, S., A. G. Kleber and J. P. Kucera. Optical recording of impulse propagation in designer cultures. Cardiac tissue architectures inducing ultra-slow conduction. *Trends in Cardiovascular Medicine*, 9(7):173–9, 1999.
48. Kucera, J. P., A. G. Kleber and S. Rohr. Slow conduction in cardiac tissue. II. Effects of branching tissue geometry. *Circulation Research*, 83(8):795–805, 1998.
49. Franzone, P., L. Guerri and S. Rovida. Wavefront propagation in an activation model of the anisotropic cardiac tissue: asymptotic analysis and numerical simulations. *Journal of Mathematical Biology*, 28(2):121–76, 1990.

50. Franzone, P. C., L. Guerri and S. Tentoni. Mathematical modeling of the excitation process in myocardial tissue: influence of fiber rotation on wavefront propagation and potential field. *Mathematical Bioscience*, 101(2):155–235, 1990.
51. Franzone, P., L. Guerri and B. Taccardi. Potential distributions generated by point stimulation in a myocardial volume: simulation studies in a model of anisotropic ventricular muscle. *Journal of Cardiovascular Electrophysiology*, 4(4):438–58, 1993.
52. Franzone, P. C., et al. Spread of excitation in 3-D models of the anisotropic cardiac tissue. II. Effects of fiber architecture and ventricular geometry. *Mathematical Bioscience*, 147(2):131–71, 1998.
53. Kanai, A. and G. Salama. Optical mapping reveals that repolarization spreads anisotropically and is guided by fiber orientation in guinea pig hearts. *Circulation Research*, 77(4):784–802, 1995.
54. Roberts, D. E., L. T. Hersh and A. M. Scher. Influence of cardiac fiber orientation on wavefront voltage, conduction velocity, and tissue resistivity in the dog. *Circulation Research*, 44(5):701–12, 1979.
55. Neunlist, M. and L. Tung. Spatial distribution of cardiac transmembrane potentials around an extracellular electrode: dependence on fiber orientation. *Biophysical Journal*, 68(6):2310–22, 1995.
56. Taccardi, B., et al. Effect of myocardial fiber direction on epicardial potentials. *Circulation*, 90(6):3076–90, 1994.
57. Colli Franzone, P., et al. Spread of excitation in 3-D models of the anisotropic cardiac tissue. III. Effects of ventricular geometry and fiber structure on the potential distribution. *Mathematical Bioscience*, 151(1):51–98, 1998.
58. Teare, R. Asymmetrical hypertrophy of the heart in young patients. *British Heart Journal*, 20:1, 1958.
59. Maron, B. J., et al. Quantitative analysis of cardiac muscle cell disorganization in the ventricular septum. Comparison of fetuses and infants with and without congenital heart disease and patients with hypertrophic cardiomyopathy. *Circulation*, 60(3):685–96, 1979.
60. Maron, B. J. and W. C. Roberts. Hypertrophic cardiomyopathy and cardiac muscle cell disorganization revisited: relation between the two and significance. *American Heart Journal*, 102(1):95–110, 1981.
61. Maron, B. J., T. J. Anan and W. C. Roberts. Quantitative analysis of the distribution of cardiac muscle cell disorganization in the left ventricular wall of patients with hypertrophic cardiomyopathy. *Circulation*, 63(4): 882–94, 1981.
62. Roberts, W. C., R. J. Siegel and B. M. McManus. Idiopathic dilated cardiomyopathy: analysis of 152 necropsy patients. *American Journal of Cardiology*, 60:1340–55, 1987.
63. Rose, A. G. and W. Beck. Dilated (congestive) cardiomyopathy: a syndrome of severe dysfunction with remarkably few morphological features of myocardial damage. *Histopathology*, 9:367–79, 1985.
64. Anderson, K. P., et al. Myocardial electrical propagation in patients with idiopathic dilated cardiomyopathy. *Journal of Clinical Investigation*, 92: 122–40, 1993.
65. Larregina, A., et al. Biventricular hypoplasia with myocardial fiber hypertrophy and disarray. *Pediatrical Pathology*, 10(6):993–9, 1990.

66. Honer, W. G., et al. Temporal lobe and ventricular anatomy in families with schizophrenia showing linkage to chromosomes 8q and 13p. *Molecular Psychiatry*, 4:16, 1999.
67. Horvath, G., et al. Electrophysiological and anatomic heterogeneity in evolving canine myocardial infarction. *Pacing Clinical Electrophysiology*, 23(7):1068–79, 2000.
68. Fujiwara, H., C. Kawai and Y. Hamashima. Myocardial fascicle and fiber disarray in 25 mu-thick sections. *Circulation*, 59(6):1293–8, 1979.
69. Chen, P. S., et al. Effects of myocardial fiber orientation on the electrical induction of ventricular fibrillation. *American Journal of Physiology*, 264(6 Pt 2):H1760–73, 1993.
70. MacCallum, J. B. On the muscular architecture and growth of the ventricles of the heart. *Johns Hopkins Hospital Reports*, 9:307–35, 1900.
71. Mall, F. On the muscular architecture of the ventricles of the human heart. *American Journal of Anatomy*, 11:211–66, 1911.
72. Fox, C. C. and G. M. Hutchins. The architecture of the human ventricular myocardium. *Johns Hopkins Medical Journal*, 25:289–99, 1972.
73. Streeter, D., et al. Fiber orientation in the canine left ventricle during diastole and systole. *Circulation Research*, 24:339–47, 1969.
74. Streeter, D. D. and W. T. Hanna. Engineering mechanics for successive states in canine left ventricular myocardium. II. Fiber angle and sarcomere length. *Circulation Research*, 33:656–64, 1973.
75. Ross, A. A. and D. D. Streeter, Jr. Myocardial fiber disarray [letter]. *Circulation*, 60(6):1425–6, 1979.
76. Streeter, D. Gross morphology and fiber geometry of the heart. In R. Berne (Ed.), *Handbook of Physiology: The Cardiovascular System I* (pp. 61–112). American Physiological Society, Bethesda, Md., 1979.
77. Streeter, D. D. and C. Ramon. Muscle pathway geometry in the heart wall. *Journal of Biomechanical Engineering*, 105(4):367–73, 1983.
78. Nielsen, P. M. F., et al. Mathematical model of geometry and fibrous structure of the heart. *American Journal of Physiology*, 260:H1365–78, 1991.
79. Basser, P., J. Mattiello and D. LeBihan. Estimation of the effective self-diffusion tensor from the NMR spin echo. *Journal of Magnetic Resonance, Series B*, 103(3):247–54, 1994.
80. Skejskal, E. O. and J. E. Tanner. Spin diffusion measurement: spin echoes in the presence of time-dependent field gradients. *Journal of Chemical Physics*, 42:288–92, 1965.
81. Holmes, A. A., D. F. Scollan and R. L. Winslow. Direct histological validation of diffusion tensor MRI in formaldehyde-fixed myocardium. *Magnetic Resonance Medicine*, 44:157–61, 2000.
82. Hsu, E. W., et al. Magnetic resonance myocardial fiber-orientation mapping with direct histological correlation. *American Journal of Physiology*, 274: H1627–34, 1998.
83. Scollan, D., et al. Histologic validation of reconstructed myocardial microstructure from high resolution MR diffusion tensor imaging. *American Journal of Physiology*, 275:H2308–18, 1998.
84. Scollan, D., et al. Reconstruction of cardiac ventricular geometry and fiber orientation using GRASS and diffusion-tensor magnetic resonance imaging. *Annals of Biomedical Engineering*, 28(8):934–44, 2000.

85. Ennis, D. B., G. Kindlman, P. A. Helm and E. R. McVeigh. Visualization of tensors data using superquadric glyphs. *Magnetic Resonance Medicine*, 53:169–76, 2005.
86. LeGrice, I. J., P. J. Hunter and B. H. Smaill. Laminar structure of the heart: a mathematical model. *American Journal of Physiology*, 272(5 Pt 2): H2466–76, 1997.
87. Pollard, A., M. J. Burgess and K. W. Spitzer. Computer simulations of three-dimensional propagation in ventricular myocardium. *Circulation Research*, 72:744–56, 1993.
88. Pollard, A. E., N. Hooke and C. S. Henriquez. Cardiac propagation simulation high performance computing. In T. C. Pilkington et al. (Eds.), *Biomedical Research*. CRC Press, Boca Raton, Fl., 1993.
89. Henriquez, C. S. Simulating the electrical behavior of cardiac tissue using the bidomain model. *Critical Review of Biomedical Engineering*, 21(1): 1–77, 1993.
90. Nichols, C. G. and W. J. Lederer. The regulation of ATP-sensitive K channel activity in intact and permeabilized rat ventricular cells. *Journal of Physiology*, 423:91–110, 1990.
91. Cortassa, S., et al. An integrated model of cardiac mitochondrial energy metabolism and calcium dynamics. *Biophysical Journal*, 84:2734–55, 2003.
92. Katz, A. M. *Physiology of the Heart*, 2nd ed. Raven Press, New York, 1992.
93. Tomaselli, G. T. and E. Marban. Electrophysiological remodeling in hypertrophy and heart failure. *Cardiovascular Research*, 42:270–83, 1999.

10

Embryonic Stem Cells as a Module for Systems Biology

Andrew M. Thomson, Paul Robson, Huck Hui Ng, Hasan H. Otu, & Bing Lim

The current trend of biological and medical research in generating massive data sets in diverse fields using multiple and varying platforms necessitates a new perspective in the way we anticipate the harvesting of new knowledge from the vast amount of accumulating data. Systems biology may be thought of as a discipline that seeks to achieve breakthrough into new understandings of biology from the cell to the organism by a computational integration of diverse large biological data sets to capture information not possible through analysis of a single or a few parameters such as transcript, protein and activity states, protein–protein interaction, protein-nucleic acid interaction, extracellular signals, and cellular transformation [1]. The goals of systems biology range from the simplest ability to predict specific functional responses, or identify networks for lineage specification at the cellular level, to grand revelations of physiological and behavioral responses at the tissue and even organism level [1].

Three advances have driven the movement toward the emergence of this discipline: (1) the technologies to acquire complete annotated genome sequences of species; (2) the development of technologies to capture parallel gene expression of transcripts and proteins on a genomic scale; and (3) the development of miniaturization and high-throughout analysis of gene function.

The overarching guiding principle of systems biology is to integrate findings from these technologies to seek connections of networks of genes in cellular and biochemical responses, thereby anticipating the discovery of new laws governing the function of networks [2–5]. Therefore, the essential requirements for systems biology to achieve robustness include: (1) digitalization of biological readouts; (2) computational hardware to absorb and process large data sets; and (3) algorithms for integration and interpretation of data.

DEVELOPMENTAL BIOLOGY AND EMBRYONIC STEM (ES) CELLS

The development of a single-celled fertilized egg to a complete organism of a few trillion cells, with all its complex connections and communications,

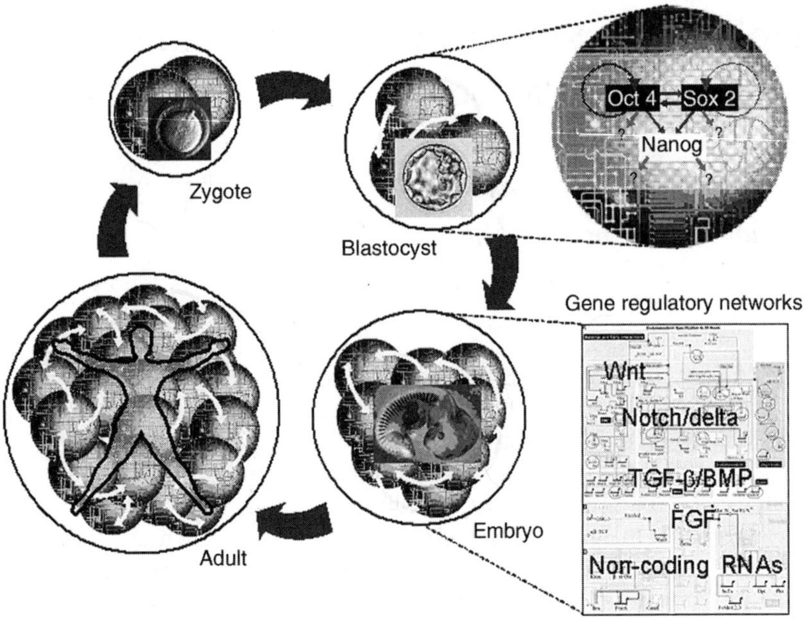

Figure 10.1 Development from a single fertilized oocyte (zygote) to adult. Early-stage blastocyst is shown with inner cell mass. At this stage, key transcriptional factors are at work in either maintaining pluripotency or directing differentiation into tissues of mesoderm, ectoderm, and endoderm lineages. A later-stage embryo shows the emergence of body shape with head, body, and limb buds and somites for the musculoskeletal system. Different transcriptional pathways are recruited, and the known dominant developmental players are shown here, including *Wnt*, *TGF-β/BMP*, *Notch*, and noncoding RNAs. At each stage, in the background, is shown the interaction of gene networks.

reflects an adherence to programmed and coordinated activation and suppression of regulatory gene networks (figure 10.1). This involves the primordial reprogramming of highly differentiated oocyte and sperm genomes into the trophectoderm and primitive endoderm that form the extraembryonic tissues, such as the placenta, required for supporting the development of the embryo proper and the pluripotent cells of the epiblast/inner cell mass (ICM). It is the ICM that develops into the newborn. The information needed to understand developmental biology therefore requires multidimensional considerations. The most elementary level of information required is that about the genes and their products. There is still much knowledge to gather and organize for all the genetic products involved, which include coding and noncoding transcripts. There is the critical dimension of precise chronological programs of genes recruited for driving the progression of differentiation and the migration of clusters of embryonic cells. The timing of activation and repression of genetic pathways is central to this process. There is also a specific spatial orientation of cellular populations during development in which signals from the environment and cues to individual cells are critical to the three-dimensional growth of the embryo. Large data sets can and are being collected for these primary components of development.

One way to discover the nature of the interaction between these components is to envisage each collection of cells in time and space as having its own set of rules guiding the genetic elements from which we can derive all possible information exhaustively. As the embryo increases in cellular and tissue complexity, new genetic elements and networks are brought into play while established genetic pathways are shut down or continue to work the same way or take on new interactive functions. The integration of these many layers of high-density genetic information to reveal natural laws governing the propelling of development by the genome can only be possible with an approach that incorporates the computational underpinnings of systems biology (figure 10.1).

Embryonic stem (ES) cells, first characterized from the mouse in 1981 [6,7], are derived from one of the three cell types present in the blastocyst. The blastocyst (figure 10.2), consisting of approximately 64 cells in the mouse, is the cellular structure that forms approximately 3.5 days after fertilization over which time the highly differentiated oocyte and sperm haploid genomes have reprogrammed to initiate the new zygotic genome's genetic regulatory program.

Within the blastocyst, the trophectoderm (the outer cell layer) and the primitive endoderm give rise to extraembryonic structures whereas the epiblast (also referred to as the inner cell mass, ICM) gives rise to all cell types of the embryo proper and of the resulting adult. In culture, in addition to replicating indefinitely in the undifferentiated state, ES cells maintain this pluripotent developmental potential of the epiblast and indeed, if injected back in the blastocyst, ES cells are capable of giving rise to all these cell types.

These characteristics of mammalian ES cells provide us with a great resource for molecular and biochemical studies aimed at understanding the earliest events in mammalian development. In practical terms, human ES cells, first derived in 1998 [8], hold enormous potential for regenerative medicine. The molecular mechanisms governing the fate decisions of ES cells are minimally understood, yet an understanding of this machinery is fundamental to the successful applications of stem cells in regenerative medicine and other potential clinical applications [9]. This is why there is great interest in expanding our knowledge to a systems level of understanding of the ES cell and its developmental derivatives (figure 10.2).

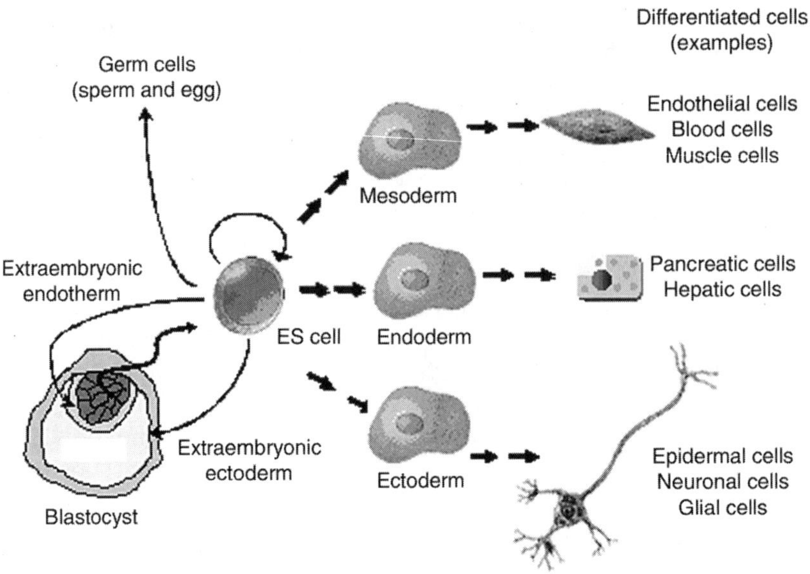

Figure 10.2 Embryonic stem cell differentiation. The origin of ES cells from the inner cell mass of the blastocyst is shown with the derivation of the three major embryonic lineages (mesoderm, endoderm, and ectoderm) and the possible tissue types into which they differentiate.

Another attractive attribute of mouse ES cells is the availability of techniques to specifically and precisely modify the genome through homologous recombination where the possibilities range from large-scale deletions (megabases) to single point mutations [10]. In addition to studying the effects of these genome alterations in an in vitro ES cell culture system, one can also generate mice from these modified ES cells to study whole animal systems. The ability to perform such genome modifications allows us to perform wet lab validations of hypotheses and predictions derived from in silico analyses.

Here we will describe studies of ES cell growth and differentiation and discuss the analyses of various large-scale data sets generated to illustrate how the ES cell is an excellent model to test the validity of a systems biology approach.

COMPUTATIONAL DISCOVERY OF GENE MODULES AND REGULATORY NETWORKS BY ANALYSIS OF GENE EXPRESSION PROFILES OF ES AND DIFFERENTIATING CELLS

One of the compelling quests in developmental biology has been to identify genes required for ES cells to maintain their pluripotency ("stemness"), and those for specific cell type differentiation. A practical value in knowing the identity of these genes and the pathways they regulate is the use of such knowledge to advance our ability to isolate, maintain, and differentiate ES cells.

Network Predictions: Transcriptome Analysis of ES and Embryonic Cells

Earlier attempts using methods such as differential screens and subtractive hybridization uncovered only a handful of genes preferentially expressed in ES cells and, by implication, important for their unique properties. With the advent of the idea of spotting multiple probes on microarray chips [11], it became possible to interrogate RNAs simultaneously with a large collection of gene probes in a multiplex fashion for the transcriptome expressed in ES cells and during differentiation. Such an approach immediately revealed that the number of genes that may be important for maintaining ES cell states could be in the hundreds [12–14]. Data sets combined from studies performed by several groups using similar arrays further provided the opportunity to intersect databases to find common elements [14]. This comparative analysis revealed a set of genes that are found by all studies to be preferentially expressed in ES cells compared to other cell types. Because the list includes genes that are known to be important for stemness properties of ES cells, such as *Oct-4/Pou5f1* and *Nanog* (figure 10.3), by implication some or many of the genes in this list would likely be involved in ES cell growth and development.

However, an integrated picture of how these genes work together to determine ES stemness has not emerged and remains to be established. To demonstrate how such lists of genes may be further utilized to reveal biological meaning, we analyzed the 332 "ES genes" shown in figure 10.3 using Ingenuity's Pathway analysis software (http://www.ingenuity.com) and Pathway Assist (http://www.ariadnegenomics.com). These programs identify and extend connections between a given set of genes using the enormous wealth of information buried in scientific literature. These two approaches differ in the way their respective knowledge bases are developed. Ingenuity's knowledge base is manually indexed, increasing its accuracy, but this means that it covers less of the scientific literature (figure 10.4a). Meanwhile, Pathway Assist (figure 10.4b) uses an automated natural language processor to gain knowledge from the literature rendering a noisy knowledge base with reduced accuracy, but gains a more exhaustive coverage [15].

Interestingly, despite differences in the method of knowledge base development between the two programs, very similar major network pathways were revealed when the 332 ES genes were analyzed (figures 10.4a and 10.4b). Each of these networks has a hub-like node and here we show the integration of the *p53* tumor suppressor, *c-myc* and *c-fos*

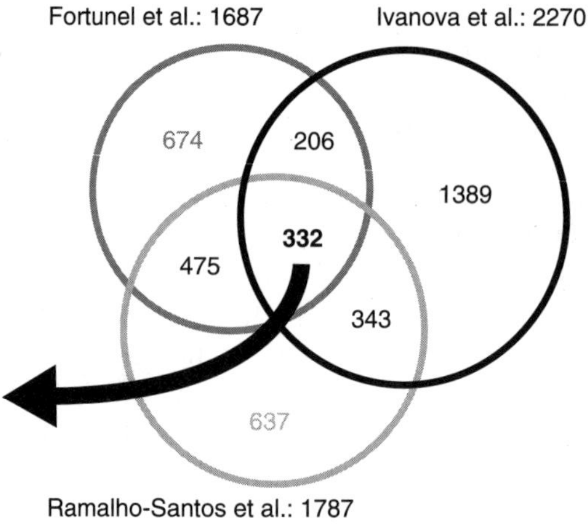

embryonal stem cell specific gene 1	*Esg1*
POU domain, class 5, transcription factor 1 (Oct-4)	*Pou5f1*
zinc finger protein 42	*Zfp42*
makorin, ring finger protein 3	*Mkm3*
testis expressed gene 20	*Tex20*
keratin complex 2, basic, gene 8	*Krt2-8*
claudin 6	*Cldn6*
dual specific phosphatase	*Dusp9*
serine protease inhibitor, Kunitz type 1	*Spint1*
RIKEN cDNA 2410002E02 gene	*Nanog*
RIKEN cDNA 2410088E07 gene	*Dppa2*
RIKEN cDNA 2410091M23 gene	*Dppa4*
undifferentiated embryonic cell transcription factor 1	*Utf1*
solute carrier family 7 (cationic amino acid transporter)	*Slc7a7*
uridine phosphorylase	*Upp*
E26 avian leukemia oncogene 2, 3′ domain	*Ets2*
coding region determinant-binding protein	*Crdbp*
f-box only protein 15	*Fbox15*

Figure 10.3 Discovery of ES stemness genes by transcriptome profiling. Venn diagram showing intersection of ES-specific genes from three studies (indicated by first author's name). The 332 genes are common to all studies and therefore represent genes most likely to play a role in either maintaining ES cell totipotency or important for ES cellular physiology. The top 20 genes on the list are shown in the table.

oncogenes, SP1 transcription factor, caspase 3 (CASP) protease, and retinoic acid receptor activator (RARA) pathways analyzed by Ingenuity's Pathway Assist. All the networks showing a high degree of connectivity are pathways related to cell cycle, apoptosis, and differentiation, which are critical functions during ES cell growth and proliferation and embryo development. When the program was interrogated further to compute

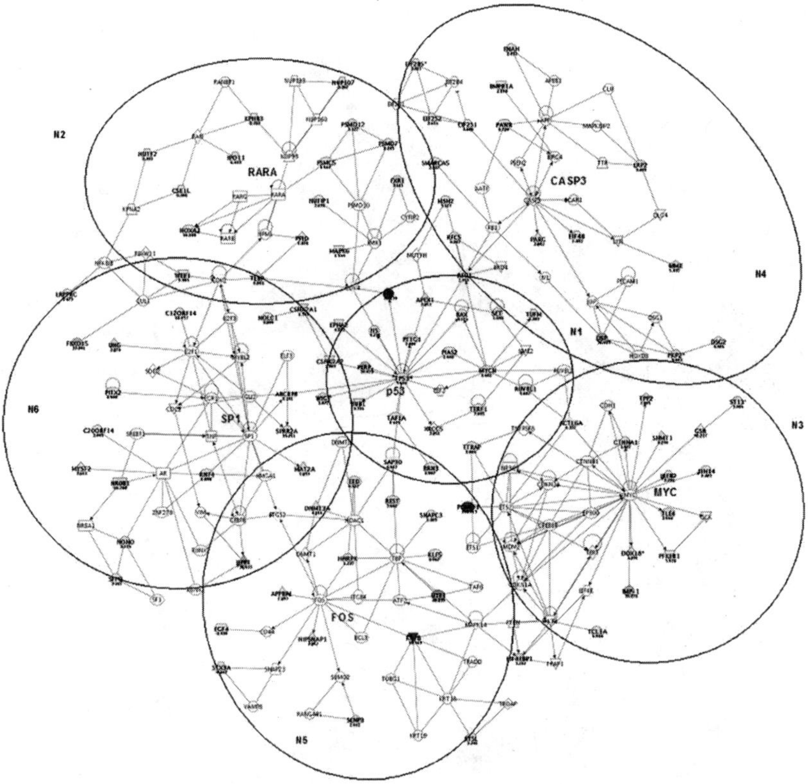

Figure 10.4a Ingenuity analysis of six networks (N1 to N6) identified from analysis of 332 ES genes. Each of the six networks is highly connected within its members and overlaps minimally with each other. We indicate the hub-like center(s) of each network in larger font (*p53*, *RARA*, *MYC*, *CASP3*, *FOS*, *SP1*). © 2000–2005 Ingenuity Systems.

an additional search for canonical pathways that may be buried within the network containing the key ES transcriptional factor, *Oct4*, the program suggested and displayed relationships between *c-myc* (cell cycle control gene), *Oct4* transcription factor, and the *Wnt* signaling pathway (figure 10.4c). This is significant as the *Wnt* pathway represents a conserved family of secreted proteins involved in various embryonic and adult tissues' function, including proliferation, survival, differentiation, and motility [16]. Defects in *Wnt* pathways have been implicated in development of human disease such as many types of cancer and bone and retinal vascular disorders [17]. Furthermore, the *Wnt* pathway appears to be critical in the early stages of ES cell self-renewal and initiation of differentiation [18,19].

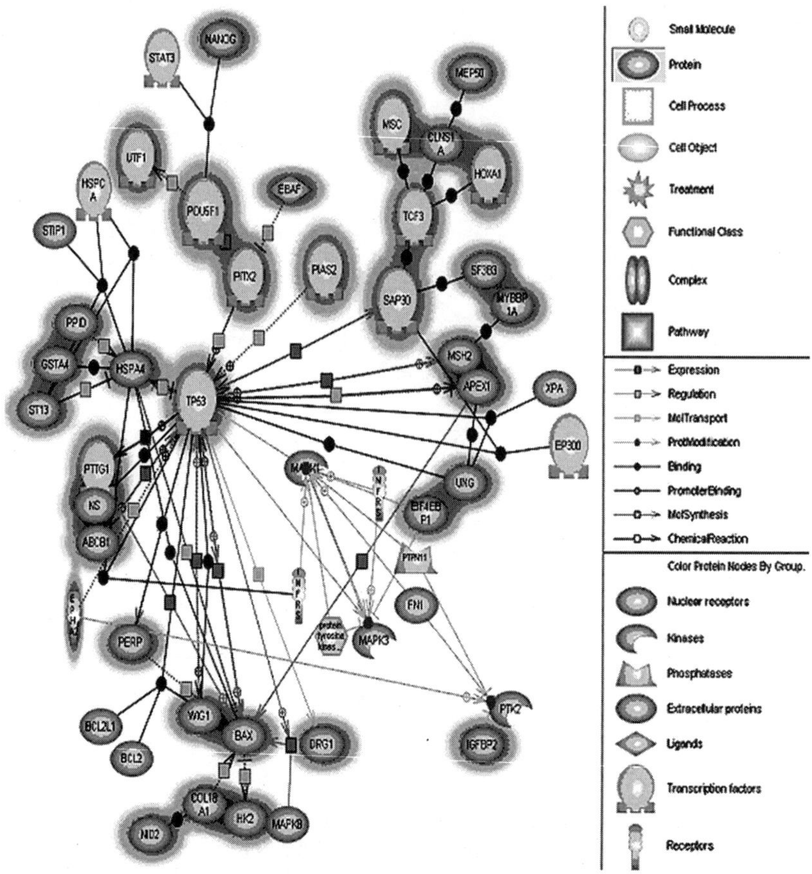

Figure 10.4b Pathway Assist network analysis of the 332 ES genes. The input genes are highlighted with a halo. The type of nodes and links used are shown in the table of symbols. © 2000–2005 Ingenuity Systems.

Whilst such an analysis demonstrates the power of a computer-based dissection of data, the approach is limited to utilizing information on known and functionally described target genes, with proof of the interactions still to be determined by functional characterizations in vivo. Additional programs in Pathway Assist can show pathways that result from direct interactions with either upstream common regulators or downstream common targets of the 332 genes originally identified from microarray analysis (data not shown).

One of the limitations of microarrays is that the information acquired is limited to known target genes or sequences of genes of unknown function in the public databases and scientific literature. Other approaches

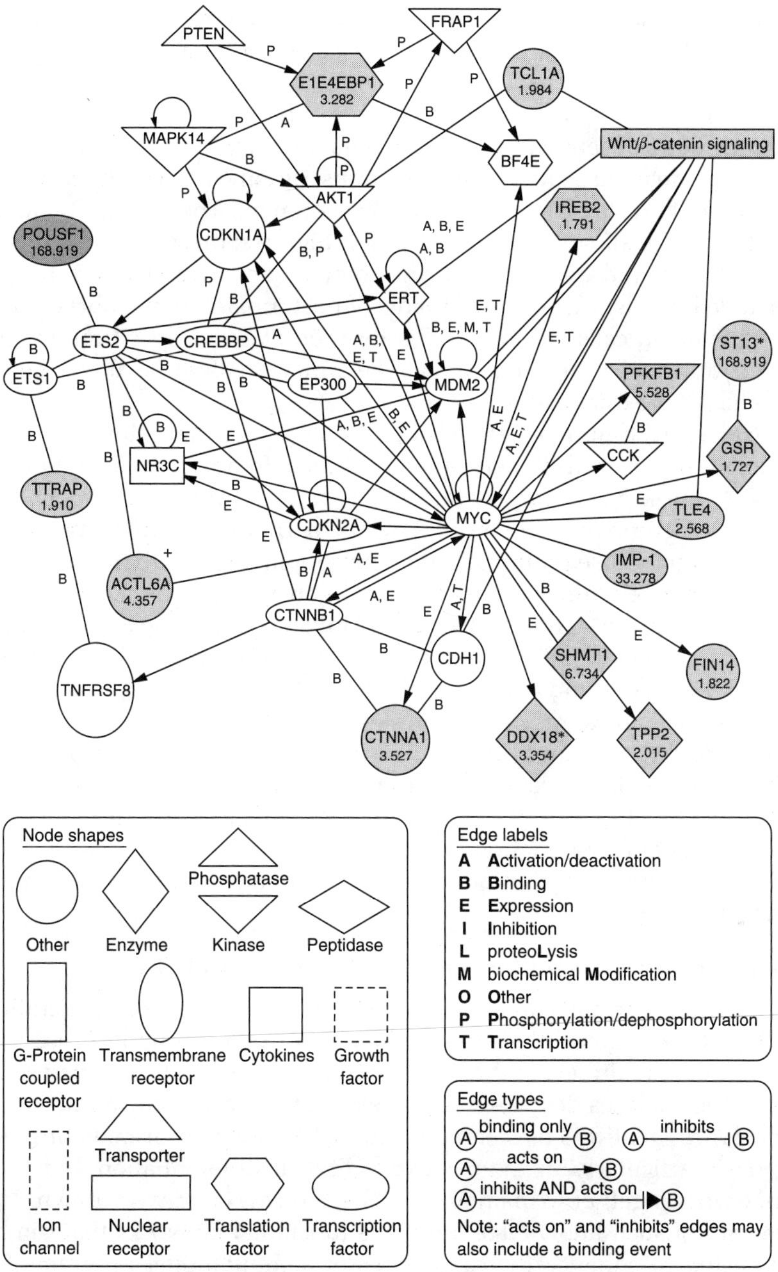

Figure 10.4c Ingenuity analysis for the network involving *Oct-4*. The canonical pathway, Wnt Signaling, is superimposed on the network by indicating the nodes shared by the two networks. The numbers indicate the fold change in ES cells compared to differentiated cells. The type of nodes and links used in the network are shown in the three boxes. © 2000–2005 Ingenuity Systems.

that are unbiased and sensitive, such as a sequence-based approach, would allow deeper analysis that reveals not only rare transcripts but completely novel transcripts [20,21]. One such method, the Massively Parallel Signature Sequencing (MPSS) technique (Lynx Technologies), has generated sequenced tagged data into millions of transcripts, allowing detailed sequencing of the transcriptome from many different cell types and stages of embryonic development. The sensitivity of the methodology allows for many new non-highly expressed transcripts to be detected and quantified. Cross-comparisons across data sets that include ES cells, various time points of postimplantation embryos, and differentiated cells led to the identification of genes specifically regulated during ES differentiation. The function of these novel genes remains to be established.

Currently, a number of websites are available that allow access to MPSS data for murine and human ES cells. In addition, an extensive EST database has been generated from mouse early development and ES cells. Gene expression profiles by microarray at multiple time points throughout mouse preimplantation development [22,23] provide additional data sets for comparison. Finally, new methods, such as that developed at the Genome Institute of Singapore based on a di-Tag technology (see below), provide the ability to map precisely the origins of novel transcripts and reveal novel and naturally occurring fusion transcripts that are a result of trans-splicing of RNA originating from different chromosomes [21,24].

Altogether, these databases represent a repository of the most comprehensive information about the mammalian transcriptome that challenges our imagination of the interrelatedness of all these genetic elements, each with a unique function, that give life to and determine the fate of ES cells.

Genome-Wide Screens for Gene Function Using ES Cells

While microarray technologies have been very efficient in generating robust lists of genes that may be involved in the common process of pluripotent maintenance or differentiation, the derivation of networks from the analyses described above takes into account only a fraction of the genes identified as stemness genes. The function of most of these genes, particularly in the context of ES cell differentiation and how they work together to maintain ES cells, remains largely unknown. The paucity of such knowledge limits the functional networks that can be extracted to enhance the capabilities of computational programs for increased accuracy of predictive models of cellular responses. Therefore, one of the major efforts in systems biology has been to seek ways to perform high-throughput screens for gene function [25]. The availability of the ES cell as a line of almost pure stem cells that can be transfected relatively easily with high efficiency provides a means to study gene responses and phenotypic changes in a high-throughput micro-well platform. Such a maneuver is technically challenging or not possible with other stem cells.

The advances made in molecular techniques that provide the ability to control endogenous levels of protein and nucleic acids has revolutionized our ability to study the over- or underexpression of a specific gene. These include improvements in nonviral and viral-mediated cell gene transfer vectors, expression modules, antisense oligonucleotides, and RNA interference (RNAi) technologies [26–29]. Current methods can achieve knockdown of RNA species that result in significant reduction of the proteins for which they code. The effect on ES cells can be monitored by a combination of changes in morphology, cell death, proliferation, or molecular markers indicative of differentiation. Profiling of transcriptome changes can be global, performed with microarray chips with high-density probes, or concentrated on monitoring embryonic lineage differentiation using focused chips or custom-designed microfluidic cards (Applied Biosystems). Transgenic ES lines can also be generated containing reporter genes, such as GFP (green fluorescent protein) or luciferase, knocked into genes of interest (such as *Oct4*).

Several repositories with large collections of ES cell lines are now available with heterozygous knockouts of single alleles using a random genome-wide targeting with highly efficient promoter trap vectors (Lexicon Genetics, http://www.lexicon-genetics.com; International Genetrap Consortium, Sanger Center: http://www.igtc.org.uk/). An investigator can browse these websites to order ES cell lines of interest to study the loss-of-function mutation of the gene in the context of a whole organism. Thus, altogether, there is an armamentarium of methodologies that are suitable for studying ES cell response that can be utilized to study multiple gene interactions and test the potential of systems biology approaches.

Expanding the Cellular Signaling Networks: Transcription Mechanisms

Gene expression in eukaryotic cells is controlled by regulatory elements that recruit transcription factors with specific DNA recognition properties. A major effort to advance understanding of gene regulation is therefore directed at studies of transcription factors, their cognate DNA binding sites, and protein-binding partners constituting coactivators and suppressors of transcription. However, given the large number of transcription factors, estimated at about 1500 in the mammalian genome, we are only at the very elementary stage in identifying and characterizing the many components, and understanding the manner in which they form interacting networks to modulate and coordinate gene expression.

Current databases can only paint a rough sketch of a few well-studied transcriptional pathways and more data needs to be generated to add to the power of predictive models. Indeed, a recent genetic screen on the *Wnt* pathway in *Drosophila* suggests that some transcriptional pathways may in fact involve many more molecular branches and connections than expected [30].

IDENTIFICATION OF KEY TRANSCRIPTION FACTOR BINDING SITES IN PROMOTER REGIONS OF ES GENES BY SEQUENCE COMPARISON

Several genes that appear to be key functional regulators of pluripotency maintenance and differentiation have been identified in ES cells. These regulators include the transcription factors *Oct4, Sox2,* and *Nanog*. *Oct4* binds to a specific DNA octamer, ATGCAAAT, within promoter regions of target genes. *Oct4* is highly expressed in ES cells and its reduction is necessary for differentiation to proceed [31–33]. *Oct4* targets and controls the expression pattern of a number of genes (*Fgf4, Utf1, Opn, Rex1/Zfp42, Fbx15*), whose regulatory elements are found to contain putative in silico *Oct4* binding sites. Mutation analysis of these putative sites has confirmed their importance in modulating gene expression via *Oct4* binding ex vivo. Interestingly, when computational genome-wide scanning was performed in several species, a conserved *Oct4* octamer binding sequence was found to be present within the promoter regions of *Oct4, Sox2,* and *Nanog* genes. The binding of *Oct4* to these sites was confirmed by chromatin immunoprecipitation (ChIP) [34]. These exciting findings (figure 10.5) demonstrated that these three key ES cell transcription factors autoregulate each other via an intricate network, where the molecular details have yet to be fully elucidated.

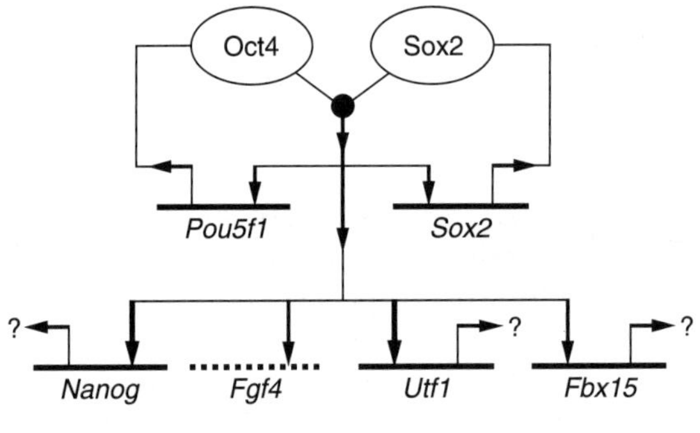

Figure 10.5 The network of genes regulated by two key ES transcriptional factors. The feedback regulation of the transcriptional factors *Oct4/Pou5f1* and *Sox-2* are shown, interconnected by each binding to the other's promoter. *Oct4* and *Sox2* colocalized right next to each other in their own promoters and in many target genes, examples of which are shown in the figure.

TRANSCRIPTION BINDING SITE DISCOVERY BY MICROARRAY AND SEQUENCE-BASED ANALYSIS OF ChIP

The miniature circuitry described above marks the activities of master regulatory genes at the top of the hierarchy of the ES gene regulatory network. An obvious and important question is how then does this circuit reach out to other genes to generate the larger networks that must be brought into play to orchestrate self-renewal, proliferation, and decisions to differentiate of ES cells into the three primary embryonic lineages of mesoderm, ectoderm, and mesoderm? Here we illustrate how once again the availability of ES as a stem cell line together with new powerful technologies driven by a systems biology approach can accelerate the clarification of these fundamental questions.

Based on the similar principle of microarray probes for a multiplex parallel screen of transcriptome, one high-throughput approach is to use chromatin-immunoprecipitated DNAs as probes to hybridize against arrayed nucleotide fragments scanning entire chromosomes. Such "ChIP on CHIP" platforms have been used successfully for whole genome localization analysis in yeast [35]. This technology is still not readily available or accessible for robust application in mammalian cells due to the large size and complexity of the mammalian genomes. Nevertheless, small-scale ChIP on CHIP analysis in mammalian cells has been applied to specific areas, such as identifying CpG islands or flanking sequences associated with transcription start sites, and small chromosome arrays [36,37]. Significantly, these partial arrays have shown that a transcription factor can bind to a large number of sites not just in the promoter but to introns, in distal locations from genes and in genome "desert" regions. However, as the amount of ChIP DNA needed for such studies is substantial, requiring a huge number of homogenous stem cell populations, only ES cells can provide such an inexhaustible supply of material.

A beautiful example of this is the recent work by Boyer et al. showing the mapping of the binding sites of *Oct4*, *Nanog*, and *Sox-2* using ChIP DNA hybridization to CHIPs containing probes for the promoters of known genes [38]. The data revealed not only the network of genes regulated by each transcriptional factor but also a core set of genes coregulated by all three factors, thereby demonstrating the power and relevance of such studies to systems biology. Another direct experimental approach to identifying DNA *cis*-elements bound by specific DNA *trans*-acting binding proteins is ChIP followed by sequencing of the protein-bound DNA. The basic idea is to make a complete library from the ChIP DNA fragments followed by high-throughput sequencing. With sufficient depth of sequencing the approach has the potential to identify all DNA segments enriched by ChIP. The scale of coverage can be enhanced by SAGE (series analysis of gene expression) tag-based sequencing strategies.

To improve the efficiency and accuracy of mapping the DNA binding domain, we have developed a method incorporating the Paired-End diTag (PET) procedure that extracts 18 nucleotide base signatures in pairs from the 5′-3′ ends of each cDNA clone, concatenating these PETs for efficient sequencing, followed by a very precise mapping of the PET sequences to the genome that demarcates clearly the protein binding site boundaries [21,24]. We have begun applying this strategy to characterize ChIP-enriched DNA fragments for the key ES cell transcription factors. *Oct-4*, *Sox-2*, and *Nanog*, for example, are found to bind to over a thousand sites in the murine genome [39]. We thus anticipate that a combination of such data sets for only a handful of key transcriptional factors should begin to yield significant clues to the chain of gene networks crucial to specifying ES cell growth and fate.

POSTTRANSCRIPTIONAL REGULATION: NONCODING (NC) RNAS

There are now increasing data showing that noncoding (nc) RNAs exert a critical role in regulating gene expression at multiple levels. From the data emerging from RIKEN and other large-scale comprehensive sequencing projects it is now estimated that over 50% of the mammalian transcriptome are ncRNAs that do not produce proteins. The various mechanisms by which these ncRNAs regulate the transcriptome include: (1) modulating transcription by functioning as transcription factor coactivators (e.g., steroid receptor RNA activator, SRA) [40]; (2) modulating mRNA decay by targeting specific *cis*-acting regions and inducing mRNA-degradation mechanisms (small inhibitory (si) RNAs) [41]; (3) modulating translation and protein production by blocking translation machinery progress upon binding to target mRNA sequences (microRNAs) [42]; and (4) small RNA directed methylation of DNA [43,44] (figures 10.6). In both (2) and (3) above, the ncRNAs are directed to targeted mRNAs through a multiple-protein complex, RISC (RNA-induced silencing complex) [45].

From our own comprehensive analysis of the ES transriptome [20] we have found that, during differentiation of ES cells, ncRNAs are differentially regulated, suggesting that the timing of expression of ncRNAs is crucial to the differentiation process. The MPSS data of murine ES cells shows dramatic changes in SRA levels between ES and differentiating cells, suggesting that this transcription factor coactivator plays a role in the differentiation pathways. Many novel ES-specific transcripts that do not contain long reading frames have been identified, although their function remains to be established.

Among the noncoding RNAs, the microRNAs are turning up rapidly in an surprising diversity of regulatory pathways, including development, tissue differentiation, apoptosis, proliferation, and organ development and, significantly, they appear to be important in carcinogenesis also [46–48].

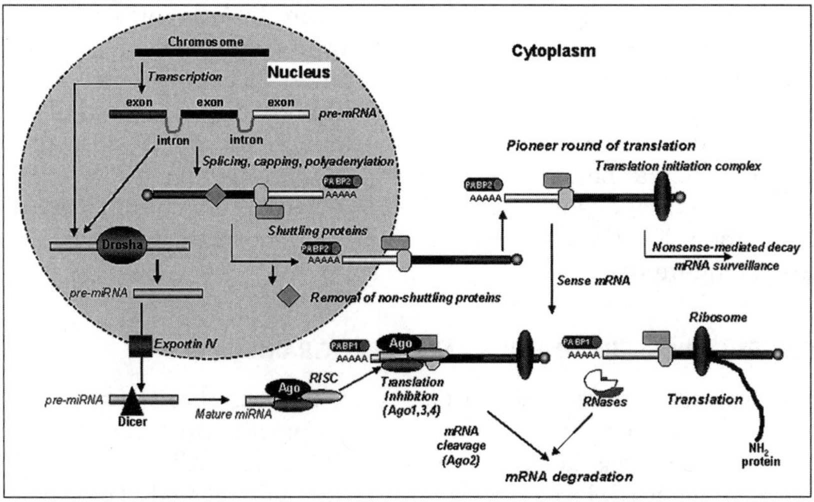

Figure 10.6 Transcription, processing, and translation of RNAs. The coding RNAs are translated, or degraded and suppressed by microRNAs. The noncoding microRNAs are processed from primary transcripts in the nucleus by Drosha and exported to the cytoplasm by Exportin where they are further cut down to small mature miRNA by Dicer and become part of the RISC complex that inhibits translation or degrades mRNAs. Ago, Argonaut; PABP, poly-A binding protein; RISC, RNA-induced silencing complex.

The small 22-nucleotide microRNAs function by inhibiting other RNAs (see chapter 10 of vol. 1). From total libraries made from size-selected small RNAs of murine and human ES cells, microRNAs have been isolated that are expressed only in ES cells and not in other tissues [49,50]. Using microRNA chips to analyze RNA from differentiation-induced ES cells, we found that certain microRNAs are invariably downregulated regardless of the method of differentiation. On the other hand, different sets of microRNAs are upregulated depending on the method of induction of differentiation, suggesting that different sets of microRNAs are recruited during development of different lineages. MicroRNAs can be overexpressed using vector-based systems or mimetics (Ambion) and, similarly, their antisense (AS) (Dharmacon), allowing for perturbation of microRNA levels for studies on a high-throughput platform using ES cells to screen for phenotypic correlates with specific microRNAs. Our own experiments show that certain microRNAs appear to exert a subtle but detectable effect in modulating pathways critical for maintenance of ES cell totipotency and their differentiation (data not shown).

Emerging data suggest that individual microRNAs regulate multiple target mRNAs, increasing the network complexity of gene regulation, especially as the number of microRNAs produced in eukaryotic cells is now estimated to be in the thousands, not hundreds as proposed previously. That a mRNA can be targeted by multiple microRNAs also adds to the elaborate network of gene regulation, implying that systems biology applications will be required to determine the multiplicity of associations between microRNAs and the gene regulatory network [51,52] (figure 10.6).

EPIGENOMICS: EPIGENETICS AND REPROGRAMMING

The analyses of transcriptomes emphasize the central role of networks of regulated genes in orchestrating development. However, there is another level of regulation determining cell fate that is directed through the remodeling of chromatin via the posttranslational methyl modification of cytosine and lysine residues of histones bound to DNA [53]. During mammalian development, the activity of the genome is dynamically regulated to allow selective activation and repression of different expression programs that specify different cell types [54]. The pattern of some of the modifications to histones can be retained in daughter cells after cell division and the information associated with the gene in the mother cell can be inherited. This phenomenon is generally known as epigenetics. Understanding the inheritance of silencing and methylation and demethylation patterns, how chromatin modification is regulated, and the subsequent coordinated changes in gene regulation are major goals in developmental biology. Several unanswered questions remain as major challenges.

Reprogramming by ES Cells

Of particular relevance to stem cell regeneration is the mechanism of reprogramming that occurs when a somatic nucleus is transferred into the oocyte. The oocyte environment contains active ingredients that are sufficient to cause differentiated nuclei to revert to the zygotic state. This is a major step in the development of therapeutic cloning.

ES cells have provided significant and relevant insights into the reprogramming phenomenon [55–57]. When somatic cell, such as lymphocytes, are fused with ES cells, the resulting fusion cells take on the phenotype and behavior of ES cells. Fusion with other cells, such as neurospheres, gives rise to the same phenotype of ES-like cells (unpublished data). When karyoplasts (nuclei-containing fraction of cells) or cytoplasts (cytoplasmic fraction of cells) are fused to neurosphere cells, it is found that only the karyoplast can reactivate ES-specific genes in the neurosphere cells [58]. Transcription profiling experiments indicate that the hybrid cells resemble ES cells more closely than the differentiated fusion partner cells (unpublished data). This switch from a differentiated-cell transcriptome to that of an ES cell involves a reactivation of ES-specific genes and requires silencing of non-ES cell genes. To date, no transcription factors have been shown to be able to convert a differentiated cell to an ES-like cell or the reactivation of specific ES genes. As discussed above, *Oct4*, *Sox2*, and *Nanog* form a key trinity of ES transcriptional factors essential for ES state and during reprogramming these must be reactivated. It is plausible that several transcription activators may be required. Interestingly, the overexpression of *Oct4* in non-ES cells does not appear to elicit reprogramming in cell lines, but does lead to tumor induction in animal models [59]. This does suggest that certain populations of uncommitted cells in different niches within an adult animal are responsive to the reintroduction of *Oct4* protein.

Reprogramming by ncRNAs

The recent demonstration that double-stranded RNA targeted against complementary DNA segments can induce methylation of CpG residues leading to a reduction in transcription or complete gene silencing further increases the scope of the role of noncoding RNAs in regulating the transcriptome [60–62] Indeed, naturally occurring microRNAs in plants have been shown to direct this mechanism toward silencing gene regions that act as transposon elements, shutting down their ability to excise and recombine. Whether naturally occurring microRNAs or other ncRNAs are involved in epigenetic regulation and modulating silencing of transcription of coding genes remains to be demonstrated. If verified, the incorporation of this network of regulatory RNAs into the conventional transcriptional network will present yet another level of complexity.

CONCLUSIONS

Major advances in genomics technology continue to drive the recovery of large biological data sets. This windfall of information presents, for the first time, possibilities of breaking new ground in biological understanding based on a comprehensive global consideration of the relation of networks of genes. The endpoint is to arrive at unifying principles underlying the dynamics of networks of molecules that will allow computational prediction of biochemical and cellular responses to specific perturbations and signals. ES cells offer a unique and versatile stem cell model system that allows derivation of transcriptome and proteome data from cells to tissues and whole organism. By generating predictive models using ES cells we should be able to test, improve and, hopefully, apply such systems strategies in research on tissue engineering, pathogenesis of developmental disorders and other diseases, and targets for therapeutics and gene therapy.

Systems biology approaches are currently in their infancy. However, given the continuing accumulation of massive new genomic and proteomic data sets in conjunction with a steady development of powerful computational hardware and algorithms to analyze these data comprehensively, it is not an unreasonable expectation that systems biology can lead to a better understanding of biological processes, revealing details at levels as complex as behavior, personality, addiction, neurological disorders, mental retardation, and sexuality. This in itself carries many ethical implications, but the potential of medical discovery is exciting, ensuring that systems biology will become a mainstream area of research in this century.

ACKNOWLEDGMENTS

This work was supported by A-Star, Singapore, NIH grants RO1 DK47636 and AI54973 (B.L.) and a grant from the Leukemia and Lymphoma Society of America (B.L.). We thank Leonard Lipovich, Lim Sai Kiang (Genome Institute Singapore), Frank McKeon, Towia Liberman, and Dan Tenen (Harvard Medical School) for helpful discussions and in some instances reading the manuscript. We thank all members of the Stem Cell and Developmental Biology Group, Transcription Group, and Cloning and Sequencing Group at GIS for their contribution to data and the ideas that drove this chapter.

REFERENCES

1. Kitano, H. Computational systems biology. *Nature*, 420:206–10, 2002.
2. Levine, M. and E. H. Davidson. Gene regulatory networks for development. *Proceedings of the National Academy of Sciences USA*, 102:4936–42, 2005.
3. Papin, J. A., T. Hunter, B. O. Palsson and S. Subramaniam. Reconstruction of cellular signaling networks and analysis of their properties. *Nature Reviews in Molecular Cellular Biology*, 6:99–111, 2005.
4. Bar-Joseph, Z., G. K. Gerber, T. I. Lee, N. J. Rinaldi, J. Y. Yoo, F. Robert, D. B. Gordon, E. Fraenkel, T. S. Jaakkola, R. A. Young, and D. K. Gifford. Computational discovery of gene modules and regulatory networks. *Nature Biotechnology*, 21:1337–42, 2003.
5. Barabasi, A. L. and Z. N. Oltvai. Network biology: understanding the cell's functional organization. *Nature Reviews in Genetics*, 5:101–13, 2004.
6. Evans, M. J. and M. H. Kaufman. Establishment in culture of pluripotential cells from mouse embryos. *Nature*, 292, 154–6, 1981.
7. Martin, G. R. Isolation of a pluripotent cell line from early mouse embryos cultured in medium conditioned by teratocarcinoma stem cells. *Proceedings of the National Academy of Sciences USA*, 78:7634–8, 1981.
8. Thomson, J. A., J. Itskovitz-Eldor, S. S. Shapiro, M. A. Waknitz, J. J. Swiergiel, V. S. Marshall and J. M. Jones. Embryonic stem cell lines derived from human blastocysts. *Science*, 282:1145–7, 1998.
9. Keller, G. Embryonic stem cell differentiation: emergence of a new era in biology and medicine. *Genes and Development*, 19:1129–55, 2005.

10. Downing, G. J. and J. F. Battey. Technical assessment of the first 20 years of research using mouse embryonic stem cell lines. *Stem Cells*, 22:1168–80, 2004.
11. Schena, M., D. Shalon, R. W. Davis and P. O. Brown. Quantitative monitoring of gene expression patterns with a complementary DNA microarray. *Science*, 270:467–70, 1995.
12. Ramalho-Santos, M., S. Yoon, Y. Matsuzaki, R. C. Mulligan and D. A. Melton. "Stemness": transcriptional profiling of embryonic and adult stem cells. *Science*, 298:597–600, 2002.
13. Ivanova, N. B., J. T. Dimos, C. Schaniel, J. A. Hackney, K. A. Moore and I. R. Lemischka. A stem cell molecular signature. *Science*, 298:601–4, 2002.
14. Fortunel, N. O., H. H. Otu, H. H. Ng, J. Chen, X. Mu, T. Chevassut, X. Li, M. Joseph, C. Bailey, J. A. Hatzfeld, A. Hatzfeld, F. Usta, V. B. Vega, P. M. Long, T. A. Libermann and B. Lim. Comment on "Stemness: transcriptional profiling of embryonic and adult stem cells" and "A stem cell molecular signature. " *Science*, 302:393, 2003.
15. Nikitin, A., S. Egorov, N. Daraselia and I. Mazo. Pathway studio—the analysis and navigation of molecular networks. *Bioinformatics*, 19:2155–7, 2003.
16. Logan, C. Y. and R. Nusse. The Wnt signaling pathway in development and disease. *Annual Reviews in Cellular and Developmental Biology*, 20: 781–810, 2004.
17. Reya, T. and H. Clevers. Wnt signaling in stem cells and cancer. *Nature*, 434:843–50, 2005.
18. Loebel, D. A., C. M. Watson, R. A. De Young and P. P. Tam. Lineage choice and differentiation in mouse embryos and embryonic stem cells. *Developmental Biology*, 264:1–14, 2003.
19. Sato, N., L. Meijer, L. Skaltsounis, P. Greengard and A. H. Brivanlou. Maintenance of pluripotency in human and mouse embryonic stem cells through activation of Wnt signaling by a pharmacological GSK-3-specific inhibitor. *Nature Medicine*, 10:55–63, 2004.
20. Wei, C. L., T. Miura, P. Robson, S. K. Lim, X. Q. Xu, M. Y. Lee, S. Gupta, L. Stanton, Y. Luo, J. Schmitt, S. Thies, W. Wang, I. Khrebtukova, D. Zhou, E. T. Liu, Y. J. Ruan, M. Rao and B. Lim. Transcriptome profiling of human and murine ESCs identifies divergent paths required to maintain the stem cell state. *Stem Cells*, 23:166–85, 2005.
21. Ng, P., C. L. Wei, W. K. Sung, K. P. Chiu, L. Lipovich, C. C. Ang, S. Gupta, A. Shahab, A. Ridwan, C. H. Wong, E. T. Liu and Y. Ruan. Gene identification signature (GIS) analysis for transcriptome characterization and genome annotation. *Nature Methods*, 2:105–11, 2005.
22. Hamatani, T., M. G. Carter, A. A. Sharov and M. S. Ko. Dynamics of global gene expression changes during mouse preimplantation development. *Developmental Cell*, 6:117–31, 2004.
23. Wang, Q. T., K. Piotrowska, M. A. Ciemerych, L. Milenkovic, M. P. Scott, R. W. Davis and M. Zernicka-Goetz. A genome-wide study of gene activity reveals developmental signaling pathways in the preimplantation mouse embryo. *Developmental Cell*, 6:133–44, 2004.
24. Wei, C. L., P. Ng, K. P. Chiu, C. H. Wong, C. C. Ang, L. Lipovich, E. T. Liu and Y. Ruan. 5′ Long serial analysis of gene expression (LongSAGE) and

3′ LongSAGE for transcriptome characterization and genome annotation. *Proceedings of the National Academy of Sciences USA*, 101:11701–6, 2004.
25. Carpenter, A. E. and D. M. Sabatini. Systematic genome-wide screens of gene function. *Nature Reviews in Genetics*, 5:11–22, 2004.
26. Cullen, L. M. and G. M. Arndt. Genome-wide screening for gene function using RNAi in mammalian cells. *Immunology and Cell Biology*, 83:217–23, 2005.
27. Salic, A., E. Lee, L. Mayer and M. W. Kirschner. Control of β-catenin stability: reconstitution of the cytoplasmic steps of the wnt pathway in *Xenopus* egg extracts. *Molecular Cell*, 5:523–32, 2002.
28. Sanchez, Y., C. Wong, R. S. Thomas, R. Richman, Z. Wu, H. Piwnica-Worms and S. J. Elledge. Conservation of the Chk1 checkpoint pathway in mammals: linkage of DNA damage to Cdk regulation through Cdc25. *Science*, 277:1497–1501, 1997.
29. Wheeler, D. B., A. E. Carpenter and D. M. Sabatini. Cell microarrays and RNA interference chip away at gene function. *Nature Genetics*, 37:S25–30, 2005.
30. DasGupta, R., A. Kaykas, R. T. Moon and N. Perrimon. Functional genomic analysis of the Wnt-wingless signaling pathway. *Science*, 308: 826–33, 2005.
31. Okumura-Nakanishi, S., M. Saito, H. Niwa, and F. Ishikawa. Oct-3/4 and Sox2 regulate Oct-3/4 gene in embryonic stem cells. *Journal of Biological Chemistry*, 280:5307–17, 2005.
32. Nichols, J., B. Zevnik, K. Anastassiadis, H. Niwa, D. Klewe-Nebenius, I. Chambers, H. Scholer and A. Smith. Formation of pluripotent stem cells in the mammalian embryo depends on the POU transcription factor Oct4. *Cell*, 95:379–91, 1998.
33. Niwa, H., J. Miyazaki and A. G. Smith. Quantitative expression of Oct-3/4 defines differentiation, dedifferentiation or self-renewal of ES cells. *Nature Genetics*, 24:372–6, 2000.
34. Rodda, D. J., J. L. Chew, L. H. Lim, Y. H. Loh, B. Wang, H. H. Ng and P. Robson. Transcriptional regulation of nanog by OCT4 and SOX2. *Journal of Biological Chemistry*, 280(26):24731–7, 2005.
35. Ren, B., F. Robert, J. J. Wyrick, O. Aparicio, E. G. Jennings, I. Simon, J. Zeitlinger, J. Schreiber, N. Hannett, E. Kanin, T. L. Volkert, C. J. Wilson, S. P. Bell and R. A. Young. Genome-wide location and function of DNA binding proteins. *Science*, 290:2306–9, 2000.
36. Cawley, S., S. Bekiranov, H. H. Ng, P. Kapranov, E. A. Sekinger, D. Kampa, A. Piccolboni, V. Sementchenko, J. Cheng, A. J. Williams, R. Wheeler, B. Wong, J. Drenkow, M. Yamanaka, S. Patel, S. Brubaker, H. Tammana, G. Helt, K. Struhl and T. R. Gingeras. Unbiased mapping of transcription factor binding sites along human chromosomes 21 and 22 points to widespread regulation of noncoding RNAs. *Cell*, 116:499–509, 2004.
37. Weinmann, A. S., P. S. Yan, M. J. Oberley, T. H. Huang and P. J. Farnham. Isolating human transcription factor targets by coupling chromatin immunoprecipitation and CpG island microarray analysis. *Genes and Development*, 16:235–44, 2002.
38. Boyer, L. A., T. I. Lee. M. F. Cole, S. E. Johnstone, S. S. Levine, J. P. Zucker, M. G. Guenther, R. M. Kumar, H. L. Murray, R. G. Jenner, D. K. Gifford,

D. A. Melton, R. Jaenisch and R. A. Young. Core transcriptional regulatory circuitry in human embryonic stem cells. *Cell*, 122:947–56, 2005.

39. Loh, Y. H., Q. Wu, J. L. Chew, V. B. Vega, W. Zhang, X. Chen, G. Bourque, J. George, B. Leong, J. Liu, K. Y. Wong, K. W. Sung, C. W. Lee, X. D. Zhao, K. P. Chiu, L. Lipovich, V. A. Kuznetsov, P. Robson, L. W. Stanton, C. L. Wei, Y. Ruan, B. Lim and H. H. Ng. The Oct4 and Nanog transcription network regulates pluripotency in mouse embryonic stem cells. *Nature Genetics*, 38:431–40, 2006.
40. Lanz, R. B., N. J. McKenna, S. A. Onate, U. Albrecht, J. Wong, S. Y. Tsai, M. J. Tsai and B. W. O'Malley. A steroid receptor coactivator, SRA, functions as a RNA and is present in a SRC-1 complex. *Cell*, 97:17–27, 1999.
41. Tomari, Y. and P. D. Zamore. Perspective: machines for RNAi. *Genes and Development*, 19:517–29, 2005.
42. Gebauer, F. and M. W. Hentze. Molecular mechanisms of translational control. *Nature Reviews in Molecular and Cellular Biology*, 5:827–35, 2004.
43. Bayne, E. H. and R. C. Allshire. RNA-directed transcriptional gene silencing in mammals. *Trends in Genetics*, 21:370–3, 2005.
44. Kawasaki, H. and K. Taira. Transcriptional gene silencing by short interfering RNAs. *Current Opinion in Molecular Therapeutics*, 7:125–31, 2005.
45. Tang, G. SiRNA and miRNA: an insight into RISCs. *Trends in Biochemical Science*, 30:106–14, 2005.
46. Lu, J., G. Getz, E. A. Miska, E. Alvarez-Saavedra, J. Lamb, D. Peck, A. Sweet-Cordero, B. L. Ebert, R. H. Mak, A. A. Ferrando, J. R. Downing, T. Jacks, H. R. Horvitz and T. R. Golub. MicroRNA expression profiles classify human cancers. *Nature*, 435:834–8, 2005.
47. Gregory, R. I. and R. Shiekhattar. MicroRNA biogenesis and cancer. *Cancer Research*, 65:3509–12, 2005.
48. Ambros, V. The functions of animal microRNAs. *Nature*, 431:350–5, 2004.
49. Suh, M. R., Y. Lee, J. Y. Kim, S. K. Kim, S. H. Moon, J. Y. Lee, K. Y. Cha, H. M. Chung, H. S. Yoon, S. Y. Moon, V. N. Kim and K. S. Kim. Human embryonic stem cells express a unique set of microRNAs. *Developmental Biology*, 270:488–98, 2004.
50. Houbaviy, H. B., M. F. Murray and P. A. Sharp. Embryonic stem cell-specific microRNAs. *Developmental Cell*, 5:351–8, 2003.
51. Cheng, L. C., M. Tavazoie and F. Doetsch. Stem cells: from epigenetics to microRNAs. *Neuron*, 46:363–7, 2005.
52. Lim, L. P., N. C. Lau, P. Garrett-Engele, A. Grimson, J. M. Schelter, J. Castle, D. P. Bartel, P. S. Linsley and J. M. Johnson. Microarray analysis shows that some microRNAs downregulate large numbers of target mRNAs. *Nature*, 433:769–73, 2005.
53. Jaenisch, R. and A. Bird. Epigenetic regulation of gene expression: how the genome integrates intrinsic and environmental signals. *Nature Genetics*, 33:S245–54, 2003.
54. Bird, A. DNA methylation patterns and epigenetic memory. *Genes and Development*, 16:6–21, 2002.
55. Tada, M., Y. Takahama, K. Abe, N. Nakatsuji and T. Tada. Nuclear reprogramming of somatic cells by in vitro hybridization with ES cells. *Current Biology*, 11:1553–8, 2001.

56. Ying, Q. L., J. Nichols, E. P. Evans and A. G. Smith. Changing potency by spontaneous fusion. *Nature*, 416:545–8, 2002.
57. Kimura, H., M. Tada, N. Nakatsuji and T. Tada. Histone code modifications on pluripotential nuclei of reprogrammed somatic cells. *Molecular and Cellular Biology*, 24:5710–20, 2004.
58. Do, J. T. and H. R. Scholer. Nuclei of embryonic stem cells reprogram somatic cells. *Stem Cells*, 22:941–9, 2004.
59. Hochedlinger, K., Y. Yamada, C. Beard and R. Jaenisch. Ectopic expression of Oct-4 blocks progenitor-cell differentiation and causes dysplasia in epithelial tissues. *Cell*, 121:465–77, 2005.
60. Morris, K. V., S. W. Chan, S. E. Jacobsen and D. J. Looney. Small interfering RNA-induced transcriptional gene silencing in human cells. *Science*, 305:1289–92, 2004.
61. Kawasaki, H. and K. Taira. Induction of DNA methylation and gene silencing by short interfering RNAs in human cells. *Nature*, 431:211–17, 2004.
62. Matzke, M. A. and J. A. Birchler. RNAi-mediated pathways in the nucleus. *Nature Reviews in Genetics*, 6:24–35, 2005.

Index

Page references to tables and figures are in *italics*.